Physics for a New Generation

H. Latal H. Mitter (Eds.)

Physics for a New Generation

Prospects for High-Energy Physics at New Accelerators

Proceedings of the
XXVIII Int. Universitätswochen für Kernphysik
Schladming, Austria, March 1989

With 226 Figures

Springer-Verlag Berlin Heidelberg New York
London Paris Tokyo Hong Kong Barcelona

Professor Dr. Heimo Latal
Professor Dr. Heinrich Mitter
Institut für Theoretische Physik, Karl-Franzens-Universität, Universitätsplatz 5, A-8010 Graz, Austria

ISBN-13: 978-3-642-75569-9 e-ISBN-13: 978-3-642-75567-5
DOI: 10.1007/978-3-642-75567-5

Library of Congress Cataloging-in-Publication Data. Internationale Universitätswochen für Kernphysik der Karl-Franzens-Universität Graz (28th : 1989 : Schladming, Austria) Physics for a new generation : prospects for high-energy physics at new accelerators : proceedings of the XXVIII Int. Universitätswochen für Kernphysik, Schladming, Austria, March 1989 / H. Latal, H. Mitter, eds. p. cm. Includes bibliographical references. ISBN-13: 978-3-642-75569-9 1. Particles (Nuclear physics) – Congresses. 2. Quantum chromodynamics – Congresses. I. Latal, H. (Heimo), 1939- II. Mitter, Heinrich. III. Title. QC793.I57 1989 539.7'2 – dc20 90-9665

Softcover reprint of the hardcover 1st edition 1990

The text was processed by the authors using the TEX macro package from Springer-Verlag.

2155/3150-543210 – Printed on acid-free paper

Preface

This volume contains the written versions of invited lectures presented at the 28th "Internationale Universitätswochen für Kernphysik" in Schladming, Austria in March 1989.

The generous support of our sponsors, the Austrian Ministry of Science and Research, the Government of Styria, and others, made it again possible to invite expert lecturers. The courses were centered on elementary particle physics to be performed with large accelerators accessible in the immediate future, including some reports on the current situation. Thanks to the efforts of the speakers it was possible to obtain excellent surveys.

After the School the lecture notes were revised and partially rewritten in TEX by the authors, whom we thank for their labour. Unfortunately the report on the situation at SLAC by M. Swartz could not be included.

Thanks are also due to the publishers for their patience.

Graz, Austria
December 1989

H. Latal
H. Mitter

Contents

The HERA Project

Physics in p$\bar{\text{p}}$ Collisions

Phenomenology of and Beyond the Standard Electroweak Model

A. Bartl, H. Pietschmann, and H. Stremnitzer

Institut für Theoretische Physik, Universität Wien
Boltzmanngasse 5, A-1090 Wien

1. The Standard Model

1.1 Introduction

It has become a ritual to begin any lecture on the Standard Model for electroweak interactions (which might better be called Quantum Flavourdynamics, QFD) with a lament about its weaknesses; i.e. there are so many free parameters, there is no genuine unification and so on. In these lectures, we shall not follow this ritual. Rather, we shall try to emphasize the tremendous achievements without forgetting that they lead naturally to new open problems and the question of what comes beyond. In this spirit, a recent reference volume by D. Haidt and one of the authors on electroweak interactions [1.1] has been concluded with the following statement:

At the end of this volume it is fair to conclude that the vast body of experimental weak and electromagnetic data, in some cases accurate to the level of 10^{-10}*, appears systematically ordered and quantitatively described within Quantum Flavourdynamics. This is one of the rare moments in the history of physics, where a theory is complete up to a certain level and when confronted with experiments does not reveal essential refutations. It is on the basis of this success that the questions leading beyond the present framework gain their relevance.*

1.2 Defining the Standard Model

Since the notion of the "Standard Model" is not always used in precisely the same way, let us give a definition for the sake of these lectures. We shall do it by stating six basic assumptions, see Table 1.1. Any extension or deviation from these assumptions shall be denoted explicitly, for instance as "modified standard model" or the like.

Table 1.1 represents the strongest set of assumptions (i.e. the most minimal model) consistent with all experiments as of today. Slight modifications can easily be made, for example assumption e) can be dropped if a neutrino mass is empirically demonstrated beyond doubt.

On the basis of these assumptions, a Lagrangian for the standard model (or QFD) can be written down. The Higgs-mechanism is to be applied for generating masses.

Physics for a New Generation Editors: H. Latal · H. Mitter

Table 1.1. Basic assumptions defining the "Standard Model" or QFD

a) The gauge group is SU(2) ⊗ U(1), i.e. there are 4 gauge bosons.
b) There is one single Higgs doublet which generates *all* mass terms spontaneously.
c) The gauge group is broken spontaneously down to $U(1)_{em}$. Thus the photon is the only massless gauge boson and it is a mixture of the gauge boson of the unbroken U(1) and the neutral gauge boson of SU(2) with the "weak mixing angle Θ_W".
d) All fermions are represented by left-handed doublets and right-handed singlets.
e) There are no right-handed neutrinos, i.e. neutrinos are massless.
f) There are no transitions between lepton generations; quark generations are mixed by the charged weak current.

In so doing, we shall choose the U-formalism because it exhibits physical properties more explicitly.

The resulting Lagrangian can be broken into 6 sectors:

$$\mathcal{L}_{QFD} = \mathcal{L}_0 + \mathcal{L}_{GB} + \mathcal{L}_{em} + \mathcal{L}_{CC} + \mathcal{L}_{NC} + \mathcal{L}_H, \tag{1.1}$$

where $\mathcal{L}_0$ is the free Lagrangian including mass terms. The gauge boson self-interaction is explicitly exhibited by

$$\begin{aligned}\mathcal{L}_{GB} = ig\cos\Theta_W\{&Z_\lambda[W_\rho^\dagger(\partial^\lambda W^\rho) - W_\rho^\dagger(\partial^\rho W^\lambda)\\ &+ W_\rho(\partial^\rho W^{\lambda\dagger}) - W_\rho(\partial^\lambda W^{\rho\dagger})]\\ &+ (\partial^\rho Z^\lambda)(W_\rho^\dagger W_\lambda - W_\lambda^\dagger W_\rho)\}\\ &- ie\{A_\lambda[W_\rho^\dagger(\partial^\lambda W^\rho) - W_\rho^\dagger(\partial^\rho W^\lambda) + W_\rho(\partial^\rho W^{\lambda\dagger}) - W_\rho(\partial^\lambda W^{\rho\dagger})]\\ &+ (\partial^\rho A^\lambda)(W_\rho^\dagger W_\lambda - W_\lambda^\dagger W_\rho)\} + O(g^2),\end{aligned} \tag{1.2}$$

where g is the triplet coupling constant related to the electric charge e and the weak mixing angle Θ_W by

$$e = g\sin\Theta_W, \tag{1.3}$$

Z, W and A are the physical fields for neutral and charged intermediate bosons and the photon.

Let us emphasize at this point that a verification of Eq. (1.2) is the ultimate test of the gauge nature of the interaction but a precision test has to wait until much higher energies (twice the W–mass) are reached experimentally.

The electromagnetic interaction is just the familiar one, found in any textbook on QED. The genuine weak interactions are

$$\mathcal{L}_{CC} + \mathcal{L}_{NC} = \frac{g}{2\sqrt{2}}(J_\lambda^{CC}W^\lambda + h.c.) + \frac{g}{4\cos\Theta_W}J_\lambda^{NC}Z^\lambda. \tag{1.4}$$

Notice that the relative strength of charged and neutral current interactions is uniquely given within the standard model based on the assumptions of Table 1.1. In order to test it, a parameter ρ is often introduced multiplying $\mathcal{L}_{NC}$. In a more general model with n Higgs multiplets of weak isospin t_i and 3rd component of its neutral member t_{3i} with vacuum expectation value λ_i, the parameter ρ becomes [1.2]

Table 1.2. ρ from different experiments

Type of experiment	ρ		
NC/CC	1.006	$\pm$	0.008
ν_μ-scattering	1.04	$\pm$	0.05
m_W, m_Z	1.009	$\pm$	0.026
ed-scattering	1.74	$\pm$	0.36

$$\rho = \frac{\sum_{i=1}^{n} [t_i(t_i+1) - t_{3i}^2] |\lambda_i|^2}{2 \sum_{i=1}^{n} t_{3i}^2 |\lambda_i|^2} \tag{1.5}$$

which reduces to $\rho = 1$ in the standard model due to assumption b) of Table 1.1. The very good experimental support is exhibited in Table 1.2, taken from Ref. 1.1.

In Eq. (1.4), the charged current is given by

$$J_\lambda^{\mathrm{CC}} = \sum_{\ell} \bar{\nu}_\ell \gamma_\lambda (1+\gamma_5) \ell + \sum_{q_1 q_2} \bar{q}_1 \gamma_\lambda (1+\gamma_5) U_{q_1 q_2} q_2 , \tag{1.6}$$

where q_1 and q_2 refer to the mass eigenstates of the quarks, (u, c, t) and (d, s, b), respectively. $U_{q_1 q_2}$ is the weak quark mixing matrix of Kobayashi and Maskawa [1.3].

The weak neutral current is given by

$$\begin{aligned} J_\lambda^{\mathrm{NC}} = & \sum_{\ell} [\bar{\nu}_\ell \gamma_\lambda (1+\gamma_5) \nu_\ell - \bar{\ell} \gamma_\lambda (C_V^\ell + \gamma_5) \ell] \\ & + \sum_{q_1} \bar{q}_1 \gamma_\lambda (C_V^1 + \gamma_5) q_1 - \sum_{q_2} \bar{q}_2 \gamma_\lambda (C_V^2 + \gamma_5) q_2 \end{aligned} \tag{1.7}$$

with

$$\begin{aligned} C_V^\ell &= 1 - 4 \sin^2 \Theta_W \\ C_V^1 &= 1 - \tfrac{8}{3} \sin^2 \Theta_W \\ C_V^2 &= 1 - \tfrac{4}{3} \sin^2 \Theta_W \end{aligned} \tag{1.8}$$

Finally, the interaction of the Higgs particle is

$$\begin{aligned} \mathcal{L}_H = & \frac{g^4}{4} \left(W_\sigma^\dagger W^\sigma + \frac{1}{2\cos^2 \Theta_W} Z_\sigma Z^\sigma \right) (\lambda^2 + 2\lambda H + H^2) \\ & + \sum_{\ell, q_1, q_2} \frac{1}{\lambda} (m_\ell \bar{\ell} \ell + m_{q_1} \bar{q}_1 q_1 + m_{q_2} \bar{q}_2 q_2) H - h \lambda H^3 - \frac{1}{4} h H^4 , \end{aligned} \tag{1.9}$$

where λ is $\sqrt{2}$ times the vacuum expectation value of the neutral Higgs component. It is related to the familiar 4 fermion coupling constant G by

$$\frac{G}{\sqrt{2}} = \frac{1}{2\lambda^2} . \tag{1.10}$$

Table 1.3. Best values for the mass parameters

Parameter	Best value or limit (90% C.L.)
m_{ν_e}	< 19 eV 95% C.L.
m_e	0.5110034(14) MeV
m_{ν_μ}	< 0.42 MeV
m_μ	105.65916(30) MeV
m_{ν_τ}	< 35 MeV
m_τ	1784.2 ± 3.2 MeV
	current quark masses at 1 GeV
m_u	3.1 – 6.6 MeV
m_d	5.5 – 11.1MeV
m_c	1.2 – 1.4 GeV
m_s	105 – 230 MeV
m_t	23 – 200 GeV
m_b	4.15 – 4.35 GeV
m_γ	$< 3 \cdot 10^{-27}$ eV
m_W	80.8 ± 0.7 GeV
m_Z	92.1 ± 0.8 GeV
m_H	4 GeV – 1 TeV

Table 1.4. Best value for the Kobayashi-Maskawa matrix elements and the coupling parameters of QFD

$\|U_{ud}\|$	0.9744 ± 0.0009
$\|U_{us}\|$	0.221 ± 0.002
$\|U_{ub}\|$	< 0.02
$\|U_{cd}\|$	0.20 ± 0.03
$\|U_{cs}\|$	> 0.66
$\|U_{cb}\|$	0.047 ± 0.006
$\|U_{td}\|$	–
$\|U_{ts}\|$	–
$\|U_{tb}\|$	–
α	1/137.03606(11)
G	$1.166344(11) \cdot 10^{-5}$ GeV^{-2}
$\sin^2 \Theta_W$	0.229 ± 0.004

Notice that the coupling of the Higgs to fermions is proportional to their masses which identifies a Higgs particle uniquely through its decay branching ratios, once it is found.

By the spontaneous symmetry breaking mechanism, all masses are generated, leading to the tree level relations (see also Eq. (1.10))

$$\frac{G}{\sqrt{2}} = \frac{e^2}{8m_W^2 \sin^2 \Theta_W} = \frac{e^2}{2m_Z^2 \sin^2 2\Theta_W} = \frac{h}{m_H^2}. \tag{1.11}$$

Since the quartic Higgs self-interaction coupling constant h does not enter in any other empirical relation, the mass of the Higgs particle remains unspecified.

Let us conclude this section by presenting the current best values for the parameters of the Standard Model taken from Ref. 1.1 with updating by new values from the Neutrino 88 Symposium in Boston, June 1988 and the HEP Conference in Munich, August 1988. Table 1.3 gives the mass values and Table 1.4 the Kobayashi-Maskawa matrix elements as well as the coupling parameters.

1.3 Testing the Standard Model

With our stringent definition of the term "Standard Model" it is also important to clearly distinguish different kinds of tests with different consequences to the Standard Model. We shall distinguish

A. Predictions
B. Open questions
C. Hypothesis
D. New frontiers.

By "predictions" we shall refer to all those consequences of Eq. (1.1) which have been beautifully corroborated by experiments as well as those whose empirical test is still pending. An unexpected refutation of such a prediction could lead to dramatic consequences, possibly even shaking the whole picture down to its roots.

By "open questions" we shall refer to problems as yet unanswered but whose eventual answer will simply enrich our knowledge, no matter what the answer shall be.

By "hypothesis" we refer to modifications and/or extensions of the standard model which wait for experimental decision as to their correctness.

Finally, "new frontiers" are all those ideas based on the standard model but reaching far into new territory. The 2 most prominent trials of today shall be described in Chaps. 2 and 3.

1.4 Predictions

It has been said in the Introduction that a wealth of experiments is correctly described by the standard model. At the present moment, there may be one single hint at a possible problem which could very well turn out not to be totally of experimental origin: The decay of the τ-lepton.

As early as 1985 it was noted by Gilman and Rhie [1.4] that the exclusive one-prong decay modes show a deficit of a few per cent with respect to the inclusive one-prong decay modes. Radiative corrections and the effect of a massive τ-neutrino are both less than 1/2% [1.5]. In spite of quite some effort, this deficit still persists [1.6]. For a time, it was thought that second class weak currents could fill the gap (through one-prong decays with η^0) but this possibility was ruled out and led to new limits on second class currents instead [1.7, 1.8]. The problem may or may not be related to the leptonic branching ratios for $\tau^- \to \ell^- \bar{\nu}_\ell \nu_\tau$ which tend to come out too low as compared to lepton universality [1.1, 1.9]:

$$\frac{G_\tau}{G_\mu} = 0.963 \pm 0.023. \tag{1.12}$$

Since we dwell into per cent effects, let us notice in passing that $\sin^2 \Theta_W$ as derived from purely leptonic processes is about $(3 \pm 2)\%$ smaller than that derived from W, Z masses.

The most crucial prediction of the standard model is the existence of the top quark. Since so-called "topless models" are ruled out by experiments for quite a while [1.10], lack of producing the top would be a disaster for the standard model. The most direct lower limit for the top mass stems from e^+e^- machines which could not produce toponium at the highest available energy. More indirectly, from $q\bar{q}$,

$$m_t > 44 \text{ GeV}. \tag{1.13}$$

Indirectly, it is also possible to give *upper* limits on the top-mass because it enters as a parameter in radiative corrections which are limited in size experimentally. Various upper limits are extracted in the literature. A conservative suggestion by Marciano at the ν 88 Symposium in Boston is 200 GeV. It may be lowered to about 120 GeV by published data [1.11].

Another crucial prediction of the standard model, the relative strength of charged and neutral currents, has already been discussed in Section 1.2, see Eq. (1.5) and Table 1.2. It is intimately related to the question of the whole Higgs-sector. The prediction of the Higgs particle is probably the most important problem of particle physics in the near future. If it is not found, we have to look for equally plausible explanations for the origin of masses which fulfill the relations (1.11) as accurately as present day experiments.

The mass limits for the Higgs boson given in Table 1.3 stem from theoretical consistency arguments. The best empirical lower limit is 3.9 GeV.

1.5 Open Questions

Within the context of the standard model as defined through Table 1.1, the most obvious open question is N_G, the number of generations (or N_ν, the number of different neutrinos) which is also 1/2 the number of flavours. As early as 1976, it has been suggested [1.12] to use the width of the Z-boson for a determination of N_G through

$$\Gamma_{\mathrm{Z}} = \Gamma(\mathrm{Z} \to \text{visible}) + N_G \cdot \Gamma(\mathrm{Z} \to \nu\bar{\nu}) \tag{1.14}$$

with

$$\Gamma(\mathrm{Z} \to \nu\bar{\nu}) = \frac{G}{12\sqrt{2}\,\pi}\, m_{\mathrm{Z}}^3 \doteq 170\ \mathrm{MeV}. \tag{1.15}$$

As soon as a direct determination of Γ_{Z} in $\mathrm{e^+e^-}$ collisions is possible, N_G will be known. Presently, only an indirect analysis [1.13] through

$$\frac{(\sigma B)_{\mathrm{W}}}{(\sigma B)_{\mathrm{Z}}} = \frac{\sigma_{\mathrm{W}}}{\sigma_{\mathrm{Z}}} \frac{\Gamma(\mathrm{W} \to \mathrm{e}\nu)}{\Gamma(\mathrm{Z} \to \mathrm{e^+e^-})} \frac{\Gamma_{\mathrm{Z}}}{\Gamma_{\mathrm{W}}} \tag{1.16}$$

is possible, where the left-hand side is determined experimentally and the right-hand side, except Γ_{Z}, computed from QCD and QFD. Present accuracy probably allows for 1 to 3 new generations, but limits as low as $N_G \leq 3$ have been published [1.14]!

Naturally, the value of all free parameters of the standard model belongs to this section. However, a detailed discussion is rather technical and the reader is thus referred to Tables 1.3 and 1.4 for the present status and to Ref. 1.1 for a more extensive explanation.

1.6 Hypotheses

If we want to test the stringent assumptions of Table 1.1 quantitatively, the only way is to design a particular modification, extension or alternative and to restrict its new

Table 1.5. Classification of processes forbidden in the standard model

$\Delta L_\ell \neq 0,\ \Delta L = \Delta B = 0$			Flavour changing neutral currents $\nu_\ell \leftrightarrow \nu_{\ell'}$ oscillations $(\mu^+ e^-) \rightarrow (\mu^- e^+)$ conversion
$\Delta L = \Delta B = -1$			Nucleon decay: $p \rightarrow e^+\pi^0;\ \mu^+ K^0; \ldots$ $n \rightarrow e^+\pi^-;\ \bar{\nu} K^0; \ldots$
$\Delta L = \Delta B = \pm 2$			$pe^- \leftrightarrow \bar{p}e^+$
$\Delta L = -\Delta B \neq 0, \Delta F = 0$*)			Nucleon decay: $n \rightarrow e^-\pi^+; \ldots$
ΔL	$\neq$	ΔB	
0		± 2	$n \leftrightarrow \bar{n}$ oscillations $nn \rightarrow 2\pi^0; \ldots$
± 2		0	ν-less $\beta\beta$ decay $\nu_\ell \leftrightarrow \bar{\nu}_\ell$ oscillations
± 3		-1	Nucleon decay: $p \rightarrow e^+\bar{\nu}\bar{\nu}; \ldots$ $n \rightarrow \nu\nu\nu;\ \nu\nu e^-\pi^+; \ldots$

*) F = # fermions

parameters by experiment. We shall select a typical example for each assumption of Table 1.1.

But before, let us collect the processes which are forbidden within the standard model; in so doing, we follow Ref. 1.15, see Table 1.5.

Assumption a. A typical extension of the standard model which goes beyond Assumption a is given by so-called "left-right-symmetric models" [1.16] based on the gauge group $SU(2)_L \otimes SU(2)_R \otimes U(1)_{B-L}$. Lower limits for the masses of right-handed gauge bosons provide a quantitative test of Assumption a. Good limits for right-handed W's of the order of 2.5 TeV stem from the mass difference of neutral kaons [1.17]. From the supernova SN87a, even better limits can be derived (reported at ν 88, Boston)

$$m_{W_R} > 16\ \text{TeV}, \qquad m_{Z_R} > 3.2\ \text{TeV}. \tag{1.17}$$

Assumption b. An extension of the standard model to more than a single Higgs doublet has already been discussed in connection with Eq. (1.5). Clearly, the experimental value of the ρ parameter does not necessitate, but neither rules out a reacher structure in the Higgs sector.

Within the standard model defined through Table 1.1, CP-violation originates in the complex Kobayashi-Maskawa matrix predicting a neutron electric dipole moment of about 10^{-30} e.cm. Should experimental hints at a possible non-vanishing effect

Table 1.6. Extraction of $\sin^2\Theta_W$ from various experiments using the on shell scheme

Experiment	Measured quantity	Extracted value of $\sin^2\Theta_W$	
UA	$m_W = (80.0 \pm 0.5 \pm 1.2)$ GeV	0.2286	$\pm^{0.0075}_{0.0072}$
	$m_Z = (91.9 \pm 0.9 \pm 1.5)$ GeV	0.230	$\pm^{0.013}_{0.012}$
CHARM, BNL	$\sigma(\nu_\mu e)/\sigma(\bar{\nu}_\mu e)$	0.212	$\pm$ 0.023
ν, $\bar{\nu}$q	$R, \bar{R}, R_p, \bar{R}_p$	0.2331	$\pm$ 0.006
νp	$\sigma(\nu p \to \nu p)$, $\sigma(\bar{\nu} p \to \bar{\nu} p)$	0.220	$\pm$ 0.031
SLAC	eD	0.218	$\pm$ 0.020
PETRA, PEP	R_q	0.236	$\pm$ 0.020
Atomic PV	Parity violation	0.234	$\pm$ 0.027
PETRA, PEP	electroweak interference	A_μ, A_ν	

Average:
$\sin^2\Theta_W = (0.229 \pm 0.004) + (0.5 \pm 1.2) \cdot 10^{-6} \left(\frac{Q^2}{\text{GeV}^2} - 3380\right)$

at 10^{-25} e.cm (reported at ν 88 in Boston) materialize, an extra mechanism for CP-violation would be necessary. The problem of CP-violation is dealt with in the lectures of K. Kleinknecht, an excellent review is found in Ref. 1.18.

Assumption c. Instead of contemplating possible modifications, let us collect the different experimental determinations of $\sin^2\Theta_W$, Table 1.6 (from [1.1]).

The beautiful agreement is even more satisfactory when we remember that experiments cover a range of Q^2 from 10^{-6} GeV^2 up to almost 10^4 GeV^2! (Notice that Table 1.6 is based on the on-shell renormalization scheme whose $\sin^2\Theta_W$ is not a running parameter).

Assumption d. A typical modification of this assumption is the scheme of "mirror fermions" [1.19]. No compelling experimental evidence exists so far.

Assumption e. It has been said in Sect. 1.2, that this assumption has to be dropped if a non-vanishing neutrino mass is found. Present limits are given in Table 1.3. The most interesting case is $\bar{\nu}_e$ from H^3 decay because a *lower* limit for its mass is repeatedly published by the ITEP-group [1.20]. We have not taken it into account in Table 1.3, because it has been questioned in the published literature [1.21].

The way to determine m_{ν_e} from H^3 decay is by a Kurie-plot, in which

$$K(E_e) = \frac{GU_{ud}\sqrt{\xi}}{\pi\sqrt{2\pi}} \sqrt{(E_0 - E_e)^2 - m_\nu^2} \tag{1.18}$$

is plotted against the electron energy E_e. E_0 is the maximal electron energy and ξ the nuclear matrix element. For an allowed transition with $m_\nu = 0$, $K(E_e)$ is a straight line. Deviations from this straight line at the end point of the spectrum hint at a non-vanishing neutrino mass but can also originate in instrumental effects (such as misidentified resolution function). The upper limit given in Table 1.3 is the average of 3 most accurate experiments (see Ref. 1.1).

Assumption f. An extension of the standard model with respect to this assumption is the introduction of a weak lepton mixing matrix also. It leads to neutrino oscillations and/or neutrinoless double β-decay. Neither has been found in the laboratory. However, the persistent solar neutrino problem leads some authors to the belief that neutrino oscillations must occur in solar matter [1.22].

1.7 Conclusion

In spite of the beautiful corroboration of the standard model by experiments, the question naturally comes up as to what is the next step. First let us recall that as a prerequisit, we have to find the top quark. Next, the Higgs mechanism waits for discovery or clarification. A further unification of forces is the natural tendency of most physicists, either by larger groups or by means of subconstituents. These two possibilities shall be discussed now.

2. Beyond the Standard Model: Supersymmetry

Supersymmetry has quite a tradition at the Schladming Winter School [2.1–2.4]. The first lectures on this subject were presented as early as 1976 [2.1] and in 1978 already we heard about how one might incorporate these new ideas into the Standard Model [2.2]. The theoretical problems still present in the Standard Model, and how they could be solved by supersymmetry, have also been treated here [2.4] and have been reviewed in several papers [2.5–2.7]. In this lecture we shall discuss the minimal supersymmetric extension of the Standard Model based on spontaneously broken $N = 1$ supergravity (for a recent review see [2.8]). We shall only briefly mention the essential steps of the formalism necessary to make the Standard Model supersymmetric and shall then concentrate on its phenomenological consequences. We shall in particular pay attention to the question of how weakly interacting supersymmetric particles could be detected experimentally provided that they do exist and their masses are in the energy range of the new accelerators Tristan, SLC, LEP or HERA.

The motivations for supersymmetry in particle physics are purely theoretical. As a rule supersymmetric quantum field theories have less virulent divergencies than "ordinary" theories. Extended supersymmetry is the only known way to combine non-trivially an internal symmetry group with the Poincaré group. Local supersymmetry, i.e. supergravity provides in a natural way the unification of gravity with the other known gauge interactions. In grand unified theories supersymmetry solves the so-called "hierarchy problem". Related to this point is the observation that supersymmetry stabilizes the mass of the scalar Higgs particle which otherwise would be unstable against radiative corrections.

These theoretical motivations have been persuasive enough so that experimental physicists started to look for experimental signals of supersymmetry. At present, searching for supersymmetric particles is already an essential part of high-energy physics [2.8–2.10] and will play an even more important role in the experimental program of the new accelerators [2.10–2.13]. The theoretical basis for this experimental search is provided by the supersymmetric extension of the Standard Model.

2.1 Supersymmetric Extension of the Standard Model

In $N = 1$ supersymmetry every spin-1/2 (matter) field of the Standard Model receives a spin-0 partner and every spin-1 (gauge field) obtains a spin-1/2 partner. To the isodoublet of spin-0 Higgs fields of the Standard Model one assigns an isodoublet of spin-1/2 fields as supersymmetric partners. As we shall show below, one needs at least two isodoublets of Higgs fields to make the Standard Model supersymmetric. The minimal particle content of the supersymmetric version of the Standard Model obtained in this manner is shown in Table 2.1 for one generation of matter fields.

In the superfield formalism one introduces chiral and vector superfields. A chiral superfield ϕ contains the component fields A, ψ, F where A is a complex scalar field, ψ is a two-component Weyl field and the auxiliary field F is also a scalar field. A vector superfield contains the components V^μ, λ, D where V^μ is a real vector field, λ is a two-component Weyl field and the scalar field D is again an auxiliary field (see e.g. [2.14]).

The supersymmetric Lagrangian for the gauge interaction of a chiral superfield with a vector superfield then reads in component form

Table 2.1. Fundamental particles of the Standard Model (one generation of matter fields), together with their supersymmetric partners

Particle	Spin	SUSY partner	Spin
Leptons ν_e, e	1/2	Sleptons $\tilde{\nu}_e$, $\tilde{e}_{L,R}$	0
Quarks u, d	1/2	Squarks $\tilde{u}_{L,R}$, $\tilde{d}_{L,R}$	0
Photon γ Weak Bosons Z^0, $W^\pm$	1	Photino $\tilde{\gamma}$ Zino $\tilde{Z}^0$ Wino $\tilde{W}^\pm$	1/2
Gluon g	1	Gluino $\tilde{g}$	1/2
Higgs particles H_1^0, H_2^0, H_3^0, $H^\pm$	0	Higgsinos $\tilde{H}_1^0, \tilde{H}_2^0$, $\tilde{H}^\pm$	1/2

$$\mathcal{L}_g = -\tfrac{1}{4} F^a_{\mu\nu} F^{a\mu\nu} + i\bar{\lambda}^a \bar{\sigma}^\mu D_\mu \lambda^a + i\bar{\chi}\bar{\sigma}^\mu D_\mu \chi + (D_\mu A)^\dagger (D^\mu A)$$
$$+ ig\sqrt{2}(A^\dagger \lambda^a T^a \chi - \bar{\chi}\bar{\lambda}^a T^{a\dagger} A) - V(A, A^\dagger)\,. \tag{2.1}$$

Here g is the gauge coupling constant, $F^a_{\mu\nu} = \partial_\mu V^a_\nu - \partial_\nu V^a_\mu + g f^{abc} V^b_\mu V^c_\nu$ is the field strength of the vector field V^a_μ, T^a and f^{abc} are the generator and the structure constant of the gauge group, respectively, and the covariant derivatives are defined as $D_\mu \lambda^a = \partial_\mu \lambda^a + g f^{abc} V^b_\mu \lambda^c$, $D_\mu \chi = \partial_\mu \chi - ig V^a_\mu T^a \chi$, etc. The potential $V(A, A^\dagger)$ contains self-interaction terms of the scalar field A stemming from the elimination of the auxiliary fields F and D^a. It will be explained below. For every gauge interaction and for every left and right chirality component of the matter fields of the Standard Model (see Eqs. (1.1–1.9)) we have to write down the appropriate Lagrangian whose structure is given by Eq. (2.1).

The Yukawa coupling terms of the Standard Model responsible for the mass terms of quarks and leptons, see Eq. (1.9), must also be made supersymmetric. This can be achieved by introducing the superpotential

$$W = h_E L E^c H_1 + h_D Q D^c H_1 + h_U Q U^c H_2 + \mu H_1 H_2, \tag{2.2}$$

L and Q denote the isodoublets of chiral superfields according to the left-handed lepton and quark fields, E, D and U are isosinglets of chiral superfields assigned to the right-handed charged lepton, down and up quark, respectively, c means charge conjugation, H_1 and H_2 are the isodoublets of chiral superfields for the two Higgs doublets. h_E, h_D and h_U are the Yukawa coupling constants. Remember that in the Standard Model one uses a term like $h_u \bar{q}_L u_R h^\dagger$ in order to create a mass term for the up quark. In the superpotential (2.2) one cannot use the hermitian adjoint of the superfield H_1, therefore one has to introduce (at least) one further superfield H_2 to give a mass also to the up quark and its superpartner [2.15].It is also necessary for a complete cancellation of anomalies.

Finally, the bilinear mixing term $\mu H_1 H_2$ in Eq. (2.2) of the two Higgs superfields with parameter μ is necessary for inducing $SU(2)_L \times U(1)$ breaking and to avoid an axion. As we shall see, it has also important consequences for phenomenology. As it stands, the superpotential (2.2) is given for one generation of quarks and leptons. For every family a copy of Eq. (2.2) has to be written down.

So far we have formulated a model which would be exactly supersymmetric. We have to break supersymmetry. Spontaneous breaking of global supersymmetry has been tried out in the past. This has led to models with some less satisfactory features (for a review, see e.g. [2.5, 2.16]). Models based on $N = 1$ supergravity spontaneously broken in the hidden sector turned out to be theoretically more attractive and phenomenologically more flexible.

To this end one has to rewrite the Lagrangians sketched in Eqs. (2.1) and (2.2) in a locally supersymmetric form and to add the supergravity Lagrangian. Spontaneous breaking of supergravity is then achieved by adding an auxiliary singlet chiral superfield whose scalar component receives a vacuum expectation value of the order of the Planck mass M_P, $M_P = 1/\sqrt{8\pi G_N} \approx 2.4 \cdot 10^{18}$ GeV, G_N being the gravitational constant. The spin-1/2 component of this auxiliary chiral superfield then becomes the helicity-1/2 component of the gravitino (super-Higgs effect). As a result of this procedure one obtains a Lagrangian which contains two pieces:

(I) the supersymmetric Lagrangian as derived above,

(II) additional soft supersymmetry breaking terms.

Among the latter there is

(i) a mass term for the gauginos, i.e. the partners of the intermediate vector bosons

$$\mathcal{L}_M = -\tfrac{1}{2}\,\tilde{M}(\lambda_a\lambda_a + \bar{\lambda}_a\bar{\lambda}_a), \tag{2.3}$$

where the sum runs over the gluinos, winos, zino and photino;

(ii) additional mass terms and interaction terms for the scalar fields.

These additonal terms can most economically be written with the help of the Kähler potential $G(A_i, A_i^\dagger)$, a real gauge invariant function of the scalar fields A_i [2.17, 2.18]. The potential $V(A_i, A_i^\dagger)$, already introduced in Eq. (2.1), contains a "chiral" part and a "gauge" part

$$V(A_i, A_i^\dagger) = V_c(A_i, A_i^\dagger) + V_g(A_i, A_i^\dagger) \tag{2.4}$$

which in terms of $G(A_i, A_i^\dagger)$ read

$$V_c = \exp\{G\}\cdot M_{\rm P}^4[G_i G^j (G_i{}^j)^{-1} - 3] \tag{2.5}$$

$$V_g = \tfrac{1}{2}\,D^a D^a \tag{2.6}$$

with the abbreviations

$$G_i = M_{\rm P}\frac{\partial G}{\partial A_i}, \qquad G^i = M_{\rm P}\frac{\partial G}{\partial A_i^\dagger}, \tag{2.7}$$

$$D^a = gM_{\rm P}G_i(T^a)^{ij}A_j. \tag{2.8}$$

In the following we shall confine ourselves to *canonical* supergravity models, i.e. those with $G_i^j = \delta_i{}^j$, or more explicitly

$$G(A_i, A_i^\dagger) = \frac{A_i A_i^\dagger}{M_{\rm P}^2} + \ln\left|\frac{f(A_i)}{M_{\rm P}^3}\right|^2, \tag{2.9}$$

where $f(A)$ is the part of the superpotential $W(\phi)$, Eq. (2.2), which contains the scalar fields A_i only[1]. We shall only mention here the class of no-scale models [2.16] which are non-canonical and specified by the Kähler potential

$$G(A_i, A_i^\dagger) = -3\ln\frac{f + f^*}{M_{\rm P}^3}. \tag{2.10}$$

[1] Sometimes the kinetic part of the Lagrangian of equation (2.1) is written in a more general form with the help of an analytic function $f_{ab}(A_i)$ of the scalar fields, e.g., $-\frac{1}{4}{\rm Re}f_{ab}(A_i)F^a_{\mu\nu}F^{b\mu\nu} + \frac{i}{4}{\rm Im}f_{ab}(A_i)F^a_{\mu\nu}\tilde{F}^{b\mu\nu}$. Canonical models are restricted to $f_{ab} = \delta_{ab}$.

Inserting (2.9) into (2.5) gives for the chiral part of the potential

$$V_c = \exp\left\{\frac{A_i A_i^\dagger}{M_P^2}\right\} \cdot \left[F_i F_i^\dagger - \frac{3|f|^2}{M_P^2}\right] \tag{2.11}$$

with

$$F_i = \frac{\partial f}{\partial A_i} + \frac{A_i^\dagger \cdot f}{M_P^2}. \tag{2.12}$$

Remember that among the scalar fields A_i there is one belonging to the auxiliary chiral superfield, introduced to spontaneously break supergravity in the hidden sector. Calling this particular scalar field A_0, its vacuum expectation value must be $\langle A_0 \rangle \propto M_P$, and we must also specify its appropriate superpotential, suitable for this breaking mechanism. One possibility is the Polonyi potential [2.19]

$$f_h(A_0) = m_s^2(A_0 + \beta), \tag{2.13}$$

where the mass parameter m_s sets the energy scale of supersymmetry breaking, and the parameter β is determined by requiring $\langle V_c \rangle = 0$ when inserted in Eq. (2.11) (vanishing of the cosmological constant to the order M_P). Such a kind of fine-tuning of parameters is necessary in this class of canonical supergravity models, whereas it is avoided in the no-scale models, Eq.(2.10).

The gravitino mass $m_{3/2}$ turns out to be given by $m_{3/2} = \langle \exp\{\frac{G}{2}\}\rangle$. Inserting (2.13) into (2.9) then leads to

$$m_{3/2} = \exp\{2 - \sqrt{3}\} \frac{m_s^2}{M_P}, \tag{2.14}$$

which is the Deser-Zumino relation [2.20], apart from a trivial numerical factor. This relation means that with a supersymmetry breaking scale as high as $m_s \approx 10^{10}$ GeV, the gravitino mass $m_{3/2}$ could be of the order of magnitude of the Z^0 mass.

As we are interested only in the low-energy limit ($E \ll M_P$) of the model we can replace the field A_0 by its vacuum expectation value in Eq. (2.11) and take the limit $M_P \to \infty$, keeping $m_{3/2}$ fixed. Thus we obtain

$$V_c = \sum_i \left|\frac{\partial f}{\partial A_i}\right|^2 + m_{3/2}^2 \sum_i |A_i|^2 + m_{3/2}(A \cdot f^{(3)} + B \cdot f^{(2)} + h.c.), \tag{2.15}$$

where the sum over i now includes all scalar fields of the "observable sector", $i \neq 0$, whose vacuum expectation values are either of the order of the weak interaction breaking scale or zero. $f^{(3)}$ and $f^{(2)}$ are the trilinear and bilinear parts of the superpotential f. The numerical constants A and B are of the order of 1, depending on the details of the supergravity breaking mechanism. For instance, when using the Polonyi potential (2.13) we would obtain $A = 3 - \sqrt{3}$ and $B = A - 1$. Furthermore, Eq. (2.15) shows that the mass splitting between the fermions of the standard model and their supersymmetric partners is also given by $m_{3/2}$.

What we have obtained now is a model based on spontaneously broken $N = 1$ supergravity, formulated near the Planck scale, which, in addition to those of the Standard Model, contains the quantities $m_{3/2}$, $\bar{M}$, μ, A, B as free parameters. Usually one assumes that things do not change when going down to the grand unification scale $M_x \approx 10^{16}$ GeV. The continuation down until we reach the scale of weak interaction breaking, characterized by $m_{\mathrm{W}} = 82$ GeV, is then performed with the help of renormalization group equations [2.21–2.24]. As a result of this procedure masses and coupling constants are changed by the radiative corrections which are different for every particle. At the scale m_{W}, the gaugino mass Lagrangian Eq. (2.3) becomes

$$\mathcal{L}_M = -\frac{1}{2}(M'\lambda'\lambda' + M\sum_{a=1}^{3}\lambda^a\lambda^a + M_{\tilde{g}}\sum_{b=1}^{8}\lambda^b\lambda^b) + h.c., \tag{2.16}$$

where λ', λ^a and λ^b denote the Weyl spinors for the U(1) and SU(2) gauginos and gluinos, respectively. M', M and $M_{\tilde{g}}$ are related by

$$\frac{M'}{M} = \frac{5}{3}\tan^2\Theta_{\mathrm{W}}, \qquad \frac{M}{M_{\tilde{g}}} = \frac{\alpha(m_{\mathrm{W}})}{\alpha_s(m_{\mathrm{W}})\sin^2\Theta_{\mathrm{W}}}, \qquad \frac{M_{\tilde{g}}}{\bar{M}} = \frac{\alpha_s(m_{\mathrm{W}})}{\alpha_{\mathrm{GUT}}} \tag{2.17}$$

(in the numerical examples discussed later we shall use $\sin^2\Theta_{\mathrm{W}} = 0.22$, $\alpha(m_{\mathrm{W}}) = 1/128$, $\alpha_s(m_{\mathrm{W}}) = 0.136$, $\alpha_{\mathrm{GUT}} = 1/24$).

Similarly, the Higgs potential following from Eqs. (2.4), (2.6) and (2.15) is

$$\begin{aligned} V_{\mathrm{H}} &= \mu_1^2|H_1|^2 + \mu_2^2|H_2^2| + B\mu m_{3/2}(H_1H_2 + h.c.) \\ &+ \frac{g^2}{8\cos^2\Theta_{\mathrm{W}}}(|H_2|^2 - |H_1|^2)^2. \end{aligned} \tag{2.18}$$

This has to be compared with the Higgs potential of the Standard Model, see Eq. (1.9), where a quartic term with undetermined coupling strength h appears. Instead of this we now have the last term in Eq. (2.18) with the weak coupling constant g. At the scale M_x we would have $\mu_1^2 = \mu_2^2 = \mu^2 + m_{3/2}^2$. When scaling down to energies near m_{W} it turns out that, in a certain range of the parameters $m_{3/2}$, $\bar{M}$, μ, A and B, the quantity μ_2^2 may become negative [2.21, 2.22, 2.25]. This sounds like a miracle, but is exactly what we need to also break the weak SU(2) × U(1) group to U(1). Spontaneous breaking of supergravity and the electro-weak symmetry group occur simultaneously. The reason why μ_2^2 may become negative is that via the superpotential (2.2) the Higgs particle H_2^0 also couples to the top quark, the top quark loop then providing a negative contribution to μ_2^2. The mass of the top quark must be large enough ($m_t \gtrsim 20$ GeV). For $m_t \lesssim 50$ GeV there are rather strong restrictions on the other parameters.

The minimum of the Higgs potential V_{H}, Eq. (2.18), then occurs at non-vanishing vacuum expectation values, $\frac{1}{\sqrt{2}}v_{1,2}$, of the neutral Higgs bosons H_1^0 and H_2^0. They are related to the W boson mass by $m_{\mathrm{W}}^2 = \frac{1}{2}(v_1^2 + v_2^2)$. Working out the masses of the Higgs particles $\mathrm{H}^{\pm}$, H_1^0, H_2^0, H_3^0, as following from Eq. (2.18), then yields [2.26]

$$m^2_{\mathrm{H}^\pm} = m^2_{\mathrm{W}} + m^2_{\mathrm{H}_3}, \tag{2.19}$$

$$m^2_{\mathrm{H}_1(\mathrm{H}_2)} = \frac{1}{2}\left[m^2_{\mathrm{H}_3} + m^2_{\mathrm{Z}} \underset{(-)}{+} \sqrt{(m^2_{\mathrm{H}_3} + m^2_{\mathrm{Z}})^2 - 4m^2_{\mathrm{Z}} m^2_{\mathrm{H}_3}\cos^2 2\beta}\right], \tag{2.20}$$

where

$$\tan\beta = \frac{v_2}{v_1}. \tag{2.21}$$

By fixing $\tan\beta$ and e.g. the mass of H^0_3, the masses of the other Higgs particles are also given. The following mass bounds are derived from Eqs. (2.19) and (2.20):

$$|m_{\mathrm{H}^\pm}| \geq m_{\mathrm{W}}, \qquad |m_{\mathrm{H}_1}| \geq m_{\mathrm{Z}}, \qquad |m_{\mathrm{H}_2}| \leq m_{\mathrm{Z}}|\cos 2\beta|. \tag{2.22}$$

No bound exists for m_{H_3}. Eq. (2.22) means that in this class of models there is at least one neutral Higgs particle lighter than the Z boson. In its couplings to other particles H^0_2 resembles to some extent the Standard Model Higgs.

For every family there are two scalar up quarks and two scalar down quarks, $\tilde{\mathrm{q}}_{\mathrm{L}}$ and $\tilde{\mathrm{q}}_{\mathrm{R}}$, two charged sleptons, $\tilde{\ell}_{\mathrm{L}}$ and $\tilde{\ell}_{\mathrm{R}}$, and one sneutrino $\tilde{\nu}$. Their masses follow from Eq. (2.15) and depend, through the renormalization group equations, on $m_{3/2}$, $\bar{M}$, μ, A and B. Written in a simplified version we have at the scale m_{W}:

$$m^2_{\tilde{\nu}} = m^2_{3/2} + 0.53\bar{M}^2 + 0.5\cos 2\beta m^2_{\mathrm{Z}}, \tag{2.23}$$

$$m^2_{\tilde{\mathrm{e}}_{\mathrm{L}}} = m^2_{3/2} + 0.15\bar{M}^2 - 0.28\cos 2\beta m^2_{\mathrm{Z}}, \tag{2.24}$$

$$m^2_{\tilde{\mathrm{e}}_{\mathrm{R}}} = m^2_{3/2} + 0.44\bar{M}^2 - 0.22\cos 2\beta m^2_{\mathrm{Z}}, \tag{2.25}$$

$$\begin{aligned} m^2_{\tilde{\mathrm{u}}_{\mathrm{L}}} &= m^2_{3/2} + 7.6\bar{M}^2 + 0.36\cos 2\beta m^2_{\mathrm{Z}}, \\ m^2_{\tilde{\mathrm{d}}_{\mathrm{L}}} &= m^2_{3/2} + 7.6\bar{M}^2 - 0.43\cos 2\beta m^2_{\mathrm{Z}}, \end{aligned} \tag{2.26}$$

$$\begin{aligned} m^2_{\tilde{\mathrm{u}}_{\mathrm{R}}} &= m^2_{3/2} + 7.1\bar{M}^2 + 0.14\cos 2\beta m^2_{\mathrm{Z}}, \\ m^2_{\tilde{\mathrm{d}}_{\mathrm{R}}} &= m^2_{3/2} + 7.1\bar{M}^2 - 0.07\cos 2\beta m^2_{\mathrm{Z}}. \end{aligned} \tag{2.27}$$

We shall now discuss the mass terms of the gauginos and higgsinos. They follow from Eq. (2.16), the last term of Eq. (2.2) and from Eq. (2.1) after inserting the vacuum expectation values $v_{1,2}$ for the neutral Higgses. In the charged sector it is convenient to introduce the basis [2.28]

$$\psi^+_j = (-i\lambda^+, \psi^1_{\mathrm{H}_2}), \qquad \psi^-_j = (-i\lambda^-, \psi^2_{\mathrm{H}_1}), \qquad j = 1,2, \tag{2.28}$$

with $\lambda^\pm$, $\psi^1_{\mathrm{H}_2}$ and $\psi^2_{\mathrm{H}_1}$ the Weyl spinors for the $\tilde{\mathrm{W}}^\pm$, $\tilde{\mathrm{H}}^+$ and $\tilde{\mathrm{H}}^-$, respectively. Then the mass terms of the charged gaugino-higgsino system can be brought to the form

$$\mathcal{L}_c = -\frac{1}{2}(\psi^+\psi^-)\begin{pmatrix} 0 & X^T \\ X & 0 \end{pmatrix}\begin{pmatrix} \psi^+ \\ \psi^- \end{pmatrix} + h.c. \qquad \text{with} \tag{2.29}$$

$$X = \begin{pmatrix} M & gv_2 \\ gv_1 & \mu \end{pmatrix}. \tag{2.30}$$

The mass matrix (2.29) can be diagonalized by two unitary 2×2 matrices U and V,

$$M_i \delta_{ij} = U^*_{im} V_{jn} X_{mn}, \tag{2.31}$$

the chargino mass eigenstates then being gaugino-higgsino mixtures:

$$\chi^+_i = V_{ij}\psi^+_j, \qquad \chi^-_i = U_{ij}\psi^-_j, \tag{2.32}$$

Similarly, in the neutral sector we choose the basis

$$\psi^0_k = (-i\lambda_\gamma, -i\lambda_Z, \cos\beta\psi^1_{H_1} - \sin\beta\psi^2_{H_2}, \sin\beta\psi^1_{H_1} + \cos\beta\psi^2_{H_2}), \tag{2.33}$$

λ_γ, λ_Z, $\psi^1_{H_1}$, and $\psi^2_{H_2}$ denoting the two-component spinor fields of the photino, zino, and the two neutral higgsinos, respectively. Then the neutral gaugino-higgsino mass matrix is [2.28]

$$\mathcal{L}_n = -\frac{1}{2}\psi^0_k Y_{k\ell}\psi^0_\ell + h.c. \qquad \text{with} \tag{2.34}$$

$$Y = \begin{pmatrix} M'c^2_W + Ms^2_W & (M - M')s_W c_W & 0 & 0 \\ (M - M')s_W c_W & M's^2_W + Mc^2_W & m_Z & 0 \\ 0 & m_Z & \mu\sin 2\beta & -\mu\cos 2\beta \\ 0 & 0 & -\mu\cos 2\beta & -\mu\sin 2\beta \end{pmatrix} \tag{2.35}$$

where we have used the abbreviations $s_W = \sin\Theta_W$ and $c_W = \cos\Theta_W$. This mass matrix can be diagonalized by a unitary 4×4 matrix N which we take to be real [2.29]:

$$\eta_k m_k \delta_{k\ell} = N_{km} N_{\ell n} Y_{mn}. \tag{2.36}$$

The neutralino eigenstates then are photino-zino-higgsino mixtures given by

$$\chi^0_k = N_{k\ell}\psi^0_\ell, \qquad k = 1, \ldots, 4. \tag{2.37}$$

The χ^0_k are ordered according to the absolute values of their masses, with χ^0_1 the lightest eigenstate. Some of the mass eigenvalues in (2.36) may be negative. We therefore write the eigenvalue in the form $\eta_k m_k$, with $m_k \geq 0$ and $\eta_k = \pm 1$ depending on the sign of the eigenvalue. η_k is then the CP quantum number of the eigenstate χ^0_k [2.30]. For convenience, we also choose the matrices U and V as real, Eq. (2.31), and write the chargino eigenvalues as $\eta_i M_i$, $M_i \geq 0$ and $\eta_i = \pm 1$ [2.31].

Whereas for the chargino mass eigenvalues analytic expressions can be derived [2.28], the diagonalization of the neutralino mass matrix has usually to be done numerically. There exist, however, several approximate analytic expressions for the neutralino mass eigenvalues, applicable in a wide range of parameter space, which allow for a quick estimate of the eigenvalues and eigenstates [2.32].

It is clear that the coupling strengths of the chargino and neutralino eigenstates to the other particles strongly depend on the matrices U, V and N. These are determined by the parameters entering the mass matrices X and Y, i.e. M', M, μ and $\tan\beta$.

2.2 Production and Decay of Supersymmetric Particles

In this section we shall discuss the most important phenomenological implications following from the supersymmetric extension of the Standard Model described previously.

Experimental lower mass bounds from e^+e^- machines exist for the charged supersymmetric particles. They can qualitatively be summarized by stating that their masses should be larger than approximately 25 GeV. Analyses of the CERN $p\bar{p}$ collider data provide lower bounds for the masses of squarks and gluinos, $m_{\tilde{q}} \geq 45$ GeV, $M_{\tilde{g}} \geq 53$ GeV, assuming that the photino is the lightest supersymmetric particle [2.9]. New data from the CERN $p\bar{p}$ collider and the Tevatron will soon give higher values for these mass bounds.

Closer inspection of the mass matrices (2.29) and (2.35) shows that one of the charginos may be lighter than the W boson, and that some of the neutralinos may have masses smaller than m_Z. Sleptons and squarks are expected to have higher masses, see Eqs. (2.23–2.27). The lightest neutralino χ_1^0 is assumed to be the lightest supersymmetric particle. The other supersymmetric particles will decay into the χ_1^0 and ordinary particles. The χ_1^0 is stable and only weakly interacting so that it will escape detection in a typical experiment, thus giving rise to missing energy. This is considered to be an important experimental signal for supersymmetry. Depending on its nature, i.e. whether it is more like a photino, zino, or higgsino, the χ_1^0 will to a large extent also determine the phenomenology of the other supesymmetric particles. The lightest supersymmetric particle may play an important role in cosmology as well. Under this assumption certain mass bounds for the neutral supersymmetric particles can also be derived within standard cosmology (see e.g. [2.7, 2.33]).

In order to illustrate the different possibilities of how supersymmetric particles would show up in an experiment we shall optimistically assume their masses are within the reach of SLC, LEP or HERA. Nature would be extremely favourable to us if we could produce and detect supersymmetric particles at the Z^0 resonance. The production of a pair of lighter charginos or neutralinos in Z^0 decay would be most promising:

$$Z^0 \to \chi_1^0\chi_1^0, \tag{2.38}$$

$$Z^0 \to \chi_1^0\chi_2^0,\ \chi_1^0\chi_3^0,\ \chi_2^0\chi_2^0, \tag{2.39}$$

$$Z^0 \to \chi^+\chi^-. \tag{2.40}$$

Reaction (2.38) would be invisible, whereas reactions (2.39) and (2.40) could be obseved by the decays of the charginos and neutralinos according to

$$\chi^\pm \to \ell^\pm + \nu + \chi_1^0, \qquad \chi^\pm \to q + \bar{q}' + \chi_1^0, \tag{2.41}$$

$$\begin{aligned} &\chi_i^0 \to \ell^+ + \ell^- + \chi_1^0, && \chi_i^0 \to q + \bar{q} + \chi_1^0, \\ &\chi_i^0 \to H_{2,3}^0 + \chi_1^0, && i = 2, 3. \end{aligned} \tag{2.42}$$

In order to see under which circumstances such a favourable situation can occur, we have to find out in which regions of supersymmetric parameter space we may

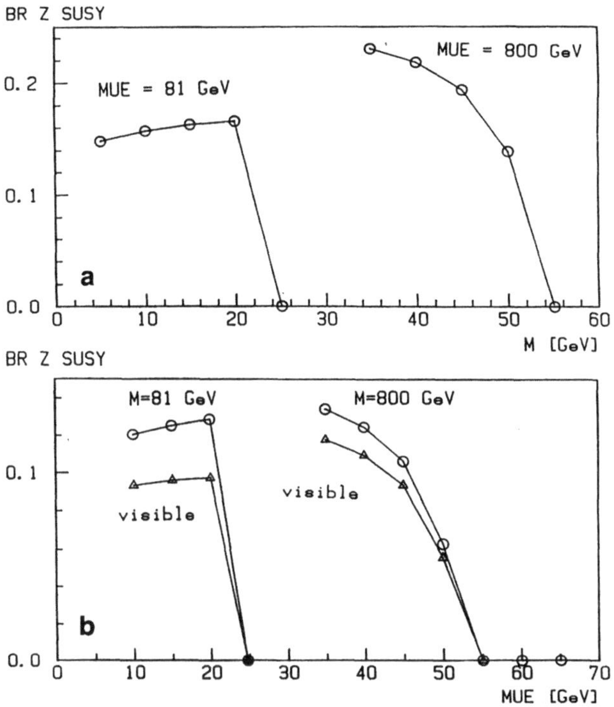

Fig. 2.1a,b. Sum of the branching ratios of Z^0 decays into chargino and neutralino pairs, Eqs. (2.38–2.40). For the non supersymmetric decays the Standard Model width $\Gamma_Z = 2.6$ GeV is taken. (**a**) As a function of the gaugino mass parameter M for $\mu = 81$ GeV ("light photino scenario") and $\mu = 800$ GeV. $\Gamma(Z^0 \to \chi_1^0\chi_1^0)$ is negligible. (**b**) As a function of supersymmetry parameter μ for $M = 81$ GeV ("light higgsino scenario") and $M = 800$ MeV. Also shown fraction of visible decays with at least one lepton, γ or jet in the final state

expect small neutralino masses, say, $m_1 < m_Z/2$, $m_2 \lesssim m_Z/2$. Examination of the mass matrix shows that there are three limiting cases leading to $m_1 \to 0$:

(I) $M', M \to 0$ (light photino scenario)

(II) $\mu \to 0$ (light higgsino scenario)

(III) $MM'\mu - (M' \cos^2 \Theta_W + M \sin^2 \Theta_W) m_Z^2 \sin 2\beta \to 0$ (light zino-higgsino scenario).

In addition, in case (I) one has to distinguish between $|\mu| < m_Z$, $|\mu| \approx m_Z$ and $|\mu| \gg m_Z$, and in case (II) between $M', M < m_Z$, $M', M \approx m_Z$ and $M', M \gg m_Z$. In case (III) the χ_1^0 is mainly a zino (higgsino) if $|\mu| > m_Z$ ($|\mu| < m_Z$).

In case (I), reaction (2.40) may give a sizable contribution to the Z^0 width, whereas the widths for the decays (2.38) and (2.40) are extremely small, because the Z^0 does not couple to the neutral gauginos. On the other hand, if case (II) is realized, there may be an appreciable decay width into neutralinos. This is illustrated in Figs. 2.1a and 2.1b which show the total and the visible supersymmetric decay branching ratios of the Z^0 in the cases (I) and (II). Here and in the following we shall always adopt $M'/M = \frac{5}{3} \tan^2 \Theta_W$.

For a discussion of the possible signatures of charginos and neutralinos we now have to consider their decays, Eqs. (2.41, 2.42). A good experimental signal would

Table 2.2. Partial widths and branching ratios for the decay modes of χ_2^0 for $M = -120$ GeV, $\mu = 20$ GeV, $m_{H_2} = 40$ GeV, $\tan\beta = 2$ and $\tan\beta = 4$, respectively, together with the masses of χ_1^0, χ_2^0, χ_2^+ (lighter chargino) and H_3^0 [2.34]. It is summed over all contributing lepton and quark flavours except the top

	$\tan\beta = 2,\ M = -120$ GeV $\mu = 20$ GeV		$\tan\beta = 4,\ M = -120$ GeV $\mu = 20$ GeV	
Particle	Mass (GeV)		Mass (GeV)	
$\tilde{\chi}_1^0$	17		12	
$\tilde{\chi}_h^0 = \tilde{\chi}_2^0$	68		65	
$\tilde{\chi}_2^+$	48		34	
$\tilde{\chi}_1^0$	40(input)		40(input)	
$\tilde{\chi}_1^0$	87		47	
Decay	Width (keV)	BR (%)	Width (keV)	BR (%)
$\tilde{\chi}_2^0 \to$				
$\nu\bar{\nu}\tilde{\chi}_1^0$	18.03	16.23	14.46	0.24
$\ell^+\ell^-\tilde{\chi}_1^0$	7.62	6.87	5.34	0.09
$u\,\bar{u}\tilde{\chi}_1^0$	19.54	17.6	14.98	0.25
$d\,\bar{d}\tilde{\chi}_1^0$	33.12	29.82	22.8	0.38
$H_2^0\tilde{\chi}_1^0$	23.25	20.94	1510	24.88
$H_3^0\tilde{\chi}_1^0$	–	–	4430	73.04
$\tilde{\chi}_2^+\ell^-\bar{\nu} + \tilde{\chi}_2^-\ell^+\nu$	3.48	3.12	25.08	0.41
$\tilde{\chi}_2^+ d\bar{u} + \tilde{\chi}_2^-\bar{d}u$	5.96	5.36	42.56	0.70
$3\tilde{\chi}_1^0$	0.07	0.06	0.63	0.01

be lepton(s) plus missing energy. As a rule, case (II) leads to the smallest leptonic branching ratios. We have illustrated this in Table 2.2 where all possible decay modes of the χ_2^0 are listed together with their branching ratios, for two sets of parameters [2.34]. One can see that in the light higgsino scenario the decay $\chi_2^0 \to H_{2,3}^0 + \chi_1^0$ dominates if it is kinematically allowed. This is further exemplified in Figs. 2.2a and 2.2b, which show how the branching ratios for the latter decay modes vary with the gaugino mass parameter M for two values of $\tan\beta$ and m_{H_2}. These decays are important in certain regions of supersymmetric parameter spaces only, which is the subject of [2.34].

At LEP II a CMS energy up to approximately 200 GeV will be available. Fig. 2.3 shows the cross section for the reaction [2.29]

$$e^+ + e^- \to \chi_1^0 + \chi_2^0 \tag{2.43}$$

as a function of the CMS energy for the four mixing cases specified in Table 2.3. These four examples were chosen in a way that they differ significantly among themselves by the nature of the lowest neutralino eigenstates χ_1^0 and χ_2^0, making

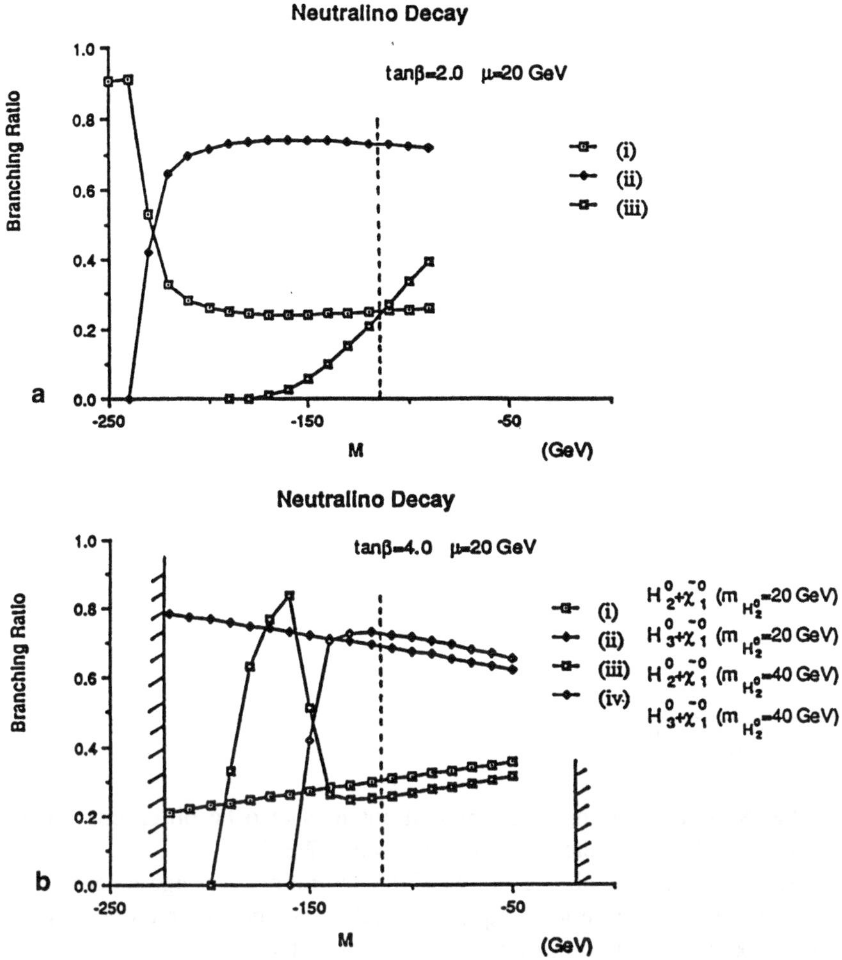

Fig. 2.2. Branching ratios for the decays of a higgsino-like neutralino $\chi_h^0 \to H_2^0\chi_1^0$ and $\chi_h^0 \to H_3^0\chi_1^0$ for $\mu = 20$ GeV as function of M [2.34], (a) for $\tan\beta = 2.0$, (b) for $\tan\beta = 4.0$. At $M = -110$ GeV a level crossing occurs: $\chi_h^0 = \chi_2^0$ for $M < -110$ GeV and $\chi_h^0 = \chi_3^0$ for $M > -110$ GeV. (i) $\chi_h^0 \to H_2^0\chi_1^0$, $m_{H_2} = 20$ GeV, (ii) $\chi_h^0 \to H_3^0\chi_1^0$, $m_{H_2} = 20$ GeV, (iii) $\chi_h^0 \to H_2^0\chi_1^0$, $m_{H_2} = 40$ GeV, (iv) $\chi_h^0 \to H_3^0\chi_1^0$, $m_{H_2} = 40$ GeV

use of the limiting cases (I), (II) and (III) given above. For the sake of comparison, however, we have tried to keep the mass values rather similar in all four cases.

Above the Z^0 resonance the t-channel selectron exchange contribution may become more important than the s-channel Z^0 exchange (in the calculation we have assumed $m_{\tilde{e}} = 40$ GeV, for $m_{\tilde{e}} = 80$ GeV the cross section for case (i) and (iv) would be reduced roughly by a factor 2 to 3). Also here the cross section strongly depends on neutralino mixing. Because there is no higgsino-lepton-slepton coupling, the cross section becomes very small above the Z^0 if one of the produced neutralinos is mainly a higgsino. We want to mention also that in neutralino pair production the forward-backward asymmetry vanishes due to the Majorana nature of the produced particles [2.35]. In neutralino production with polarized e^+ and e^- beams other con-

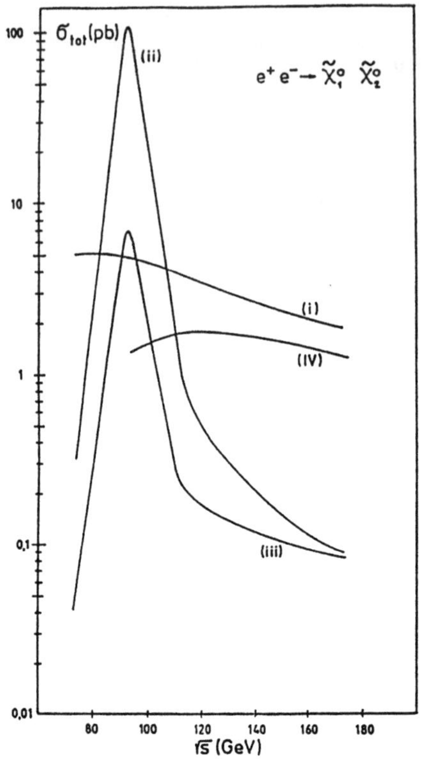

Fig. 2.3. Total cross section for $e^+ + e^- \to \chi_1^0 + \chi_2^0$ as a function of $\sqrt{s}$ in the mixing scenarios (*i*), (*ii*), (*iii*), (*iv*) of Table 2.3, with $m_{\tilde{e}} = 40$ GeV [2.29]

sequences of the Majorana nature of the neutralinos may also be observed, which differs from the behaviour expected for Dirac particles [2.36].

For a discussion of the possible signatures of the neutralinos it is again necessary to consider their various decay modes, Eq. (2.42). Table 2.4 shows which branching ratios one might expect for the mixing possibilities (i) ... (iv).
Here we have assumed that the Higgs particles are heavy so that the decay mode $\chi_2^0 \to H_{2,3}^0 \chi_1^0$ is absent. As one can see, case (ii) (χ_1^0 almost a higgsino, χ_2^0 almost a photino) and case (iii) (χ_1^0 almost a photino, χ_2^0 almost a higgsino) would give rather small leptonic branching ratios. Here, however, the decay $\chi_2^0 \to \chi_1^0 + \gamma$ (being of higher order in the weak coupling constant g) might lead to an interesting signature for the χ_2^0. It is obvious that the possible decay patterns of the χ_2^0 also strongly depend on whether $m_{\chi_2^0} < m_{\tilde{\nu}}$ or $m_{\chi_2^0} > m_{\tilde{\nu}}$ and whether $m_{\chi_2^0} < m_{\tilde{e}}$ or $m_{\chi_2^0} > m_{\tilde{e}}$.

If chargino pair production is kinematically allowed,

$$e^+ + e^- \to \chi^+ + \chi^- \ , \tag{2.44}$$

its cross section is usually an order of magnitude higher than for neutralino production, making this process particularly attractive [2.31]. Fig. 2.4 shows the total cross section for the two mixing possibilities (i) and (ii) of Table 2.3. As a consequence of the decay modes (2.41) chargino pair production would give rise to "two-sided events" with leptons and/or jets in the final states. On the other hand, "one-sided events" would be characteristic for the neutralino production process (2.43). Addi-

Table 2.3. Masses m_i and M_i of the neutralinos and charginos, together with the diagonalizing matrices N_{ki}, V_{ki} and U_{ki}, for the four mixing cases (i–iv) [2.29]. η_i are the signs of the mass eigenvalues. M, μ and $\tan\beta$ are the parameters of the gaugino-higgsino mass matrices, $\tan\beta = 1.1$ in cases (i), (ii) and (iv), $\tan\beta = 1.25$ in case (iii). $m_W = 81$ GeV and a Weinberg angle $\sin^2\Theta_W = 0.22$ are used. The basis used in the neutral sector is given in Eq. (2.33)

Scenario	(i)	(ii)	(iii)	(iv)
M, μ (GeV)	16.2, 81	81, 12.2	24.3, −48.6	97.2, 145.8
m_1, η_1	9.7 GeV, +1	12.1 GeV , −1	14.1 GeV, +1	12.7 GeV, +1
m_2, η_2	49.5 GeV, −1	46.6 GeV , +1	46.9 GeV, +1	63 GeV, +1
m_3, η_3	81.5 GeV, −1	55.7 GeV , −1	86.3 GeV, +1	145.9 GeV,−1
m_4, η_4	145.2 GeV, +1	140.5 GeV , +1	111.5 GeV, −1	213.3 GeV,+1
N_{11}, N_{12}	.999, .032	.005, −.018	.998, −.039	−.366, .766
N_{13}, N_{14}	−.040, −.004	.015, 1	−.055, −.018	−.526, −.051
N_{21}, N_{22}	−.049, .811	.983, −.065	.026, .158	.927, .251
N_{23}, N_{24}	−.562, −.154	−.172, −.004	.042, .986	−.277, −.020
N_{31}, N_{32}	−.004, .109	−.101, .591	−.060, −.807	−.003, .024
N_{33}, N_{34}	−.113, .988	−.800, .023	−.566, .156	−.060, .998
N_{41}, N_{42}	−.015, −.574	−.154, −.804	−.024, .567	−.081, −.591
N_{43}, N_{44}	−.818, −.031	−.574, −.005	−.821, −.055	−.802, −.034
M_1, η_1	135.9 GeV, +1	134.7 GeV, +1	103.5 GeV, +1	206.1 GeV,+1
M_2, η_2	38.4 GeV, +1	41.1 GeV, +1	73.3 GeV, +1	37.1 GeV,−1
$V_{11} = -V_{22}$	.524	.808	−.253	.583
$V_{12} = V_{21}$	.852	.589	.968	.813
$U_{11} = U_{22}$	.596	.858	.777	.611
$U_{12} = -U_{21}$	.803	.512	−.629	.792

tional information about neutralino and chargino pair production in e^+e^- collisions can be found in [2.37–2.40].

At the e-p collider HERA the most promising reaction for producing supersymmetric particles would be

$$e + p \to \tilde{e} + \overset{(-)}{\tilde{q}} + X, \tag{2.45}$$

$$e + p \to \tilde{\nu} + \overset{(-)}{\tilde{q}} + X, \tag{2.46}$$

The cross sections to be expected for these reactions have been calculated in [2.41–2.43]. Figs. 2.5 and 2.6 exhibit the predictions for the total cross sections of (2.45) and (2.46) at $\sqrt{s} = 314$ GeV, the nominal CMS energy of HERA, for various slepton and squark masses, for the mixing possibilities (i), (ii), (iv), and (i), (ii) of Table 2.3, respectively [2.42, 2.43]. Scenario (i) leads to the smallest cross sections. This result is somewhat surprising but can be explained by a destructive interference between the t-channel exchange contributions of the partonic reactions $e + q \to \tilde{e}(\tilde{\nu}) + \tilde{q}$.

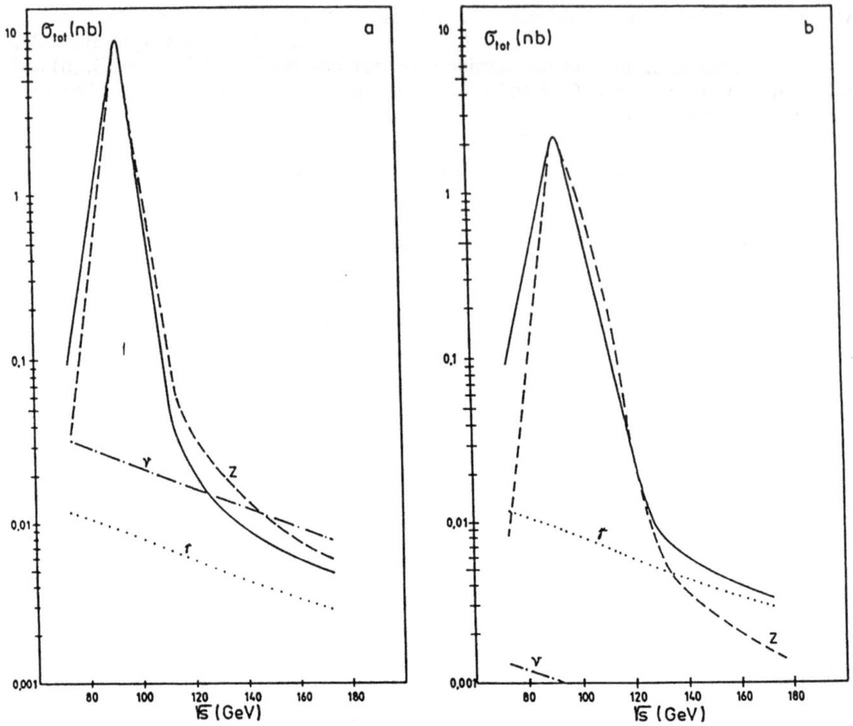

Fig. 2.4. Total cross section (*full line*) for $e^+e^- \rightarrow \chi^+ + \chi^-$ as a function of $\sqrt{s}$, together with the contributions of Z^0 exchange (*dashed line*), γ exchange (dotted line), and $\tilde{\nu}$ exchange (*dashed-dotted line*), with $m_{\tilde{\nu}}$ = 25 GeV [2.31], (**a**) for mixing case (i), "light photino scenario", (**b**) for mixing case (ii), "light higgsino scenario"

For a discussion of the possible experimental signals of reactions (2.45) and (2.46), it is again necessary to carefully treat all important decays of the produced supersymmetric particles. The selectron can decay according to

$$\tilde{e}^{\pm}_{L,R} \rightarrow e^{\pm} + \chi^0_k, \tag{2.47}$$

$$\tilde{e}^{\pm}_{L} \rightarrow \overset{(-)}{\tilde{\nu}} + \chi^{\pm}. \tag{2.48}$$

Usually it is assumed that the most important selectron decay mode is the one into the lightest neutralino, $\tilde{e}^{\pm} \rightarrow e^{\pm} + \chi^0_1$. If phase space allows, also the decays into higher neutralinos and charginos may occur, Eqs. (2.47) and (2.48). These charginos and higher neutralinos will subsequently undergo transitions to lighter neutralinos or charginos until the lightest neutralino χ^0_1 is reached (see Eqs. (2.41) and (2.42)). These cascade decays would lead to events with several leptons and/or jets plus missing energy in the lepton hemisphere of reaction (2.45).

A good experimental signature for selectron production would be electron plus missing energy. Table 2.5 shows the branching ratios for this decay mode as a function of selectron mass for the mixing cases (i), (ii) and (iv). With increasing selectron mass the branching ratio for $\tilde{e}^{\pm}_L \rightarrow e^{\pm} + \chi^0_1$ may become as low as 9%,

Table 2.4. Partial widths for the various decay channels of χ_2^0 [2.29] in the four mixing cases (i–iv) of Table 2.3. Total width and leptonic branching ratio (for one lepton species) are shown for the four different cases, (a) $m_{\chi_2^0} > m_{\tilde{\nu}}, m_{\chi_2^0} > m_{\tilde{e}}$, (b) $m_{\chi_2^0} < m_{\tilde{\nu}}, m_{\chi_2^0} > m_{\tilde{e}}$, (c) $m_{\chi_2^0} > m_{\tilde{\nu}}, m_{\chi_2^0} < m_{\tilde{e}}$, (d) $m_{\chi_2^0} < m_{\tilde{\nu}}, m_{\chi_2^0} < m_{\tilde{e}}$. Hadronic widths are summed over all contributing quark flavours. For $\Gamma(\chi_2^0 \to \nu\tilde{\nu})$ and $\Gamma(\chi_2^0 \to \ell\tilde{\ell})$ $m_{\tilde{\nu}} = 25$ GeV and $m_{\tilde{\ell}} = 40$ GeV has been used, for the other partial widths $m_{\tilde{\nu}} = m_{\tilde{\ell}} = 80$ GeV and $m_{\tilde{q}} = 100$ GeV has been taken

Scenario		(i)	(ii)	(iii)	(iv)
$\Gamma(\tilde{\chi}_2^0 \to \nu\,\bar{\tilde{\nu}} + \bar{\nu}\tilde{\nu})$		95 MeV	0.60 MeV	3.2 MeV	14 MeV
$\Gamma(\tilde{\chi}_2^0 \to \ell^+\tilde{\ell}_L^- + \ell^+\tilde{\ell}_R^- + \ell^-\tilde{\ell}_L^+ + \ell^-\tilde{\ell}_R^+)$		10 MeV	22 MeV	0.27 MeV	150 MeV
$\Gamma(\tilde{\chi}_2^0 \to \tilde{\chi}_1^0\ell^+\ell^-)$		1.0 keV	13 eV	32 eV	6.3 keV
$\Gamma(\tilde{\chi}_2^0 \to \tilde{\chi}_1^0 q\bar{q})$		1.4 keV	0.21 keV	0.13 keV	9.6 keV
$\Gamma(\tilde{\chi}_2^0 \to \tilde{\chi}_1^0\nu\bar{\nu})$		negligible	25 eV	63 eV	0.57 keV
$\Gamma(\tilde{\chi}_2^0 \to \tilde{\chi}_2^+\ell^-\nu + \tilde{\chi}_2^-\ell^+\bar{\nu})$		negligible	2.6 eV	–	negligible
$\Gamma(\tilde{\chi}_2^0 \to \tilde{\chi}_2^+ q\bar{q}')$		negligible	11 eV	–	negligible
$\Gamma(\tilde{\chi}_2^0 \to \tilde{\chi}_1^0\gamma)$		negligible	0.17 keV	0.16 keV	negligible
case (a)	Γ_{tot}	315 MeV	68 MeV	10 MeV	492 MeV
	B_ℓ	3.2 %	32 %	2.7 %	30 %
case (b)	Γ_{tot}	30 MeV	66 MeV	0.8 MeV	450 MeV
	B_ℓ	33 %	33 %	33 %	33 %
case (c)	Γ_{tot}	285 MeV	1.8 MeV	9.6 MeV	42 MeV
	B_ℓ	$3.5 \cdot 10^{-6}$	$9 \cdot 10^{-6}$	$3 \cdot 10^{-6}$	$1.5 \cdot 10^{-4}$
case (d)	Γ_{tot}	4.4 keV	0.5 keV	0.58 keV	30 keV
	B_ℓ	22 %	2.6 %	5.5 %	21.5 %

indicating the opening of new decay channels. The branching ratios for the most important other decay modes of the $\tilde{e}_L$ and $\tilde{e}_R$ are shown in Tables 2.6a and 2.6b for a selectron mass of 80 GeV. Combining the results for the cross sections, Fig. 2.5, with the branching ratios for $\tilde{e}^{\pm}_{L,R} \to e + p_{miss}$ in Table 2.5, one can conclude that selectron and squark production could be observed at HERA if $m_{\tilde{e}} + m_{\tilde{q}} \lesssim 160$ GeV.

In the mass range accessible at HERA the first steps in the decay chain of the sneutrino are

$$\tilde{\nu} \to \nu + \chi_k^0, \tag{2.49}$$

$$\tilde{\nu} \to e^- + \chi^+. \tag{2.50}$$

Subsequent decays of the higher neutralinos and the chargino would again lead to cascade decays, which would have as a result leptons and/or jets plus missing energy in the final state. A good experimental signature for the sneutrino would be $e^- + e^+ + p_{miss}$ or $e^- + \mu^+ + p_{miss}$. The branching ratios for these decay modes depend on the mass of the sneutrino and on neutralino mixing, with numerical values around 10 per cent (for more details see [2.43]). Here one can conclude that sneutrinos and

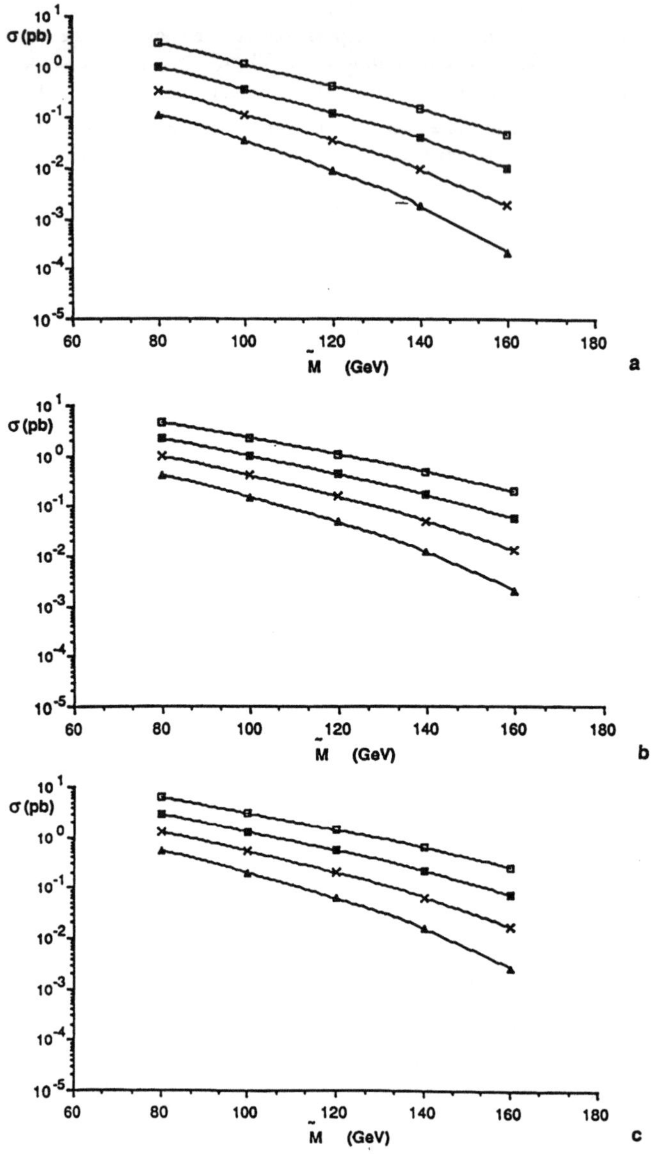

Fig. 2.5a–c. Total cross section for $e^- + p \to \tilde{e}^- + \overset{(-)}{\tilde{q}} + X$ at $\sqrt{s} = 314$ GeV, as a function of squark mass $\tilde{M}$ for different selectron masses $\tilde{m}$, summed over left and right selectron and squark final states [2.42]. □ $\tilde{m}$ = 40 GeV, ■ $\tilde{m}$ = 60 GeV, × $\tilde{m}$ = 80 GeV, ▲ $\tilde{m}$ = 100 GeV; (**a**) for mixing case (i), "light photino scenario", (**b**) for mixing case (ii), "light higgsino scenario", (**c**) for mixing case (iv), "light zino-higgsino scenario"

squarks could be detected at HERA if the sum of their masses is smaller than roughly 140 GeV.

Recently, the idea of constructing a linear e^+e^- collider with a CMS energy of 2 TeV has been proposed: NLC in the US and CLIC in Europe. With such a machine supersymmetric particles with masses up to almost 1 TeV could be produced. In

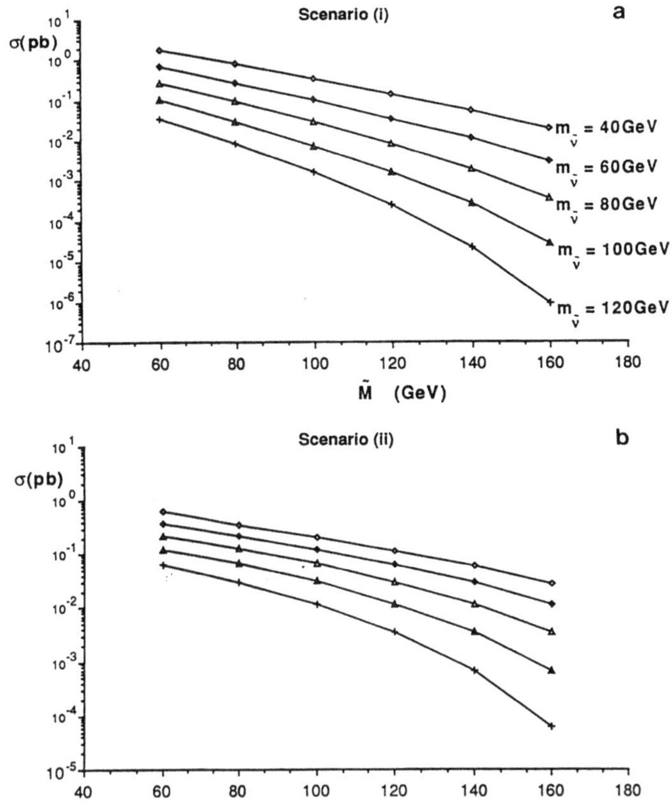

Fig. 2.6a,b. Total cross section for $e^- + p \to \tilde{\nu} + \tilde{q} + X$ at $\sqrt{s} = 314$ GeV, as a function of squark mass $\tilde{M}$ for different sneutrino masses $m_{\tilde{\nu}}$ [2.43]. ◊ $m_{\tilde{\nu}} = 40$ GeV, ♦ $m_{\tilde{\nu}} = 60$ GeV, △ $m_{\tilde{\nu}} = 80$ GeV, ▲ $m_{\tilde{\nu}} = 100$ GeV, + $m_{\tilde{\nu}} = 120$ GeV, (**a**) for mixing case (i), "light photino scenario", (**b**) for mixing case (ii), "light higgsino scenario"

Table 2.5. Branching ratios (in per cent) for $\tilde{e}_L \to e + p_{miss}$ and $\tilde{e}_R \to e + p_{miss}$ [2.42], as a function of selectron mass (in GeV), for the mixing cases (i), (ii) and (iv) of Table 2.3

	$B(\tilde{e}_L \to e + p_{miss})$			$B(\tilde{e}_R \to e + p_{miss})$		
$\tilde{m}$	(i)	(ii)	(iv)	(i)	(ii)	(iv)
40	99	100	44	100	100	100
60	72	13	12	97	13	100
80	58	9	10	91	12	86
100	52	13	9	87	11	72

order to judge the prospects of finding supersymmetry at such a supercollider, a detailed study of the signals resulting from the production and subsequent decays of sleptons, sneutrinos, charginos and neutralinos in this mass range has recently been performed [2.44]. We shall shortly sketch the most important results of this paper.

Table 2.6. Branching ratios (in per cent) for additional selectron decay modes in the mixing cases (i), (ii) and (iv), for a selectron mass of 80 GeV [2.42]

(a) for $\tilde{e}_L$

Decay mode	(i)	(ii)	(iv)
$\tilde{e}_L^- \to \ell^- + p_{miss}$ *)	6	6	7
$\tilde{e}_L^- \to e^- + \ell^+ + \ell'^- + p_{miss}$ **)	3	5	11
$\tilde{e}_L^- \to$ jets $+ p_{miss}$	26	30	59
$\tilde{e}_L^- \to e^- +$ jets $+ p_{miss}$	1	25	6
$\tilde{e}_L^- \to e^- + \ell +$ jets $+ p_{miss}$ **)	0	1	0
$\tilde{e}_L^- \to e^- + \gamma + p_{miss}$	0	18	0

(b) for $\tilde{e}_R$

Decay mode	(i)	(ii)	(iv)
$\tilde{e}_R^- \to e^- + \ell^+ + \ell'^- + p_{miss}$ **)	6	9	9
$\tilde{e}_R^- \to e^- +$ jets $+ p_{miss}$	3	45	5
$\tilde{e}_R^- \to e^- + \ell +$ jets $+ p_{miss}$ **)	0	2	0
$\tilde{e}_R^- \to e^- + \gamma + p_{miss}$	0	32	0

*) $\ell = \mu$ or τ, not summed over.
**) $\ell, \ell' = e$, μ or τ, summed over all lepton families and charge modes.

The cross sections for pair production of selectrons and electron sneutrinos at $\sqrt{s}$ = 2 TeV vary from 20 to 200 fb and from 40 to 400 fb, depending on the mass of the produced particles. They are quite insensitive to neutralino mixing. The chargino pair production cross section is also rather independent of the model parameters, being in the range of 10 to 50 fb. Neutralino pair production depends quite strongly on mixing, being of the order of 1 fb for each individual channel. This means that a luminosity of at least 10^{33} cm^{-2}s^{-1} would be necessary to obtain reasonable counting rates.

In the decay of the selectrons and sneutrinos the first steps are again the transitions (2.47–2.50). If the lighter neutralinos are mainly higgsinos, then the decays into higher neutralinos and the heavier chargino will dominate. Because these particles will also have higher masses, they will undergo transitions of the type of

$$\chi_k^0 \to Z^0 + \chi_n^0, \qquad \chi_k^0 \to W^\pm + \chi_n^\mp, \qquad \chi_k^0 \to H_{2,3}^0 + \chi_n^0, \tag{2.51}$$

and

$$\chi_1^\pm \to Z^0 + \chi_2^\pm, \qquad \chi_k^\pm \to W^\pm + \chi_n^0, \qquad \chi_1^\pm \to H_{2,3}^0 + \chi_2^\pm, \tag{2.52}$$

where $\chi_1^\pm$ is the heavier and $\chi_2^\pm$ the lighter chargino.

If the lighter neutralinos are mainly gauginos, the lighter neutralinos and the lighter chargino will predominantly be produced in the transitions (2.47–2.50). These particles then decay either according to the two-body modes (2.51) and (2.52), or according to the three-body modes (2.41) and (2.42). This will in any case lead to events with leptons, jets, $W^\pm$'s, Z^0's, $H_{2,3}^0$'s and missing energy. It depends on the details of gaugino-higgsino mixing and on the phase space available in each single decay transition, whether there are more events with multileptons and missing energy or with multi-$W^\pm$'s and Z^0's. For more information we refer to [2.44], where a detailed numerical study of the expected cross sections, decay probabilities and event topologies and rates is given for various mixing cases and mass ratios.

3. Beyond the Standard Model: Composite Models

3.1 Introduction

A review on the status of Composite Models has nowadays a fixed place at any conference or workshop which deals with elementary particle phenomenology. Correspondingly, a large number of reports can be found in the literature [3.1] and we will mention the essential ideas only briefly.

The idea that the standard model of strong and electroweak interactions itself is an effective theory of another, more fundamental interaction and quarks, leptons and bosons are composites of more fundamental fields is almost as old as the standard model itself. Since quarks carry flavor and color from a purely spectroscopical point of view, one can think of them as being glued together from entities having flavor only ("flavons") and others having color only ("chromons"). To get a similar picture for leptons, Pati and Salam [3.2], who invented this idea as early as 1974, viewed the lepton number as a fourth color. A new QCD-like gauge interaction ("metacolor", "hypercolor") binds the constituents (preons) together and its residual Van-der-Waals-type interactions are the interactions of the standard model.

But there are other reasons of taking the existence of strong fundamental gauge interactions seriously. They come from the – up to now – unexplored Higgs-sector of the standard model. Actually, the only job of the Higgs doublet in the standard model is to provide symmetry breaking. This requires a certain form of the Higgs potential, so the ground state of the system is no longer invariant under $SU(2)_L \times U(1)$. Although there is nothing wrong with this mechanism, we have not observed such a pattern for elementary boson fields, but what we have observed is spontaneous symmetry breaking in QCD, where the global chiral symmetry of the quarks is broken by condensates $\langle \bar{q}_L q_R \rangle \neq 0$ as effect of the strong color force. In fact, technicolor theories [3.3] have been invented only for this purpose.

Once familiar with the idea, one will address the whole flavor problem to composite models. In fact, any underlying strong interaction theory must yield the Yukawa couplings, the condensate values and therefore the masses of fermions as a result rather than as fundamental parameters. It is therefore a challenging task for any model builder to provide at least a gross understanding of the mass hierarchies, as well as to give an argument how many families we should expect.

Some composite models [3.4, 3.5] also address problems usually associated with Grand Unified Theories, like the existence and incorporation of additional baryon-number (or even B-L)-violating interactions. Here the experimental evidence is the existence of the baryon asymmetry in the universe, a measurement to be taken seriously by any particle physicist. This, however, means having the compositeness scale very high $\Lambda > 10^{11}$ GeV, which usually makes it difficult but not impossible [3.6] to make low energy predictions.

3.2 The Size of Composite Quarks

If a particle is composite, its most important parameter is its size. The inverse size of a particle shows up as a scale parameter in all couplings with dimension, like magnetic moments. The easiest way to define the size of a fermion [3.4] is by the strong four-fermi-interaction

$$\mathcal{L}_{\mathrm{ST}} = \frac{k^2}{M^2} \cdot \bar{\psi}_\alpha \psi_\alpha \cdot \bar{\psi}_\alpha \psi_\alpha \, . \tag{3.1}$$

For a strong interaction, we expect $\frac{k^2}{4\pi} \sim \mathcal{O}(1)$ and define

$$\Lambda_\alpha^2 = \frac{4\pi}{k^2} \cdot M^2 \, . \tag{3.2}$$

For instance, from low-energy proton-proton scattering, we find for the nucleon

$$\Lambda_{\mathrm{p}} \sim m_\rho \sim 700 \,\mathrm{MeV} \, . \tag{3.3}$$

In a preonic theory, all the inverse sizes of the composites should be related to a fundamental scale Λ. This scale is usually defined by the value of the condensates

$$\langle \bar{f}_L f_R \rangle \simeq \Lambda^3 \tag{3.4}$$

If we assume that the composite fermion ψ_α acquires its mass through dynamical breaking of chiral symmetry, and if in addition the fermion f is a constituent of ψ, we find the relation [3.7]

$$\Lambda_\alpha = \frac{1}{r_\alpha} = k \left(\frac{\Lambda^3}{m_\alpha} \right)^{1/2} \, . \tag{3.5}$$

This relation is well known for QCD, where $\Lambda = \Lambda_{QCD} \simeq 250$ MeV, $m_\alpha \equiv m_p \simeq 1$ GeV, $k \sim g_{QCD} = g_{\rho NN} = 5.5$. We find $\Lambda_p \sim 700$ MeV in agreement with (3.3).

Now let us impose Eq. (3.5) on a preonic-type theory, yielding composite quarks. In this case the condensate in (3.4) is the one relevant for $SU(2)_L \times U(1)$-breaking, ψ_α is a quark, and f is one of its constituents. Since, however, the condensates as well as the masses form a matrix in flavor space, we can only give a statement about the heaviest eigenvalue. This yields,

$$(\Lambda_q)_H = k\left[\Lambda_{HC}^3/m(q_H)\right]^{1/2} . \tag{3.6}$$

Assuming $m(q_H) = m_{top} > 40$ GeV, $k^2 < 10$, $\Lambda_{HC} \sim (G_F)^{1/2} \sim 300$ GeV from electroweak symmetry breaking, we arrive at $(\Lambda_q)_H < 3$ TeV. If there exists a fourth family, with $m_{top} \sim 350$ GeV, one finds $(\Lambda_q)_H \lesssim 1$ TeV. We would like to stress the point that this result is quite general, as long as the *same* constituents are used making up the quark as well as the condensate.

Based on this reasoning, a series of tests have been proposed [3.8] for the third (heaviest) family of quarks and leptons assuming an inverse size $\Lambda_\tau \sim 1$ TeV. One has to consider processes inside the third family only, and it turns out that especially toponium physics is a valuable research ground. In particular, one should look for

(i) toponium $(t\bar{t}) \to \tau^+\tau^-$, $b\bar{b}$
(ii) open top decay $t \to be^+\nu$
(iii) Structure of the τ-lepton

For LEP or LEP200, the effects are expected to be fairly large, particularly due to the high top mass. The structure (formfactor) of τ could be seen at SLC as a $\sim 4\%$ effect. Contrary to this, the inverse sizes of the light quarks and leptons are known to be small. The best compilation of data can be found in Ref. [3.9]. One generally can say

$$\Lambda_e, \Lambda_\mu > 150 \text{ TeV} . \tag{3.7}$$

This means that either the light quarks do not share the constituents whose condensates are responsible for $SU(2)_L \times U(1)$ breaking, or they are not composite at all. This makes an understanding of the replication of families to a central issue of composite models [3.10].

3.3 Quantum Preon Dynamics

If we take the more radical point of view and assume *all* particles are composite, even gluons, photons and even the constituents discussed in the previous chapter, the compositeness scale Λ is presumably very high, $\Lambda = \Lambda_M \gtrsim 10^{11}$ GeV. Such a theory must have a spectrum of massless composites, and therefore a residual chiral symmetry to protect fermion masses and a residual gauge symmetry to prevent vector meson masses. Let us outline a few general theorems which are important in such a case.

Assume we start with a preonic theory (QPD) based on fundamental fermions (with or without fundamental bosons), which has an invariance group $G_M(\text{local}) \times G_{fc}$

(global or local). At a certain scale Λ_M the local G_M interactions become strong and form bound states (composites), which are G_M singlets. In addition, part of G_{fc} is spontaneously broken by condensates $\langle\bar{f}f\rangle_0 \neq 0$, yielding an effective low energy symmetry group $G_{LE} \subset G_{fc}$ for the composites. This effective symmetry group is local, if the composite vector mesons carrying the quantum numbers of G_{LE} are born massless at Λ_M, otherwise it is global. Can G_{LE} contain the familiar $SU(3)_C \times SU(2)_L \times U(1)_Y$ gauge group?

Before going to certain details, we should remark the following: Certainly any effective theory among composite objects is nonrenormalizable, because one can write down arbitrarily high dimensional operators in terms of the fundamental fields. However, in terms of composite fields, d-dimensional operators are suppressed by a scale factor $\Lambda_M^{-(d-4)}$ $(d > 4)$, and their contribution is of order $(m/\Lambda_M)^{d-4}$, where m is the mass of the composites. Therefore, if we restrict ourselves to the mass zero composites (i.e. $m/\Lambda_M \ll 1$), the remaining low-dimensional operators have to be renormalizable and therefore are assumed to be a gauge theory (with certain Yukawa couplings).

One has further to distinguish whether G_M is vectorlike or chiral [3.11]. In the latter case, the preons have to be in a complex representation R alone without being accompanied by another set of preons belonging to $\bar{R}$. For vectorlike theories the following constraints are given for the spectrum of composites:

(i) The Weingarten [3.12], Nussinov [3.13], and Witten [3.14] mass inequalities require, among other things, that the lightest composite fermion be heavier than the light composite boson:

$$m_F > \frac{N}{2} m_B \tag{3.8}$$

(assuming G_M = SU(N));

(ii) The Weinberg-Witten theorem [3.15] shows that in a vectorlike theory with a Lorentz-covariant conserved current (energy-momentum tensor), no composite particle with spin $j > 1/2$ ($j > 1$) with mass zero and corresponding charge unequal zero can be formed.

(iii) The Vafa-Witten theorem [3.16] shows that vectorlike *global* symmetries U_{L+R} (N) are not broken by condensates, whereas

(iv) chiral symmetries $U_L(N) \times U_R(N)$ are broken. This has been proved by Weingarten [3.12] using lattice field theory calculations.

(v) Furthermore, even in chiral QPD, one has to fulfill 't Hooft's anomaly-matching conditions [3.17] which state that the Adler-Bell-Jackiw anomalies [3.18] associated with background gauge fields of G_{fc} have to be the same at the preon as at the composite level. (It should be noted that G_M, as a gauge theory, has to be anomaly-free anyway in order to be consistent.) This puts certain severe restrictions on the preon representation of G_M in chiral QPD. One example of a preon theory of the latter type is given in Ref. [3.19].

We see that the mass inequalities [3.12–3.14] and the Vafa-Witten constraint prohibit the appearance of a residual $SU(2)_L$ protecting the masses of composite quarks and leptons (without having lighter composite bosons) in a vectorlike QPD.

Fortunately, these constraints can be avoided in a *supersymmetric* QPD for the following reasons:

(i) Both the Weingarten inequality and the Vafa-Witten theorem use the positivity of the fermion determinant in QCD-type theories in their proof. However, in supersymmetric QCD-theories, there exist Yukawa-interactions between gauginos, matter fermions and spin-zero matter bosons which destroy the positivity [3.20].

(ii) The proof of the breakdown of chiral symmetry in lattice QCD also ignores the presence of Yukawa interactions.

(iii) In a local supersymmetric Yang-Mills theory there is no Lorentz-covariant conserved current and energy-momentum tensor.

(iv) The anomaly matching conditions are easily fulfilled in a rather economical way, if quarks and leptons are constructed as fermion-boson composites (as is the case in supersymmetric theories) [3.21].

The necessity for supersymmetry at the preon level opens two new aspects. First, one is led to the intriguing possibility that the supersymmetric QPD itself is the result of a supergravity or superstring theory at a higher scale, presumably the Planck scale. Secondly, one has to answer how the supersymmetry is broken. The most promising way is to assume the breakdown of local supersymmetry through gaugino condensates [3.22] formed by the G_M-gauge force itself at Λ_M. This in turn implies the appearance of a vacuum expectation value for the auxiliary fields and therefore a breakdown of global supersymmetry. There exists however, certain restrictions on global SUSY-breaking, depending on the representation and mass of your fundamental preon fields [3.23]. For such cases, one finds that the condensate values have to be damped by powers of Λ_M/M_{Pl}:

$$\langle\lambda_M\lambda_M\rangle_0 \simeq \Lambda_M^3 \cdot \left(\Lambda_M/M_{Pl}\right)^{n_\lambda} \quad . \tag{3.9}$$

Here, λ_M is the (metacolor)-gaugino, and n_λ some (unknown) exponent. From this, it follows that the relevant scale for supersymmetric mass splitting, $\delta m_S \equiv m_{\text{boson}} - m_{\text{fermion}}$, to be of the order

$$\delta m_S \sim \Lambda_M \cdot \left(\Lambda_M/M_{Pl}\right)^{n_\lambda} \quad . \tag{3.10}$$

If one takes $n_\lambda = 1$, $\Lambda_M \sim 10^{11}$ GeV the supersymmetric mass splitting can be as low as ~ 1 TeV.

This peculiar behavior of a certain class of supersymmetric preon models allows one to construct theories with a very high scale parameter, $\Lambda_M > 10^{11}$ GeV, (that are therefore capable of the successes of Grand Unified Theories), but with a very low mass spectrum and low SUSY-breaking mass differences, according to the appearance of factors $\Lambda_M/M_{Pl} \sim 10^{-8}$. One is therefore capable of constructing supersymmetric extensions of the standard model, as discussed in Chap. 2 of these lectures, from high scale composite theories [3.6]. In addition, some composite mesons with unusual behavior required by cosmological reasons may also be based on such models [3.24].

We have outlined here that is it indeed possible to push the more radical point of compositeness consistently by constructing preon models at a high scale. Supersymmetry comes in naturally as the easiest way to avoid the no-go-theorems. This favors strongly a certain hierarchy of new physics from the Fermi scale $(G_F)^{-1/2}$ up to the Planck scale M_{Pl}, and we have to await more experimental numbers from future accelerators as well as from cosmology to sharpen our point of view.

3.4 Other Features of Composite Models

Up to now, we have focussed on theoretical aspects of composite models and bounds on fermion substructures. Since the fermion sector of the standard model is fairly well known, additional, more delicate tests can be presented. Furthermore, a few words for boson substructure are in order.

(i) Mixing problems. If a composite model presents a mass matrix for the fermions, the parameters are usually adjusted in such a way to yield small sizes for the light quarks and larger sizes for the heavy quarks. However, this in turn induces mixing of the current eigenstates and therefore a small amount of flavor changing neutral currents, $K_0 - \bar{K}_0$ as well as $B_0 - \bar{B}_0$ mixings, etc. For the mass matrix of the "minimal" model given by Ref. [3.4], a lower limit for $K_L \to \bar{\mu}e$ has been given [3.25] by

$$B\left(K_L \to \mu^+ e^-\right) > \tfrac{1}{3}\, 10^{-10}$$

in agreement with the recent Yale-Brookhaven experiment [3.26]. Furthermore, a large effect of mixing, either in $B_d - \bar{B}_d$, or $B_s - \bar{B}_s$, has been predicted [3.27].

(ii) Composite Gauge Bosons. Several models treat the weak gauge bosons $W^\pm$ and Z as composite objects, like the vectormesons of QCD. A particular example of that type is the "Strongly coupled Standard Model", (SCSM) of Ref. [3.28]. This model starts exactly from the QFD-Lagrangian of the Standard model as given in Chap. 1.1, however, with an *unbroken* gauge group $SU(2)_L \times U(1)$. Instead, it is assumed that $SU(2)_L$ becomes strong at a scale Λ_2 far higher then the Higgs vacuum expectation value λ, just like the running coupling constant of QCD becomes large at Λ_{QCD}. In this case, the usual discussion of spontaneous symmetry breaking does not apply, $SU(2)_L$ becomes confining and only singlet bound states survive at low energies. The theory, however, possesses an approximate global symmetry $SU(2)_W$, which characterizes the composites. Lefthanded quarks and leptons, scalars and vectormesons appear as composite dubletts, singlets, and tripletts of $SU(2)_W$ which plays phenomenologically the same role as the spontaneously broken $SU(2)_L$ of the usual Standard Model. Therefore the interesting feature arises that the W and Z are only the lowest lying states of a whole spectrum of vectormesons.

The SCSM reproduces the usual low energy weak interaction data if one additionally implies the principles of vectormeson dominance of the weak currents, and $W_3 - \gamma$ mixing. One arrives at an effective "massive Yang-Mills" theory, as one was used to a long time ago from the field theoretic descriptons of ρ and A_1-mesons. But since vectormeson dominance is only an *approximate* principle, small deviations of the structure in the WWZ or WWWW vertices as well as in their masses from the standard model prediction may be a hint of such a picture. In addition, excited W' and Z' as well as their isoscalar partners are predicted [3.29] in this model.

(iii) Exotic Composite Mesons. Leptoquarks ($B = 1/3\,, L = -1$) may appear in almost all of the composite models, however, their mass may be quite different. In their most likely realization [3.30], they may be as light as 40 GeV, whereas in others [3.5] they are expected around $\sim$ 1 TeV. Somewhat higher are the mass limits for diquarks and dileptons, since they may be linked to Baryon number violating processes.

In concluding this chapter, we would like to say that the motivation for substructure of quarks, leptons, and bosons is still very good. Most of them may have structure effects at the TeV scale. Unfortunately, the predictive power of most models concerning the open questions in the standard model – the number of families, hierarchy of masses, details of the KM-matrix – is up to now very weak. It is generally hoped that future accelerator-experiments, and in particular the discovery of the Higgs meson may give us some hint how to think about composite models in more detail.

References

Chapter 1

1.1 D. Haidt and H. Pietschmann: *Electroweak Interactions, Experimental Facts and Theoretical Foundation*, Landolt-Börnstein New Series, Group I, Vol.10 (Springer-Verlag Berlin, Heidelberg, 1988)

1.2 D.A. Ross and M. Veltman: Nucl. Phys. **B95** (1975) 135

1.3 M. Kobayashi and T. Maskawa: Progr. Theor. Phys. **49** (1973) 652

1.4 F.J.Gilman and S.H. Rhie: Phys. Rev. **D31** (1985) 1066

1.5 H. Rupertsberger: private communication

1.6 M.L. Perl: SLAC-PUB-4632,May 1988

1.7 S. Fajfer and R.J. Oakes: Phys. Lett. **B213** (1988) 376

1.8 ARGUS-collaboration: DESY 88-113, Aug. 1988

1.9 Particle Data Group: Phys. Lett. **170B** (1988)

1.10 A. Chen et al.: Phys. Lett. **122B** (1983) 317

1.11 D. Haidt: private communication

1.12 R. Bertlmann and H.Pietschmann: Phys. Rev. **D15** (1977) 683

1.13 F. Halzen and K. Mursula: Phys. Rev. Lett. **51** (1983) 857;
K. Hikasa: Phys. Rev. **D29** (1984) 1939;
N.G. Deshpande et al.: Phys. Rev. Lett. **54** (1985) 1757

1.14 R. Ansari et al.: Phys. Lett. **186B** (1987) 440; erratum **190B** (1987) 238

1.15 S. Weinberg: Phys. Rev. **D22** (1980) 1694;
J.D. Vergados: Phys. Rep. **133** (1986) 1

1.16 J.C. Pati and A. Salam: Phys. Rev. **D10** (1975) 275;
R.N. Mohapatra and J.D. Vergados: Phys. Rev. Lett. **47** (1981) 1713

1.17 G. Ecker and W. Grimus: Z. Phys. **C30** (1986) 293

1.18 W. Grimus: Fortschr. d. Phys. **36** (1988) 201

1.19 K. Engvist, K. Mursula, M. Roos: Nucl. Phys. **B226** (1983) 121;
J. Maalampi and M. Roos: Phys. Rep. (1989)

1.20 V.A. Lubimov et al.: Phys. Lett. **94B** (1980) 266; S.A. Boris et al.: Phys. Lett. **159B** (1985) 217

1.21 K.E. Bergkvist: Phys. Lett. **154B** (1985) 224

1.22 S.P. Mikheyev, A. Smirnov: Nuovo Cim. **9C** (1986) 17;
S.T. Petcov: Phys. Lett. **214B** (1988) 139

Chapter 2

2.1 J. Wess: Acta Phys. Austr. Suppl. XV, 475 (1976); *Concepts and Trends in Particle Physics, Proc. XXV Int. Universitätswochen für Kernphysik, Schladming 1986*, eds. H. Latal, H. Mitter (Springer-Verlag 1987) p. 29

2.2 W. Kummer: Acta Phys. Austr. Suppl. XIX, 399 (1978); *Concepts and Trends in Particle Physics, Proc. XXV Int. Universitätswochen für Kernphysik, Schladming 1986*, eds. H. Latal, H. Mitter (Springer-Verlag 1987) p. 59

2.3 H. Nicolai: Acta Phys. Austr. Suppl. XXV, 71 (1983)

2.4 R. Barbieri: Acta Phys. Austr. Suppl. XXIV, 363 (1982); G.G. Ross: Acta Phys. Austr. Suppl. XXV, 145, 1983

2.5 H.P. Nilles: Phys. Rep. **110**, 1 (1984)

2.6 P. Nath, R. Arnowitt, A.H. Chamseddine: *Applied N = 1 Supergravity* (World Scientific, Singapore 1984); N. Dragon, U. Ellwanger, M.G. Schmidt: *Supersymmetry and Supergravity*, Progr. in Particle and Nuclear Physics, Vol. 18 (Pergamon Press 1987)

2.7 R. Barbieri: "Looking beyond the standard model: the supersymmetric option", University of Pisa preprint IFUP-TH33/87, 1987

2.8 E. Reya: *Proc. 23rd Int. Conf. on High Energy Physics, Berkeley 1986*, vol. I (World Scientific) p. 285

2.9 Talks by T. Devlin, J.D. Dowell, R.A. Flores, J.F Gunion and H.E. Haber, L.J. Hall, T. Kamae, F. Pauss, A.J. Sanda, M.J. Shochet, M. Yamauchi: *Proc. 24th Intern. Conf. on High Energy Physics, München 1988* (Springer-Verlag)

2.10 F. Pauss: these Proceedings

2.11 M. Swartz: these Proceedings

2.12 H. Schopper: these Proceedings

2.13 R. Felst: these Proceedings

2.14 J. Wess, J. Bagger: *Supersymmetry and Supergravity* (Princeton University Press, Princeton 1983)

2.15 P. Fayet: Nucl. Phys. **B90**, 104 (1975)

2.16 A.B. Lahanas, D.V. Nanopoulos: Phys. Rep. **145**, 1 (1987)

2.17 E. Cremmer, B. Julia, J. Scherk, S. Ferrara, L. Girardello, P. van Nieuwenhuizen: Nucl. Phys. **B147**, 105 (1979)

2.18 E.Cremmer, S.Ferrara, L. Girardello, A. van Proyen: Nucl. Phys. **B212**, 413 (1983)

2.19 J. Polonyi: Budapest preprint KFKI-93 (1977)

2.20 S. Deser, B. Zumino: Phys. Rev. Lett. **38**, 1433 (1977)

2.21 K. Inoue, A. Kakuto, H.Komatsu, S. Takeshita: Prog. Theor. Phys. **68**, 927 (1982); **71**, 413 (1984)

2.22 L. Alvarez-Gaumé, J. Polchinski, M.B. Wise: Nucl.Phys. **B221**, 495 (1983); J.Ellis, J.S. Hagelin, D.V. Nanopoulos, K. Tamvakis: Phys. Lett. **125B**, 275(1983); C. Kounnas, A.B. Lahanas, D.V. Nanopoulos, M. Quirós: Nucl. Phys. **B236**, 438 (1984)

2.23 L.E. Ibañez, C. Lopez: Nucl. Phys. **B233**, 511 (1984); L.E. Ibañez, C. Lopez, C. Muñoz: Nucl. Phys. **B256**, 218 (1985); A. Bouquet, J. Kaplan, C.A. Savoy: Nucl. Phys. **B262**, 299 (1985)

2.24 B. Gato, J. León, J. Pérez-Mercader, M. Quirós: Nucl. Phys. **B253**, 285 (1985)

2.25 L.E. Ibañez, G.G. Ross: Phys. Lett. **110B**, 227 (1982); L.E. Ibañez: Nucl. Phys. **B218**, 514 (1983)

2.26 J.F. Gunion, H.E. Haber: Nucl. Phys. **B272**, 1 (1986); **B278**, 449 (1986); E. Reya: Phys. Rev. **D33**, 733 (1986)

2.27 L.J. Hall, J. Polchinski: Phys. Lett. **152B**, 335 (1985)

2.28 H.E. Haber, G.L. Kane: Phys. Rep. **117**, 75 (1985)

2.29 A. Bartl, H. Fraas, W. Majerotto: Nucl. Phys. **B278**, 1 (1986)

2.30 S.M. Bilenky, N.P. Nedelcheva, S.T. Petcov: Nucl. Phys. **247**, 61 (1984)

2.31 A. Bartl, H. Fraas, W. Majerotto: Z. Phys. **C30**, 441 (1986)

2.32 A. Bartl, H. Fraas, W. Majerotto, N. Oshimo: University of Vienna preprint UWThPh-1989-13

2.33 S. Raby: Los Alamos preprint LA-UR-84-2693, 1984

2.34 A. Bartl, W. Majerotto, N. Oshimo: Phys. Lett. **216B**, 233 (1989)

2.35 S.T. Petcov: Phys. Lett. **139B**, 421 (1984)

2.36 E.Ch. Christova, N.P. Nedelcheva: Phys. Lett. **208B**, 525 (1988)

2.37 H. Baer et al.: In *Physics at LEP*, Yellow Report, CERN 86-02, Vol. 1, ed. J. Ellis, R. Peccei, p. 297; M. Chen, C. Dionisi, M. Martinez, X. Tata: Phys. Rep. **159**, 201 (1988)

2.38 D.A. Dicus, S. Nandi, W. Repko, X. Tata: Phys. Rev. Lett. **51**, 1030 (1983);
V. Barger, R.W. Robinett, W.Y. Keung, R.J.N. Phillips: Phys. Lett. **131B**, 372 (1983);
S. Dawson, E. Eichten, C. Quigg: Phys. Rev. **D31**, 1581 (1985);
X. Tata, D.A. Dicus: Phys. Rev. **D35**, 2110 (1987)

2.39 J.M. Frère, G.L. Kane: Nucl. Phys. **B223**, 331 (1983);
T. Schimert, C. Burgess, X. Tata: Phys. Rev. **D32**, 707 (1985);
P. Chiapetta, J. Soffer, P. Taxil, F.M. Renard, P. Sorba: Nucl. Phys. **B262**, 495 (1985);
R. Arnowitt, P. Nath: Phys. Rev. **D35**, 1085 (1987);
B.F.L. Ward: Phys. Rev. **D35**, 2092 (1987);
N. Oshimo, Y. Kizukuri: Phys. Lett. **186B**, 217 (1987);
T. Fukai, Y. Kizukuri, N. Oshimo, Y. Otake, N. Sugiyama: Prog. Theor. Phys. **78**, 395 (1987)

2.40 J. Ellis, J.S. Hagelin, D.V. Nanopoulos, M. Srednicki: Phys. Lett. **127B**, 233 (1983);
J. Ellis, J.M. Frère, J.S. Hagelin, G.L. Kane, S.T. Petcov: Phys. Lett. **132B**, 436 (1983);
E. Reya: Phys. Lett. **133B**, 245 (1983);
D. Dicus, S. Nandi, W. Repko, X. Tata: Phys. Rev. **D29**, 1317 (1984); Phys. Rev. **D30**, 1112 (1984);
H. Komatsu, J. Kubo: Nucl. Phys. **B263**, 265 (1986);
H. Komatsu: Phys. Lett. **177B**, 201 (1986);
R. Barbieri, G. Gamberini, G.F. Giudice, G. Ridolfi: Phys. Lett. **195B**, 500 (1987); Nucl. Phys. **B296**, 75 (1988);
K. Griest, H.E. Haber: Phys. Rev. **D37**, 719 (1988);
M. Drees, C.S. Kim, X. Tata: Phys. Rev. **D37**, 784 (1988);
E.Ch. Christova, N.P. Nedelcheva: Phys. Lett. **208B**, 525 (1988)

2.41 S.K. Jones, C.H. Llewellyn-Smith: Nucl. Phys. **B217**, 145 (1983);
P.R. Harrison: Nucl. Phys. **B249**, 704 (1985);
J.A. Bagger, M.E. Peskin: Phys. Rev. **D311**, 2211 (1985); Erratum **D32**, 1260 (1985);
H. Komatsu, R. Rückl: Nucl. Phys. **B299**, 407 (1988);
J. Bartels, W. Hollik: Z. Phys. **C39**, 433 (1988)

2.42 A. Bartl, H. Fraas, W. Majerotto: Nucl. Phys. **B297**, 479 (1988)

2.43 A. Bartl, H. Fraas, W. Majerotto: Z. Phys. **C41**, 475 (1988)

2.44 H. Baer, A. Bartl, D. Karatas, W. Majerotto, X. Tata: University of Wisconsin preprint MAD/PH/422, to be published in Int. Journal Mod. Phys. A

Chapter 3

3.1 See, e.g. W. Buchmüller: Acta Phys. Austr. Suppl. XXVII, 517 (1985);
I. Bars: *Proc. International Workshop on Superstrings, Cosmology and Composite Structures, Univ. Maryland 1987*, eds. S.J. Gates and R.N. Mohapatra, p. 433

3.2 J.C. Pati, A. Salam: Phys. Rev. **D10**, 275 (1974)

3.3 E. Farhi, L. Susskind: Phys. Rep. **74**, 277 (1981)

3.4 J.C. Pati: Phys. Lett. **144B**, 375 (1984)

3.5 P. Mohapatra, J.C. Pati, H. Stremnitzer: Phys. Lett. **166B**, 407 (1986);
A. Masiero, R.N. Mohapatra, R.D. Peccei: Nucl. Phys. **B192**, 66 (1981)

3.6 J.C. Pati: *Proc. XXIV Int. Conf on High Energy Physics, Munich 1988*, p. 1504

3.7 J.C. Pati: Phys. Rev. **D30**, 1144 (1984)

3.8 J.C. Pati, H. Stremnitzer: Phys. Rev. Lett. **56**, 2152 (1986);
J.C. Pati, H. Stremnitzer: Phys. Lett. **196B**, 240 (1987)

3.9 W. Buchmüller, D. Wyler: Nucl. Phys. **B268**, 621 (1986)

3.10 H. Harari: Phys. Rep. **104**, 159 (1984)

3.11 J.M. Gipson, Y. Tosa, R.E. Marshak: Phys. Rev. **D32**, 284 (1985)

3.12 D. Weingarten: Phys. Rev. Lett. **51**, 1830 (1983)

3.13 S. Nussinov: Phys. Rev. Lett. **51**, 2081 (1983)

3.14 E. Witten: Phys. Rev. Lett. **51**, 2081 (1983)

3.15 S. Weinberg, E. Witten: Phys. Lett. **96B**, 59 (1980)

3.16 C. Vafa, E. Witten: Nucl. Phys. **B234**, 173 (1984)

3.17 G. 't Hooft: In *Recent Developments in Gauge Theories*, ed. G. 't Hooft et al. (Plenum Press, New York 1980)

3.18 S.L. Adler: Phys. Rev. **177**, 2426 (1969);
J.S. Bell, R. Jackiw: Nuovo Cim. **60**, 147 (1969);
W.A. Bardeen: Phys. Rev. **184**, 1848 (1969);
J. Wess, B. Zumino: Phys. Lett. **B37**, 95 (1971)
3.19 Y. Tosa, R.E. Marshak: Phys. Rev. **D27**, 616 (1983)
3.20 J.C. Pati: Proc. 3rd Annual ICTP Workshop on Particle Physics, IC/83/221
3.21 W. Buchmüller, R.D. Peccei, T. Yanagida: Phys. Lett. **124B**, 67 (1983);
R. Barbieri, R.N. Mohapatra, A. Masiero: Phys. Lett. **105B**, 369 (1981)
3.22 S. Ferrara, L. Girardello, H.P. Nilles: Phys. Lett. **125B**, 457 (1983)
3.23 J.C. Pati, M. Cvetič, H.S. Sharatchandra: Phys. Rev. Lett. **58**, 851 (1987)
3.24 M. Cvetič, T. Hübsch, J.C. Pati, H. Stremnitzer: Univ. Maryland preprint #89-078
3.25 J.C. Pati, H. Stremnitzer: Phys. Lett. **172B**, 661 (1986)
3.26 H.B.Greenlee et al.: Phys. Rev. Lett. **60**,893 (1988)
3.27 A. Datta, J.C. Pati: Phys. Lett. **172B**, 441 (1986)
3.28 M. Claudson, E. Farhi, R.L. Jaffe: Phys. Rev. **D34**, 873 (1986), and references therein
3.29 See, e.g. B. Schrempp: *Proceedings of the XXIII Int. Conference on High Energy Physics, Berkeley 1986*, p. 454
3.30 B. Schrempp and F. Schrempp: Phys. Lett. **153B**, 101 (1985)

Physics at LEP

Herwig Schopper

CERN, CH-11211 Genève, Switzerland

1. Introduction

At the time this lecture was given the construction of LEP was approaching its end and first collisions were expected in Summer 1989. The four LEP experiments were very busy installing the final components and getting ready to take the first data. This exciting situation will be reflected in this report by concentrating on the physics that will be done in the near future. Therefore emphasis will be given to experiments that can be done with LEP 100 and in particular to the results which could be expected in 1989/90 [1]. In the meantime a first beam was injected into LEP as foreseen on 14 July 1989, a pilot run took place from 13 to 18 August 1989 with all four experiments taking data and within a few hours more than 50 Z^0-events were accumulated as will be described later. The first physics run is foreseen to start in September 1989.

In the past the approach of most theoretical work was mainly to investigate the consequences of physics beyond the Standard Model for various observables. Here preferentially we shall start from the experimental accuracies which might be obtained and discuss how these results can be interpreted. An enormous amount of work is presently being done in all LEP groups in order to establish procedures on how to derive the interesting physics parameters from measured quantities including the effects of calibration and acceptance. For this purpose semi-analytical formulae or Monte Carlo programmes are used. The situation is changing continuously and therefore it does not seem useful to enter here into the details, in particular no attempt will be made to compare the different programmes. The main emphasis will rather be on general arguments which do not change so quickly.

Problems which will be attacked with LEP 200 will only be considered insofar as they are direct continuations of the experiments at lower energies [2]. Hence W-pair production will not be treated. Also physics with polarized beams will not be a topic of this report [3]. It seems that the chances to obtain transverse beam polarization at LEP are quite good but a polarization rotator to obtain longitudinally polarized beams is not compatible with the copper-rf-cavities and hence such a rotator can only be implemented after the installation of superconducting cavities.

Most of the calculations which have been done in the past assume a mass of the top quark in the region 40 to 50 GeV. The most recent results from CDF at TEVATRON show that the top mass is higher and perhaps well above the Z mass.

Physics for a New Generation Editors: H. Latal · H. Mitter

This implies that toponium physics will hardly be possible at LEP and therefore we shall not treat it here. On the other hand, top decays were considered to be a serious background for other interesting searches and this is not true anymore – a positive aspect of the high top mass.

For some time in the past it was thought that some experiments indicated a Higgs mass higher than about 7 GeV. Recently, however, it became evident that the QCD corrections on which this limit was based are not reliable and hence a very light Higgs is not excluded anymore. The Higgs search will be one of the most exciting fields at LEP, in particular since hadron colliders have practically no chance to find it. Possible ways of detecting the Higgs at LEP will be discussed.

We shall consider three classes of experiments with different values of integrated luminosities:

1) low statistics : 10 to 50 pb^{-1} ; such luminosities could be achieved during the runs 1989/90.
2) normal statistics : 100 to 200 pb^{-1}; after the running-in of LEP these could be luminosities obtained in one "normal" year.
3) high statistics : > 400 pb^{-1} will require several years of running.

It should also be mentioned that at the beginning of LEP operation a big effort will be necessary to calibrate the detectors and determine the efficiencies and acceptances.

Of course, in this report we can consider only new physics which is predicted by theory. In the past e^+e^--colliders have very often produced unexpected results and hence it will be very important to keep an open eye for surprises.

2. LEP Machine

2.1 LEP Parameters

LEP has been described at many occasions and therefore only the most important parameters, in particular those which are relevant for the operation of the machine, will be reviewed here.

LEP is designed and optimized for a beam energy of about 100 GeV (Fig. 2.1). For an electron machine the maximum energy is not limited by the bending power of the dipole magnets but by the accelerating system which has to compensate the synchrotron radiation losses. These losses are given by

$$E_{\mathrm{loss}} \sim E^4/R$$

where E is the electron energy and R the radius of the machine. To compensate these losses an acceleration field U has to be applied for which one must require

$$E_{\mathrm{loss}} = e \cdot U.$$

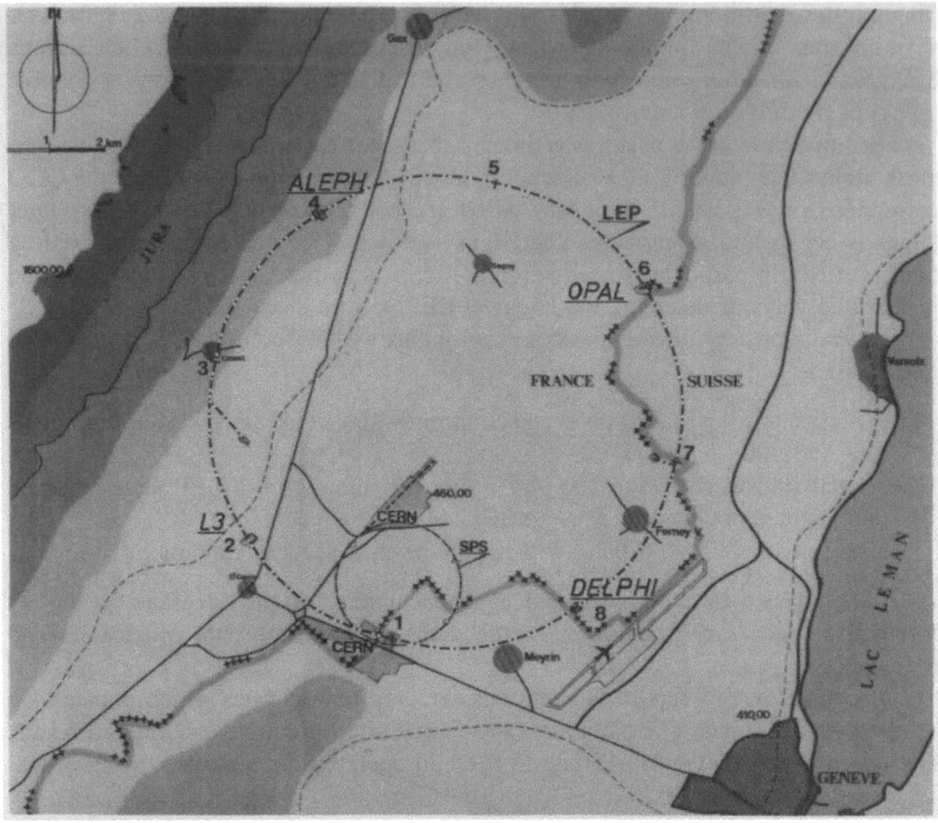

Fig. 2.1. Plan of LEP. The experiments are located in the even intersections, the rf accelerating structures in straight sections 2 and 6

The accelerating field is produced in rf-cavities which have a shunt impedance Z (which is proportional to the number of cavities) and are fed by the rf-power P. One has

$$P = U^2/Z \sim E^8/R.$$

Hence for a given radius the energy increases only with the 8th root of the number of cavities or the rf-power. Pushing the energy up requires therefore a very non-linear effort.

The optimization for a beam energy of about 100 GeV gave a circumference of 26.7 km with 8 straight sections for the experiments, the rf-cavities and other components (Fig. 2.1). With a radius of about 4.2 km the required magnetic field is rather low. The 3300 dipole magnets are designed for a field corresponding to an energy of 125 GeV and even then the field is only 0.12 Tesla. In order to economize a new type of magnet was developed which consists of iron laminations separated by about one cm and the space in-between is filled by mortar in order to give the magnet yokes the necessary mechanical rigidity ("concrete magnets").

Table 2.1. General LEP parameters

Reference circumference (including sagitta in dipoles)	26658.883376 m
Reference average radius	4242.893066 m
Diameter P1–P5 = P3–P7	8409.45 m
Diameter P2–P6 = P4–P8	8409.73 m
Diameter from arc-centre to arc-centre	8524.91 m
Maximum slope of the median plane of the machine	1.42%
Position of the lowest point of the circumference (North-East of P8)	1348.5 m (from P8)
Lowest altidude	328.30 m
Highest altidude	450.23 m
Lattice type	FODO
Phase advance/period	60 or 90 degrees
No. of bunches per beam	4
No. of interaction points	4 + 4
Equipped experimental areas (P2, P4, P6, P8)	4
Ratio horizontal/vertical β-values at I.P.	25
RF frequency	352.209042 MHz
Revolution time	88.92446 μs
Harmonic number	31320
Nominal klystron output power (total)	16 MW
Active RF structure length	272.377 m
Injection energy	20 GeV
Beam energy with maximum luminosity in phase 1	55 GeV
Maximum energy in phase 1 (zero luminosity)	$\sim$ 60 GeV

Because of the interest in Z-physics and also because of financial limitations LEP will be realized in two stages. For LEP 100 the beam energy will be 55 GeV at a maximum luminosity of $1.7 \times 10^{31}\,\mathrm{cm}^{-2}\mathrm{s}^{-1}$ (at 60 GeV with zero luminosity). If by fighting the single beam instabilities or by increasing the injection energy the beam current could be increased relative to the design value of 3 mA (5 mA might not seem impossible) at the Z-peak (E = 46 GeV) a luminosity of $2.8 \times 10^{31}\,\mathrm{cm}^{-2}\mathrm{s}^{-1}$ could be obtained.

The main parameters for LEP 100 are shown in Tab. 2.1.

In the first stage 128 copper cavitites with a total length of 272 m will be used. They are powered by 16 klystrons, each providing 1 MW cw power. Because of the size of LEP the time between bunches traversing the cavities is relatively long (about 11 microsec) and therefore low-loss storage cavities have been added to the accelerating cavities. The two cavitites perform like two coupled pendula: when a bunch passes through the cavities it feels the full accelerating field, whereas in-between passes the energy is transferred to the storage cavity. In this way the losses in the cavity walls and hence the power consumption are considerably reduced.

For some of the experiments the energy resolution is very important. The energy spread is mainly determined by synchrotron radiation and can be calculated by [4]

$$\Delta W(\mathrm{MeV}) = \frac{4.4 \times 10^{-3} \cdot W^2(\mathrm{GeV})^2}{\sqrt{1 - 0.41 \times 10^{-6} \cdot W^3(\mathrm{GeV})^3}} \tag{2.1}$$

where W is the center-of-mass energy. At the Z-peak the energy spread is about 45 MeV.

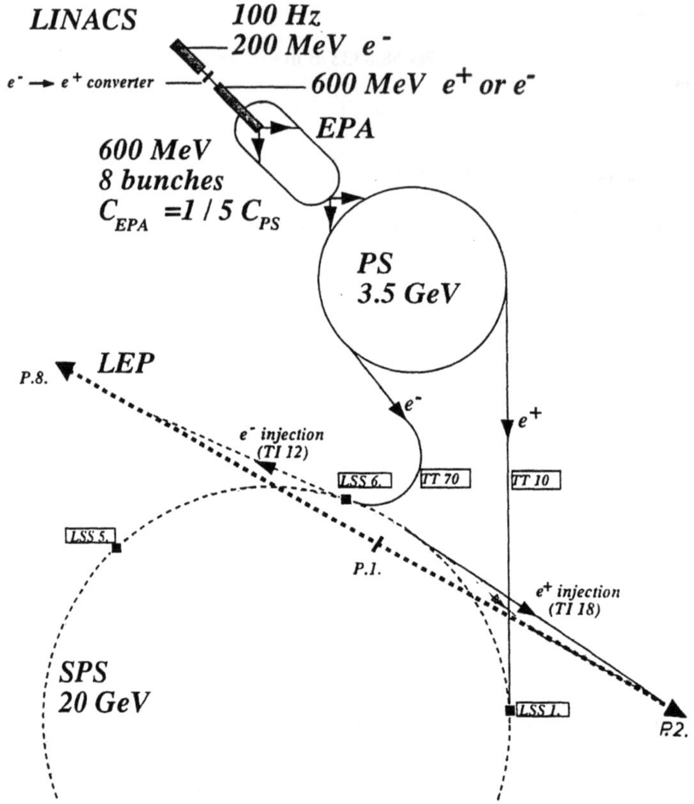

Fig. 2.2. The injection system of LEP consisting of 2 linear accelerators, the storage ring EPA, the PS and SPS. The final energies reached after each step are indicated

2.2 Injection System

The injection system is shown in Fig. 2.2. Positrons are produced by a high current 200 MeV linac. Both electrons and positrons are accelerated to about 600 MeV by a second linac. After this the particles are bunched in a small storage ring EPA (Electron-Positron Accumulator) from where they are injected into the PS, accelerated to 3.5 GeV and transferred to the SPS. Both machines had to be modified to be able to accelerate electrons. From the SPS the particles are transferred at an energy of about 20 GeV to LEP.

While particle bunches are circulating in LEP the SPS can, of course, be used for fixed target physics. In order to gain even more time for this physics, proton and electron injections can be interleaved so that even during injection into LEP protons are available. The SPS cycle foreseen for this operation is shown in Fig. 2.3. The only effect on the proton operation is a lengthening of the supercycle from about 12 to 15 seconds.

Originally it was thought that LEP and $p\bar{p}$ operation would not be compatible since proton-antiproton lifetimes would be much longer than electron-positron lifetimes. However, because of a number of improvements in the design of LEP beam

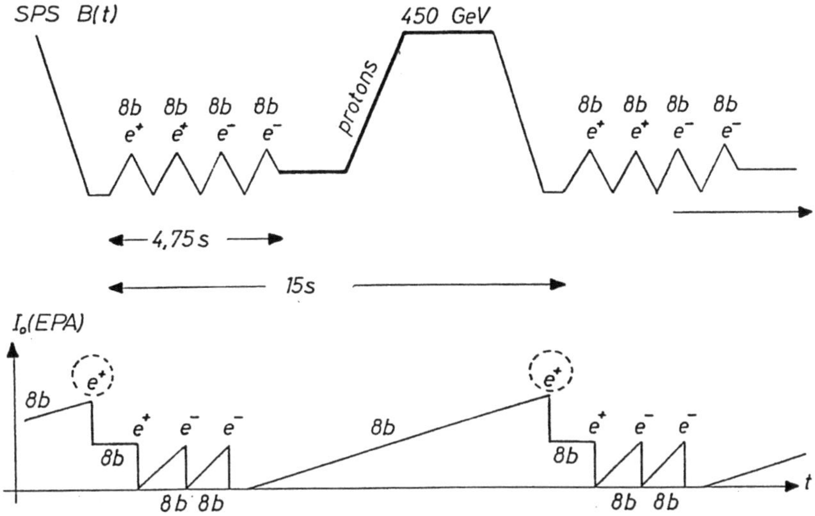

Fig. 2.3. The injection cycle of LEP. In the upper part the SPS cycle (magnetic field), in the lower part the EPA cycle (circulating current) are shown. After positrons have been accumulated in EPA they are extracted in two steps (8 bunches each) to form two positron pulses. They are followed by two electron pulses. These are accelerated in the SPS in 4 short cycles followed by the normal proton cycle

lifetimes up to 18 hours might be expected which implies that less frequent filling is needed. Since such run times are comparable to the $p\bar{p}$ runs, parallel running of LEP and the collider may now seem feasible (cf. LEPCC 28.2.1989).

2.3 Luminosity

The relevant formula giving the luminosity is

$$L = \frac{\gamma \cdot I \cdot \xi}{2\pi e \cdot r_e \cdot \beta_z^*} \tag{2.2}$$

Here the letters stand for:

γ energy of the circulating electrons
ξ tune shift due to beam-beam interaction
I beam current
β_z^* beta function at the beam crossing (determined by the power to focus the beam at the crossing)
e, r_e electron charge and classical radius

Remarks:

- the luminosity increases with energy
- the limiting tune shift can be estimated by calculations and by the experience of previous electron machines (quite different from proton accelerators because of synchrotron radiation!). A safe estimate is $\xi = 0.04$ (at PETRA 0.05, PEP 0.06)

- the beam current I is limited
  - at injection energy by single beam instabilities. This instability might limit the bunch current to $i < 0.75$ mA or equivalently the beam current to $I <$ 3 mA (for 4 bunches).
  - at high energies by the total rf power. The higher the energy the more power is lost in the cavity walls and therefore less power is available for the beam.
- the lowest beta function which can be achieved is determined by
  - the focussing power of the quadrupoles nearest to the crossing point and by the free space between quads at opposite sides;
  - chromatic effects which have to be compensated by sextupoles;
  - the bunch length (about 16 mm at LEP). If the beam sizes become comparable to the bunch length some collisions are lost.
    For LEP 100 one expects $\beta_z^* = 7$ cm.

With these assumptions one obtains the luminosity as function of the energy as shown in Fig. 2.4.

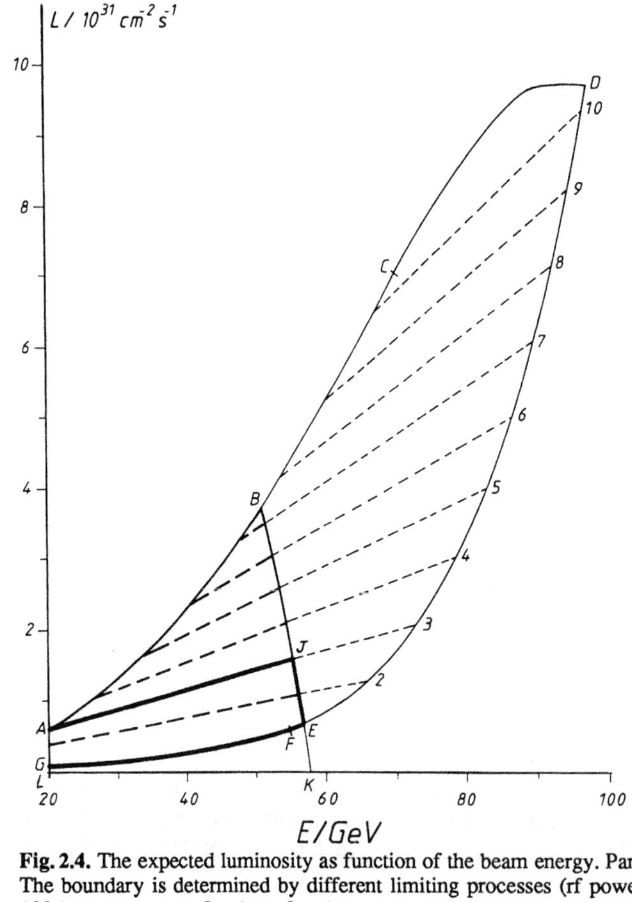

Fig. 2.4. The expected luminosity as function of the beam energy. Parameter is the beam current (in mA). The boundary is determined by different limiting processes (rf power, maximum current, etc). For LEP 100 beam currents of at least 3 mA are expected

The design value for the peak luminosity is

for LEP 100 $\quad L_{peak} = 1.6 \times 10^{31}\,\text{cm}^{-2}\text{s}^{-1}$ at 55 GeV,
for LEP 200 $\quad L_{peak} = 2.8 \times 10^{31}\,\text{cm}^{-2}\text{s}^{-1}$ at 95 GeV.

However, for the experiments the integrated luminosity is more important than the peak luminosities. It is determined apart from the peak luminosity by the filling time, beam life time, reliability. The average luminosity L_{aver} is estimated to be about a factor 2 lower than the peak luminosity. Taking into account the operating time of the machine one obtains the integrated luminosity

$$L_{integ} = 8.6 \times \frac{L_{aver}}{10^{30}} \times \frac{\text{days}}{100}\,\text{pb}^{-1} \tag{2.3}$$

In a "normal" year one might conservatively expect to run 150 to 200 days with an $L_{aver} = 10^{31}\,\text{cm}^{-2}\text{s}^{-1}$ which should give an integrated luminosity of close to 200 pb^{-1}. With a peak cross section for Z production of about 10 nb one expects about 10^6 Z / 100 pb^{-1}. Hence about 2 million Z might be obtained per year. These figures will be used when discussing the statistical accuracy of experiments. After some running-in and improvements higher rates may become available.

2.4 Status of LEP. Tests and Plans for the Commissioning

By the end of 1988 most of the civil engineering was finished and 7 of 8 octants were installed. The last octant between points 3 and 4 (Fig. 2.1) was delayed because of water breaking into the tunnel 2 years ago and hence the installation had to be done in 1989. Also most of the rf equipment was already in place.

Tests of the installed components were going on in the first part of 1989 including magnets, vacuum equipment, accelerating cavities etc.

Already in July 1988 the whole injection systems including the long and complicated transfer tunnels were tested and positrons were injected into the octant 1 to 2. The design parameters of the components of the injection chain could be obtained in a very short time and the transport of particles through the LEP octant was achieved on the "first shot".

Subsequently many beam tests were performed on various LEP hardware and measurements were made on the optics parameters both for the LEP 100 and LEP 200 lattice. All components (power supplies, vacuum, beam monitors, magnetic reproducibility, etc.) operated to specification or better. The only error found was a slight deviation in the phase advance per lattice cell due to the presence of a minute amount of ferromagnetic material in the bonding between the aluminium vacuum chamber and the lead shielding.

Positrons were injected into LEP on 14 July 1989 at 15:35 h, one day earlier than foreseen. Using fluorescence screens the beam could be steered consecutively through the 8 arcs where beam stoppers to protect the experiments were removed one after the other. Within one hour the beam made its first turn around the 27 km circumference, demonstrating that no major errors were made in the alignment. Soon many turns were observed, the rf system was switched on and trapped the particles. The next step was to accelerate the beams to 47.5 GeV which also presented no problem. In general the machine behaved as expected. The only surprise was an unexpected coupling between the horizontal and vertical beam motion. Some coupling had been expected due to the experiments' solenoids but the observed coupling was an order of magnitude higher. It was found that the origin of this coupling was uniformly distributed around the arcs. The source is not yet definitely identified but it could be the nickel layer between the vacuum chamber and the lead shield mentioned above. Due to the coupling the vertical dimensions of the beams are blown up leading to particle losses and a lower luminosity. By changing the optics and using skew quadrupoles this coupling effect could be reduced to such a size that it does not effect anymore the beam dynamics in a detrimental way.

From 13 to 18 August 1989 a pilot run took place during which the experiments could test for the first time their detectors. During this run the solenoids of the experiments were switched on whereas the superconducting low-beta-quadrupoles were not yet powered. Several runs were performed and peak luminosities of about 2.6×10^{28} $cm^{-2}s^{-1}$ could be achieved. Within one 9 hours' run 15 Z^0 were found against an expected number of 13. All experiments observed Z^0 events and a total of about 50 was accumulated. A very encouraging result within the first few hours of operation!

Note added in proof. The first physics run started by the middle of September 1989. Within a few days a peak luminosity of about 8.4×10^{29} $cm^{-2}s^{-1}$ was obtained with a total beam current of 1.8 mA. The latest record for the beam current is 3.8 mA. The longest coasting beams lasted for 12h45' and the time between fills was reduced to less than 2 hours. These results obtained after a very short period of operation are extremely encouraging. Some of the experiments have collected more than 700 Z^0 events and the total for all 4 experiments is quickly approaching about 2000.

In 1990/91 single beam currents of 3 mA or more might be obtained and luminosities of 1.7×10^{31} $cm^{-2}s^{-1}$ can be expected. With more bunches, requiring however a beam separation scheme ("Pretzel scheme"), after some years of operation, the luminosity could be further increased to values beyond 10^{32} $cm^{-2}s^{-1}$.

The beam energy was known from the start to a precision of $5 \cdot 10^{-4}$ and in the case of an energy scan the relative steps should have a precision of about $5 \cdot 10^{-5}$.

2.5 Up-grading of LEP

To achieve the design energies of about 100 GeV per beam superconducting cavities are necessary since the synchrotron radiation losses increase very rapidly with energy (see Sect. 2.1). However, for the first stage of LEP copper cavities are used since the development of superconducting cavities was not sufficiently far advanced. But all precautions were taken to make a change from copper to s.c. cavities as easy as possible. The same frequency (350 MHz) was chosen for both types of cavities which allows the use of the same klystrons, wave guides and many other components of the rf infrastructure. Essentially one has only to replace the cavities themselves and, of course, a cooling system has to be installed.

The advantage of an electron machine is the fact that the energy upgrading can be done in stages, which not only allows the best timing from the physics point of view but at each step a decision can also be taken as to the best technical solution. In the LEP straight sections 2 and 6 only about 2/3 of the available space is filled by copper cavities. As a first step this space will be filled by s.c. cavities. As a further upgrading the copper cavities could be gradually replaced by s.c. cavities. However, due to the synchrotron radiation losses the orbits of electrons and positrons become somewhat different at higher energies. Having only two places where these losses are replaced may consequently reduce the luminosities. Therefore it might turn out that opening up also the straight sections 4 and 8 for accelerating stations could turn out as necessary. This would require civil engineering work to build two more klystron galleries. Studies of the beam dynamics will be necessary to demonstrate the necessity of such a scheme.

The development of s.c. cavitites has been pushed during the past years at CERN, in laboratories and in industry. The successes are very encouraging and permit the use of such cavities in the near future. Two types of cavities have been developed at CERN: cavities fabricated of solid niobium and copper cavities which are plated with niobium at the inside. With both types good results have been obtained but Nb-Cu cavities seem even more promising. Field gradients of about 5 MeV/m were foreseen in earlier plans but it seems now that values of about 7 MeV/m could be achieved in technically built and operated cavities.

It is foreseen to install already 4 Nb-cavities and 4 Nb-Cu-cavities in the LEP tunnel by the end of 1989. These would allow to add between 68 and 91 MeV accelerating voltage to the 400 MeV produced by the 128 Cu-cavities and thus to permit to reach a beam energy of 59 to 60 GeV at maximum luminosity. It is further foreseen to order 20 Nb- and 4 Nb-Cu-cavities in Summer 1989 which could become operational in Summer 1991. The beam energy would then reach 64 to 67 GeV. Further upgrading could then be done in the following years, and if financial resources can be found, physics above the W-pair production threshold could be done after the end of 1993, about one year earlier than previously foreseen.

However, at that stage a decision concerning the number of rf-stations will have to be taken. The synchrotron radiation losses are distributed uniformly around the ring whereas the losses are replaced only by 2 localized rf-stations. As a consequence even for a perfect machine the orbits of electrons and positrons are different (Bassetti effect). The maximum separation is approximately given by

$$\Delta x \propto E^3/n_{\mathrm{rf}}$$

where n_{rf} is the number of rf-stations. It might become necessary to open up rf-stations in straight sections 4 and 8 which would require the excavation of tunnels for the klystrons and the other rf equipment. A more attractive solution would be to replace all the copper cavities by superconducting cavities. Only after the beam dynamics has been studied a decision will be possible and it should be taken in 1990.

Another option for LEP is physics with polarized beams. The question as to what degree the beams in LEP will be transversely polarized by the emission of synchrotron radiation is still not completely settled. However, recent calculations [3] done by simulation of beam tracking have given rather favorable results. It seems that even at 46 GeV a beam polarization of more than 50% could be obtained. For this purpose the existing wigglers should be complemented by dedicated wigglers. At any rate a Compton polarimeter has been installed and will measure the beam polarization. But some dedicated running time will be needed to optimize the machine parameters.

From the physics point of view a longitudinal beam polarization is necessary which requires a spin rotator. The possibility for such a rotator has been studied intensively and a feasible solution has been found. However, it requires the rebuilding of about half a km of the LEP structure in each of the straight sections and the beam has to go above and below (cuts in the floor) the present beam line. It also interferes with the copper storage cavities and hence it seems reasonable to install such a rotator only together with s.c. cavities. Since the realization of such a spin rotator seems to be several years away physics with polarized beams will not be discussed in this report.

Right from the beginning of the LEP project long-range options have been foreseen. The LEP tunnel was made wide enough and the LEP ring was installed sufficiently low in order to provide space for a hadron ring (called Large Hadron Collider – LHC) in the same tunnel. However, it is not the task of this report to deal with these very exciting possibilities.

3. LEP Detectors

Four large detectors, ALEPH, DELPHI, L 3 and OPAL, have been approved for installation in the even straight sections (Fig. 2.1). Three of them are mobile on rails. As a consequence they could be assembled outside the interaction region and afterwards rolled into the beams (Fig. 3.1). On the other hand L 3 is so big that it had to be permanently installed in the interaction area (Fig. 3.5).

The experiments have been designed and constructed by four big international collaborations involving more than 1300 physicists. They come not only from CERN member states but also from the USA, USSR, Japan, China, Israel, Canada etc. The total cost of the four detectors amounts to about 400 Million SFr. It is very remarkable that about 2/3 of this cost has been provided by the participating institutions and only 1/3 came from CERN. More remarkable is even the fact that these very sophisticated and large technical projects could be realized within the short time of 5 years inspite of the large number of components which had to be developed and constructed in so many institutes dispersed all over the world.

The LEP detectors have been described at many occasions and therefore I shall restrict myself here to a few remarks comparing the advantages and disadvantages of the various experiments. Often the question has been raised why 4 detectors have been approved. Although superficially some look very similar in detail they show differences which might become essential since it is impossible to design the best detector to discover the unknown. Of course, the "easy physics" at the beginning will be done by all experiments but later when particular and rare processes will be investigated these differences could well be important.

Fig. 3.1. Schematic plan of the intersection hall for the three mobile experiments

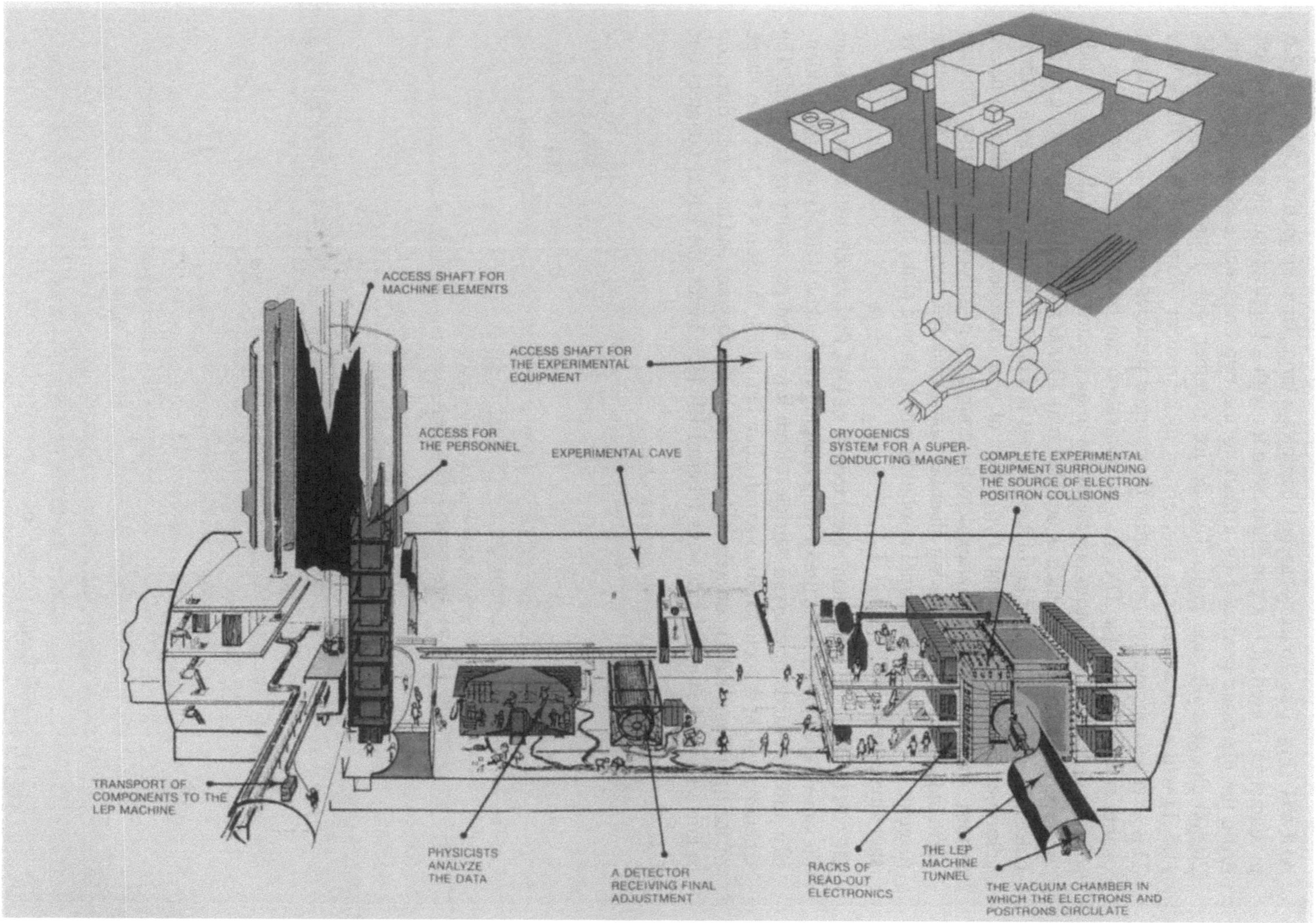

The individual detectors can be characterised in the following way:

ALEPH (Fig. 3.2). To track charged particles a large Time Projection Chamber (3.6 m diameter) is used. It is a gas-filled cylinder with a longitudinal elctric field which drifts the electrons produced by a track to the two endplates each equipped with 20,500 charge collection pads. This device together with a strong magnetic field (1.5 T) will allow a magnetic momentum measurement of about 1% at 10 GeV/c. This excellent momentum resolution for charged particles is one of the advantages of this experiment. In the TPC also the ionisation can be measured which helps to identify the particles. Photons are measured in an electromagnetic calorimeter inside the superconducting coil which is devided into 70,000 radial sectors pointing to the collision point. The iron yoke of the magnet is sandwiched to be used as a hadron calorimeter. Ouside the magnet two planes of wire chambers are installed to detect muons. ALEPH has also a vertex detector composed of silicon microstrips which is surrounded by an inner tracking chamber containing 960 sensing wires. The vertex detector has a very good spatial resolution which should make it possible to observe short-lived particles.

DELPHI (Fig. 3.3). The main characteristic of DELPHI which distinguishes it from the other experiments is the use of Ring Imaging Cherenkov (RICH) counters to identify different kinds of charged particles. Both liquid and gas radiators are used. The Cherenkov light is focussed onto a drift chamber which measures the size and shape of the light rings. RICH counters are installed in the forward and backward direction relative to the beam and sideways ("barrel RICH"). Inside the RICH counters one

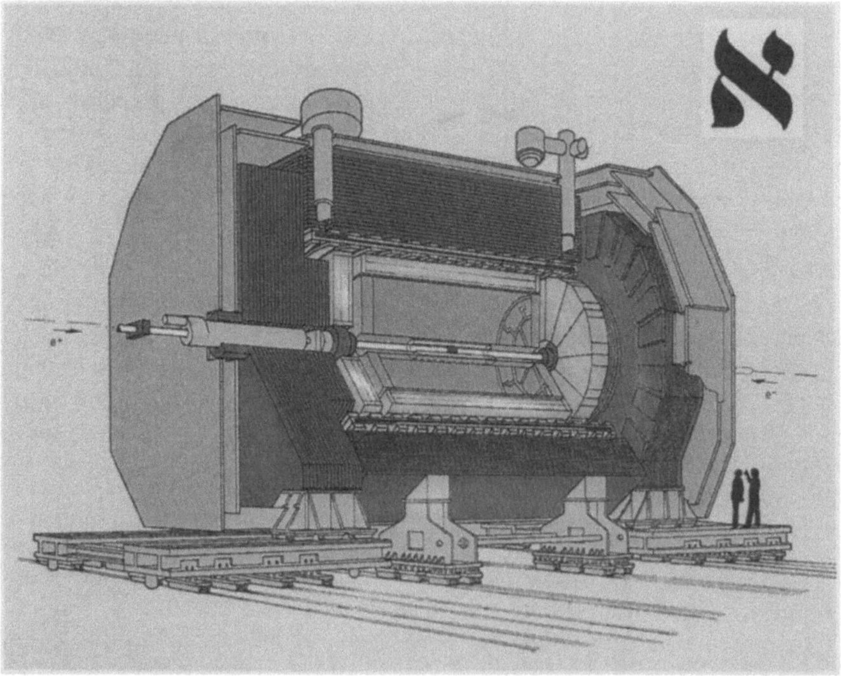

Fig. 3.2. ALEPH detector

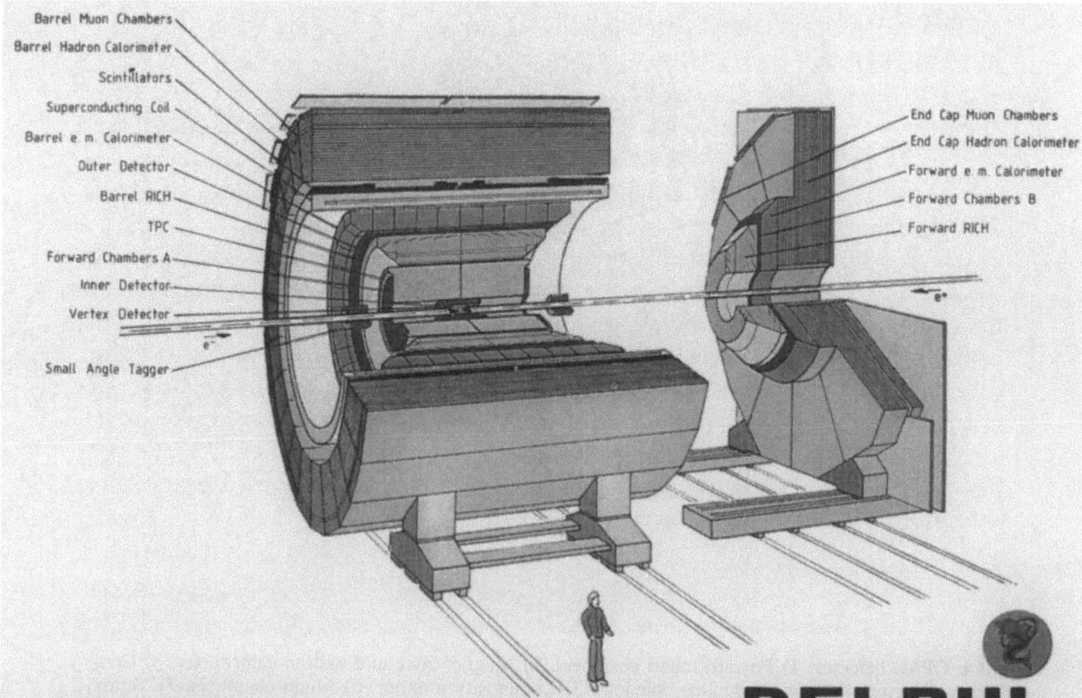

Fig. 3.3. DELPHI detector

finds a TPC, somewhat smaller than that of ALEPH, and close to the beam a microvertex detector consisting of siliconstrips. Another special feature of DELPHI is its electromagnetic calorimeter (HDPC). It is composed of long lead bands folded up and thus forming gaps in which the produced electrons are drifted to the detectors. This device provides an extraordinarily good granularity. The magnet yoke is used as hadron calorimeter. The magnet is powered by the largest superconducting solenoid ever built. Outside the magnet 2 layers of wire chambers to detect the muons are installed.

OPAL (Fig. 3.4). This detector is in a certain sense an upgrade of the JADE detector at PETRA. It uses a warm copper coil. Its main component is the so-called jet chamber which does not only allow a precise tracking of charged particles but also a measurement of their ionization. Photons and electrons will be detected by an array of lead glass crystals arranged in such a way that they point to the interaction region. The iron yoke of the magnet is sandwiched in order to be used as a hadron calorimeter and at the same time it serves as a filter for muons which are detected by chambers mounted outside the magnet.

L 3 (Fig. 3.5). This experiment differs from the others in many respects. Its magnet yoke and coils are very big in order to provide a "magnetic hall" unobstructed by the magnet. One of the objectives is to measure muon momenta with the highest possible precision. This is done with three layers of muon chambers positioned with high precision (wire position known to about 50 microns) inside the magnet. Another aim is to determine with high accuracy the energies of electrons and photons down to

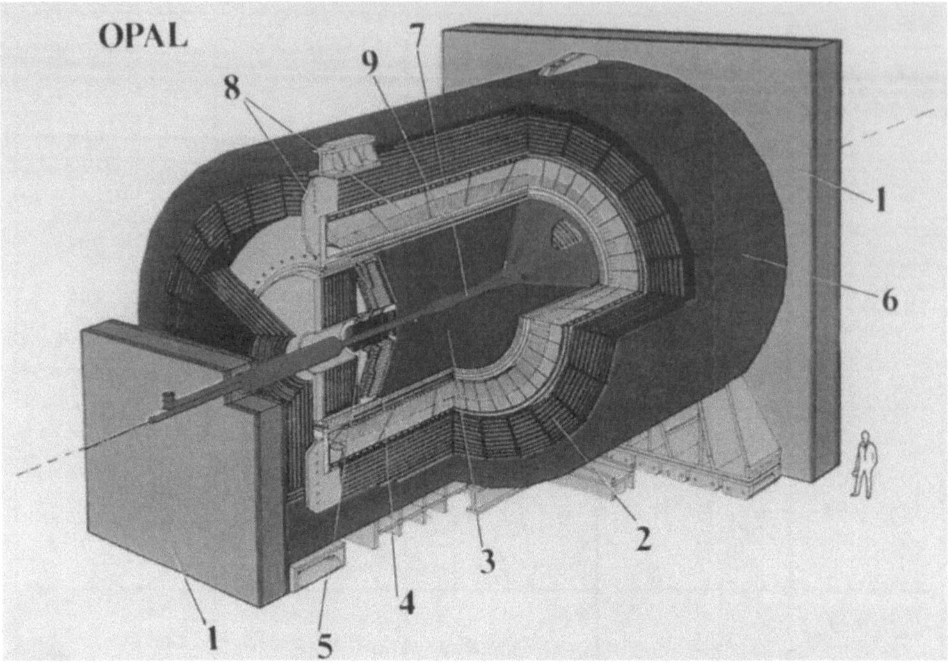

Fig. 3.4. OPAL detector. 1) Foward muon chambers. 2) Magnet yoke and hadron calorimeter. 3) Central track detector. 4) Superconducting solenoid. 5) Luminosity monitor. 6) Muon chambers. 7) Vertex detector. 8) Electromagnetic calorimeter (lead glas). 9) Time-of-flight counters

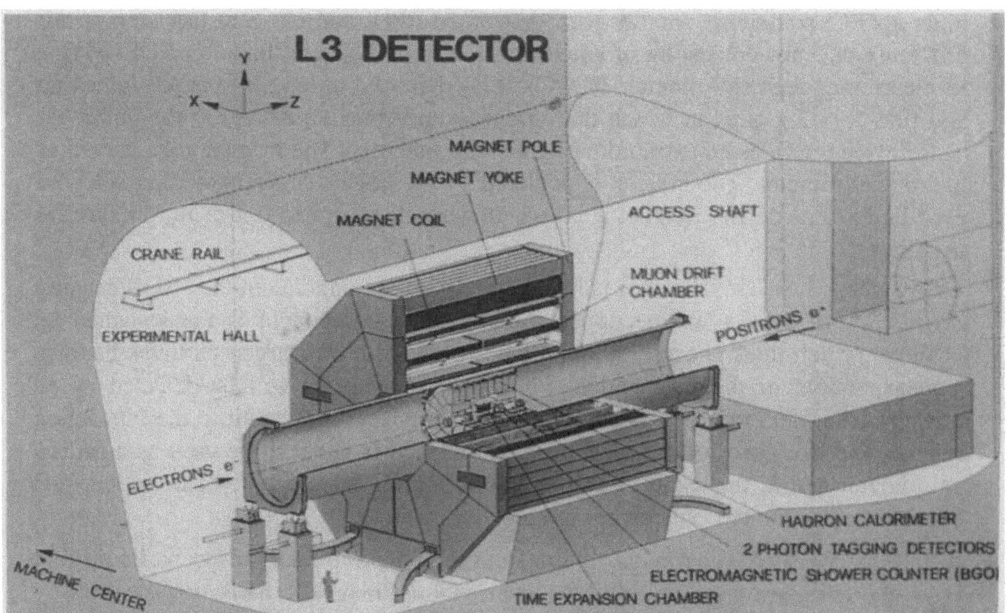

Fig. 3.5. L 3 detector

Table 3.1.

	ALEPH	DELPHI	OPAL	L 3
Vertex Detector	chamber+ Si-micro-strips $\sigma_{r\Phi} \cong 10\ \mu$	Si-micro-strips $5\ \mu$	wire-chamber	TEC Time Expansion Chamber
Inner-Tracking	wire-chamber R= 28 cm L= 2.1 m $\sigma_{r\Phi} = 100\ \mu$	wire-chamber R= 22.3 cm L= 80 cm $\sigma_{r\Phi} = 100\mu$	R= 23.5 cm L= 1 m $\sigma_{r\Phi} = 50\ \mu$	R= 50 cm L= 90 cm $\sigma_{r\Phi} = 40\ \mu$
Central Tracking	TPC 1 atm R=1.8 m L=4.4 m $\sigma_{r\Phi}$	TPC 1 atm R= 1.2 m L= 2.7 m $\sigma_{r\Phi} = 250\ \mu$	JET-Chamber 4 atm abs. R = 1.8 m L = 4.0 m $\sigma_{r\Phi} = 120\ \mu$	Muon-Chambers inside coil
Tracking Resolution	13% p	15% p	15% p	2% p μ only
Particle Identific.	dE/dx= 4.5%	dE/dx=5.5% RICH	dE/dx= 3.5%	——
EM Calorim. Resolution	Pb/Gas 3x3cm towers $18\%/\sqrt{E}$	HDPC Pb/Gas 40 slots 90cm drift inside coil $18\%/\sqrt{E}$	Pb-Glas 12300 blocks pointed $6.5\%/\sqrt{E}$	BGO 12000 crystals $2\%/\sqrt{E}$
Hadron Calorim. Resolution	Fe - Yoke Streamer tubes behind coil 5000 towers $85\%/\sqrt{E}$	Fe - Yoke Streamer tubes behind coil 20 layers $100\%/\sqrt{E}$	Fe - Yoke Streamer tubes behind coil $120\%/\sqrt{E}$	Uranium with tubes inside coil $(55/\sqrt{E} + 5)\%$
Muon Detection	2 layers of chambers	2 layers of drift chambers	4 layers of drift chambers	3 layers of precision chambers inside coil
Magnet	supercond. R=2.5 m L=6m 1.5 T	supercond. R=2.6 m L=7.4 m 1.2 T	copper coil R= 2 m L= 6.5 m 0.4 T	Al coil R = 8 m L = 14 m 0.5 T

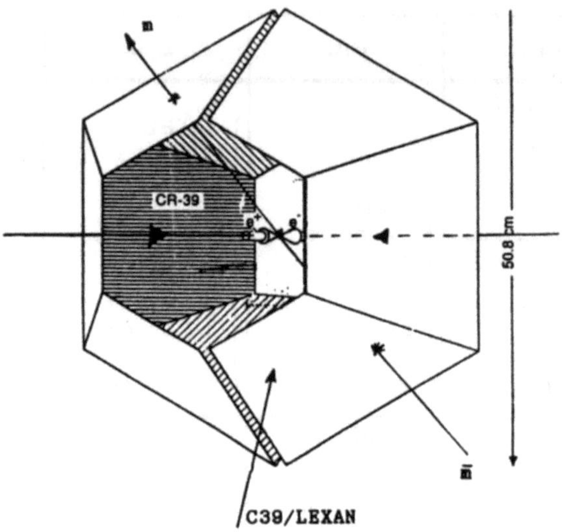

Fig. 3.6. Plastic foil arrangement to look for monopoles

about 100 MeV. For this purpose 10,000 BGO crystals pointing toward the collision point are used. Hadron energies will be measured with a specifically designed hadron calorimeter (not using the magnet yoke but depleted uranium plates) providing a very good energy resolution. L 3 has on the other hand only a small tracking device, the Time Projection Chamber.

The properties of the different components of each experiment are shown in Table 3.1 which is an up-dated version of a similar table prepared by Dydak [5].

Recently a **fifth experiment** was approved which will look for highly ionizing particles, in particular for magnetic monopoles [6]. For this purpose the interaction point 5 will be surrounded by a system of plastic detectors (Fig. 3.6). Each sheet is a sandwich of two different kinds of plastic, CR39 and Lexan, covering about 90% of the solid angle. After irradiation these plastic sheets are etched and holes of conical shape are produced. The depth of the etched pits is an increasing function of the the quantity Z/β of the ionizing particles (Z is the charge and $\beta = v/c$). CR39 is sensitive for $Z/\beta > 5$ and Lexan for $Z/\beta > 60$, whereas most of the particles produced have Z/β close to 1. The range that can be covered is $3\,e < Z < 180\,e$ or a Dirac magnetic charge of $20\,e < g < 200\,e$. With an integrated luminosity of 10^{35} $cm^{-2}s^{-1}$ (corresponding to a 12 day run) the present mass limit of monopoles (about 22 GeV, from TRISTAN) could be extended to approximately 50 GeV. The experiment could start in 1990.

4. Physics of the Z^0

The first measurements at LEP will concern the mass, width and decays of the Z^0. Before discussing the different experiments a short summary will be given of the properties of the Z^0 as expected from the Standard Model and of the radiative corrections which have to be applied.

4.1 Z^0 Parameters in the Standard Model

The Standard Model contains 3 fundamental parameters [7, 8]: the two gauge couplings g_1 and g_2 and the Higgs vacuum expectation value v. Alternatively one can use the three parameters

$$\alpha, \qquad G_\mu \qquad \text{and} \qquad M_Z$$

which can be experimentally determined: the finestructure constant α e.g. from the Josephson effect, the Fermi constant G_μ from the muon lifetime and the Z^0 mass from e^+e^- collisions.

To test the SM one additional parameter is needed, e.g. the mass of the W, or $\sin^2\theta_W$ (ratio of the gauge couplings) or the fermionic couplings V_f and A_f related to the neutral currents (can be measured by asymmetries).

The SM gives in lowest order the following relations:

$$1 - \sin^2\theta_W = M_W^2/M_Z^2 \tag{4.1}$$

$$M_W^2 = \frac{\pi\alpha}{\sqrt{2}G_F} \cdot \frac{1}{\sin^2\theta_W} \tag{4.2}$$

$$V_f = I_3^f - 2Q_f \sin^2\theta_W, \quad A_f = I_3^f \tag{4.3}$$

where V_f and A_f are the neutral current coupling constants of fermion f to the boson and

$$I_3^f = \begin{cases} +1/2 & \text{for neutrinos and u-like quarks} \\ -1/2 & \text{for charged leptons and d-like quarks} \end{cases}$$

Q_f = quark charge

Equation (4.2) relates the SM to the Fermi theory at low energies characterized by the effective coupling G_F. The equation can be rewritten for the Z-mass as

$$M_Z^2 = \frac{\pi\alpha}{\sqrt{2}G_F} \cdot \frac{1}{\sin^2\theta_W \cos_W^2} \tag{4.2a}$$

Instead of the coupling constants of Eq. (4.3) one can use the definition for the neutral current

$$J_{NC} = 2(J_3^m - \sin^2\theta_W J_{em}^m) \tag{4.3a}$$

or what is also equivalent the ratio of the neutral current cross sections

$$\frac{\sigma(\nu_\mu e - \nu_\mu e)}{\sigma(\bar{\nu}_\mu e - \bar{\nu}_\mu e)} = \frac{3 - 12\sin^2\theta_W + 16\sin^4\theta_W}{1 - 4\sin^2\theta_W + 16\sin^4\theta_W} \tag{4.3b}$$

All these equations can be interpreted as definitions of $\sin^2\theta_W$ and of course they are compatible in the lowest order. In the case of radiative corrections, however, one can choose one equation to define $\sin^2\theta_W$ with the consequence that the others will have to be changed (see Sect. 4.2.2). Various definitions can be found in the literature which makes it sometimes difficult to compare the results of different authors.

Deviations from the minimal SM in lowest order, i.e. from the above relations, could be due to:

- Radiative corrections (QED, QCD, weak)
- New particles, e.g. a 4th fermion generation, more than one Higgs doublet, SUSY particles. They would contribute to virtual loops and produce unexpected radiative corrections.
- Different structure of the SM, e.g. Higgs triplets, a Z' etc.

Deviations from the structure of the minimal SM would imply modifications already on the tree level. Instead of (4.1) one would have

$$1 - \sin^2 \theta_W = M_W^2 / \rho M_Z^2 \tag{4.4}$$

with $\rho \neq 1$. This would for instance be the case for more than one Higgs doublet.

A new Z-particle would change the relation between $\sin^2 \theta_W$ and the coupling constants V_f, A_f.

All these possibilities would not lead to large deviations. Hence precise experiments will be required and the known radiative corretions need a very careful treatment.

Before we discuss radiative corrections a summary of the Z^0 parameters will be given.

4.1.1 Branching Ratios. Total Width

Neglecting higher order effects the partial decay width of the Z^0 into two fermions is given by [7, 8]

$$\Gamma(Z^0 \to f\bar{f}) = 8\Gamma_0 c_f (V_f + A_f) \sqrt{1 - (M_f/M_Z)^2} \tag{4.5}$$

with the V_f and A_f defined in (4.3), the colour factor $c_f = 1$ or 3 for leptons and quarks respectively, and

$$\Gamma_0 = G_F M_Z^3 / 24\pi\sqrt{2}.$$

If we take for M_Z the most recent value found [9]

at	MarkII/SLC	$M_Z = (91.17 \pm 0.18)$ GeV
and	CDF/FERMILAB	$M_Z = (90.9 \pm 0.3)$ GeV

one finds $\Gamma_0 \approx 82$ MeV. For the known quarks the square root $\sqrt{1 - (M_f/M_Z)^2}$ is close to 1 and will be neglected.

For the different decay channels and using $\sin^2 \theta_W = 0.232$ [10] one obtains the following table.

For the top quark the expression for the u-like quarks would also be valid except for threshold factors [7]. Since the top mass seems to be well above the Z^0 mass we shall not take it into account here.

These new values for the partial decay widths are somewhat lower than those used so far in the literature because of the lower Z mass which enters with the 3rd power.

Table 4.1

Decay	$\Gamma(Z^0 \to f\bar{f})/\Gamma_0$	$\Gamma(Z^0 \to f\bar{f})$ (MeV)
$\nu\bar{\nu}$	2	165
e^+e^- $\mu^+\mu^-$ $\tau^+\tau^-$	$1 + (1 - 4\sin^2\theta_W)^2 \approx 1.005$	83
$u\bar{u}$, $c\bar{c}$	$3[1 + (1 - 8/3\sin^2\theta_W)^2] \approx 3.15$	259 (269)
$d\bar{d}$, $s\bar{s}$, $b\bar{b}$	$3[1 + (1 - 4/3\sin^2\theta_W)^2] \approx 4.43$	364 (379)

The total width (without the top quark) results as a sum of the partial widths and is found to be $\Gamma = 2354$ MeV. As we shall see later the color factor for the quarks has to be changed because of QCD corrections. In first order c_f should be multiplied by a factor $[1 + \alpha_s(M_Z)/\pi + \ldots] \approx 1.04 \pm 0.01$. With this correction one obtains the values in brackets in Table 4.1. Summing up one arrives at a total width of

$$\Gamma = 2419\,\text{MeV} \pm 20\,\text{MeV}$$

(without top, but with QCD corrections)

which is noticeably lower than the value 2560 MeV used so far without top and the value of 2641 MeV with a top of 40 GeV mass [7]. The error is only the one associated to the QCD correction. Other corrections will be discussed later.

4.1.2 Z^0 Line Shape

The decay channel which is easiest to treat is the decay into charged leptons, in particular into muon pairs since there are no QCD corrections and only the two diagrams of Fig. 4.1 contribute in lowest order.

Neglecting radiative corrections one finds for the cross section integrated over all angles (for the full expression giving angular and polarization dependence see Sect. 4.3 and [8, 11])

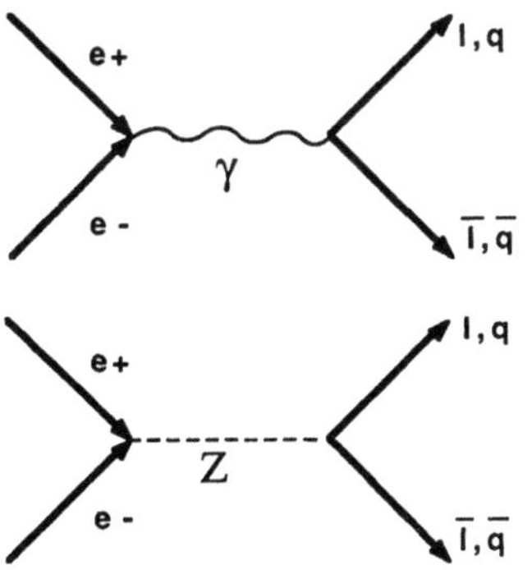

Fig. 4.1. The two diagrams contributing to e^+e^- annihilation in lowest order

$$\sigma(e^+e^- \to \mu^+\mu^-) = \frac{4\pi\alpha^2}{3s}\left\{1 + \frac{2s(s-M_Z^2)}{\sin^2\theta_W\cos^2\theta_W}\cdot\frac{V_e^2}{D^2}\right.$$
$$\left. + \frac{s^2}{\sin^4\theta_W\cos^4\theta_W}\cdot\frac{(V_e^2+A_e^2)^2}{D^2}\right\} \tag{4.6}$$

where $D^2 = |s - M_Z^2 + iM_Z\Gamma|^2 = (s-M_Z^2)^2 + M_Z^2\Gamma^2$ is the resonance denominator.

The general structure of this formula is easy to understand. The 1 corresponds to the photon graph (Fig. 4.1), the last term to the Z-graph and the middle term is the interference between the two.

Sometimes this cross section is written in the following form [12]

$$\sigma(e^+e^- \to \mu^+\mu^-) = \frac{4\pi\alpha^2}{3s}\left\{1 + 2\nu^2\cdot\frac{K\,s(s-M_Z^2)}{D^2}\right.$$
$$\left. +(\nu^2+1)^2\cdot\frac{K^2s^2}{D^2}\right\} \tag{4.6a}$$

with

$$\nu = 1 - 4\sin^2\theta_W \quad \text{and} \quad K = 1/(16\sin^2\theta_W\cos^2\theta_W).$$

This form is sometimes useful to implement the radiative corrections (see Sect. 4.2.2).

Fitting the measured line shape to this formula one can in principle determine the 3 parameters M_Z, the total width Γ and $\sin^2\theta_W$. This is not so easy in practice since radiative corrections have to be taken into account even for experiments with limited accuracy. This will be discussed in the next paragraph.

Still another form of Eq. (4.6) is sometimes useful. Using the partial widths as given in (4.5) one can write

$$\sigma(e^+e^- \to \mu^+\mu^-) = 12\pi\Gamma(Z^0 \to e^+e^-)\Gamma(Z^0 \to \mu^+\mu^-)/D^2 \tag{4.6b}$$

Integrating over the resonance and assuming that $\Gamma(Z^0 \to e^+e^-) = \Gamma(Z^0 \to \mu^+\mu^-)$ one obtains

$$\int d\sqrt{s}\,\sigma(e^+e^- \to \mu^+\mu^-) = 6\pi^2\Gamma^2(Z^0 \to \ell^+\ell^-)/M_Z^2\Gamma \tag{4.7}$$

From the two measured quantities, the integral and the total width Γ, one obtains the leptonic width $\Gamma(Z^0 \to \ell^+\ell^-)$. Experimentally the main error in the measurement of the integral will come from the absolute normalization of the cross section. The intrinsic errors in the total width are less important. Blondel et al. [13] estimate that the overall uncertainty of the integral will be about 3%, leading to an error in the leptonic width of about 2%.

By observing the decays of the Z into electron and tau pairs in addition to muon pairs it is possible to determine from (4.7) the individual leptonic widths which allow to check universality.

For the decay into quark pairs more or less the same arguments hold, but of course the appropiate color coefficient has to be used (see Eq. (4.5)) and QCD corrections have to be applied (see Sect. 4.2.3).

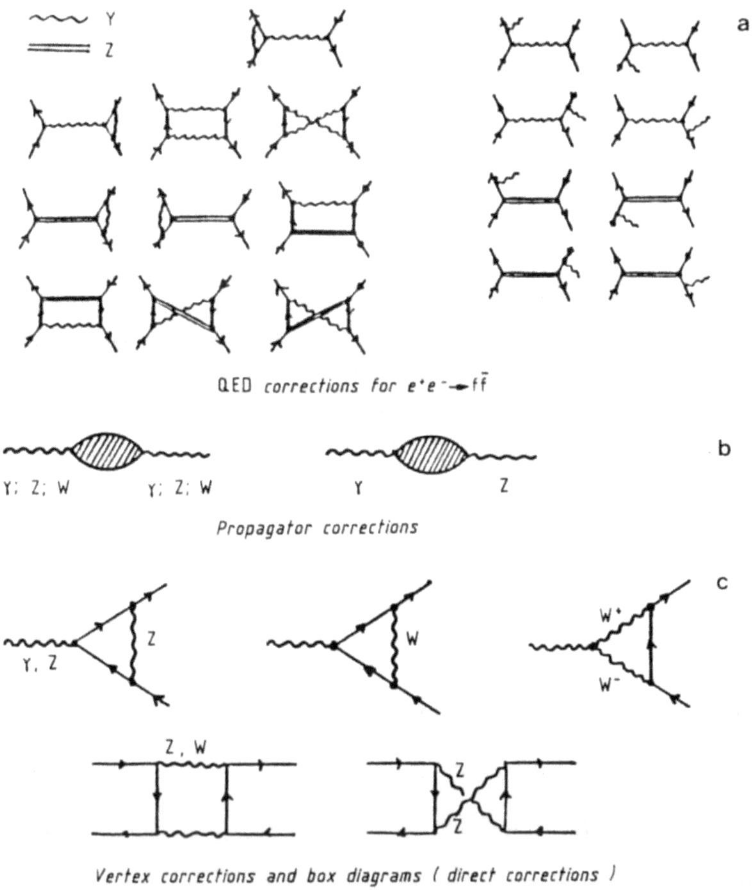

Fig. 4.2. Diagrams of electroweak radiative corrections

4.2 Radiative Corrections to the Z^0 Mass and Width

Radiative corrections can be classified into three categories:

i) QED corrections related to diagrams with one extra photon added to the Born diagram, either as a real bremsstrahlung photon or a virtual photon loop (Fig. 4.2a). They are not very interesting from the physics point of view but they can be large and hence require a careful treatment in particular for precision experiments. Above all bremsstrahlung in the initial state turns out to have quite sizable effects.

ii) Weak corrections which can be subdivided into two categories:
   - Those involving loops in the propagators (Fig. 4.2b), sometimes called "oblique" corrections. Apart from "dressing" the propagators one has to take into account also interference effects between the photon and the Z-propagator which do not propagate independently.
   - Those involving vertex corrections and box diagrams (Fig. 4.2c) which are sometimes called "direct" corrections.

iii) QCD corrections for processes in which quarks are involved.

These corrections have been investigated in numerous papers (see for instance the reports by Dydak [14] or Hollik [8] and references therein). Here the results will be summarized only to the extent that they are relevant for the analysis of the experimental data.

4.2.1 QED Corrections

These corrections are of little interest as far as new physics is concerned and they can be computed [7, 8, 14] whereby higher order graphs must be taken into account since they are important even for experiments with moderate precision. The major difficulty is given by the fact that these corrections depend on the acceptance of the detector and therefore semi-analytical or Monte Carlo methods have to be used. It is not the purpose to discuss these calculations here, only the main characteristcs of the QED corrections will be considered.

The principal modification arises from the emission of bremsstrahlung photons in the initial state (Fig. 4.2a). Kinematically this shifts the peak of the process $e^+e^- \to \mu^+\mu^-$ to higher energies and gives a long radiative tail at the right wing of the resonance [7, 8]. This is demonstrated in Fig. 4.3. The size of the effect can be estimated roughly in first order $O(\alpha)$ to give a mass shift and increase of the width by

$$\delta M_Z \approx \delta \Gamma_Z \approx (\alpha/\pi)\ln(M_Z^2/M_e^2) \cdot \Gamma \approx 150\,\mathrm{MeV}$$

and the peak cross section is reduced by a factor

$$1-(2\alpha/\pi)\ln(M_Z^2/M_e^2)\ln(M_Z/\Gamma) = 0.6 \tag{4.8}$$

which means by about 40%.

However, it turns out that these first order corrections overshoot. A partial summation of multiple photon emission is obtained by exponentiation of the leading log term which means to replace (4.8) by

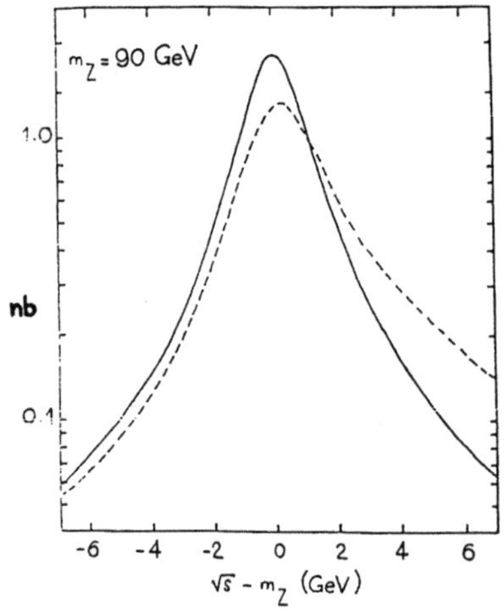

Fig. 4.3. Total cross section for $e^+e^- \to \mu^+\mu^- + \gamma$ as function of energy [1]. The solid line is the Born approximation, the broken line includes first-order QED corrections. Parameters were taken as M_Z = 90 GeV, $\sin^2\theta_W$ = 0.2435, Γ = 2.26 GeV.

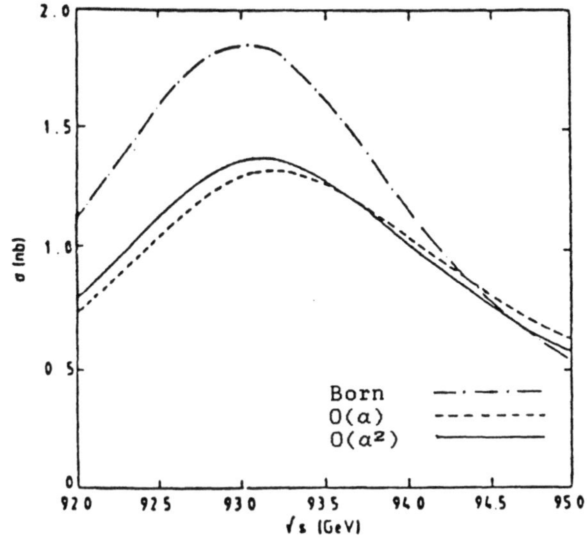

Fig. 4.4. Total cross section for $e^+e^- \to \mu^+\mu^- + \gamma$ as function of energy [27]. The influence of second order QED corrections (*full line*) are shown relative to first order (*dashed line*) and Born approximation (*point-dashed line*)

$$(M_Z/\Gamma)^{-(2\alpha/\pi)\ln(M_Z^2/M_e^2)} \,. \tag{4.9}$$

Comparing this to the first order correction one notices that the original displacement of the resonance peak of +184 MeV is reduced by −88 MeV (Fig. 4.4). The higher order corrections have been reviewed recently again (for a summary see Altarelli [15]) and agreement was reached that the results of various procedures differ by less than 0.1% and the remaining uncertainty of the mass shift amounts to only a few MeV. An approximate analytic formula for the QED correction was also derived which allows a folding with the experimental acceptances. It is rather lengthy and hence will not be reproduced here (see also [16]).

Several authors have tried to develop a strategy how to determine the Z^0 parameters, in particular folding the experimental acceptance with the QED correction and investigating the sensitivity of the experimental results (e.g. [17, 14, 18]). They have shown that the statistical errors which could be obtained with even moderate luminosities are smaller than the QED corrections. A scan with about 6 points (step width about 0.5 GeV) around the Z^0-peak with about 2 pb^{-1} per point could give [13, 19] already a statistical accuracy of about $\delta M_Z \approx 14$ MeV and $\delta\Gamma_Z \approx 40$ MeV. However, other errors will also have to be taken into account.

4.2.2 Weak Corrections

In contrast to the QED corrections electroweak higher order corrections are of strong physical interest since they allow a test of the model and might give information on the top and Higgs mass or even reveal some other new physics.

As mentioned above weak corrections can be grouped into loops affecting the boson propagator and loops involving vertices (Fig. 4.2). In calculating these corrections one must agree on a definition of $\sin^2\theta_W$. For this purpose any of the relations

(4.1) to (4.3) can be used. For various purposes of calculations different definitions are more adequate and therefore a considerable variety of procedures exists in the literature making a comparison sometimes difficult.

There is a certain consensus that the definition by Eq. (4.1) introduced by Sirlin [20] should enjoy a certain preference since it connects $\sin^2\theta_W$ to directly measurable quantities M_Z and M_W, is independent of a particular process and it also allows to estimate the radiative corrections by means of the renormalization group. It is sometimes called the on-mass-shell approach (OMS) [21] and will be used preferentially in this report.

However, it turns out that the radiative corrections for this definition turn out to be rather large and depend strongly on the top mass and therefore another definition involving the neutral current (4.3) is sometimes used. We shall discuss these two cases.

4.2.2.1 The On-Mass-Shell Definition of $\sin^2\theta_W$

Adopting the definition for $\sin^2\theta_W$ as given by (4.1) this relation will not change when applying radiative corrections whereas the other relations have to be modified. Equation (4.2) has to be changed to [20]

$$M_W^2 = A \cdot \frac{1}{\sin^2\theta_W} \cdot \left(\frac{1}{1-\Delta r}\right) \tag{4.10}$$

with

$$A = \pi\alpha/\sqrt{2}G_F = (37.281\,\text{GeV})^2.$$

The quantity $\Delta r = \Delta r(\alpha, M_Z, M_W, M_H, M_t)$ is a function of the fermion and Higgs masses. It can be calculated [22, 23] as a function of the unknown top and Higgs mass. It is displayed in Fig. 4.5. As one notices it shows a rather large dependence on M_t whereas the dependence on M_H is much smaller. In the past a value for the top mass around 40 GeV has been assumed and for instance a value $\Delta r = 0.0713 \pm 0.013$ has been derived for M_t = 45 GeV and M_H = 100 GeV.

The error of Δr is due to uncertainties in the value of the low energy ($e^+e^- \to$ hadron) cross section which is needed to estimate the vacuum polarization contributions of the light quarks.

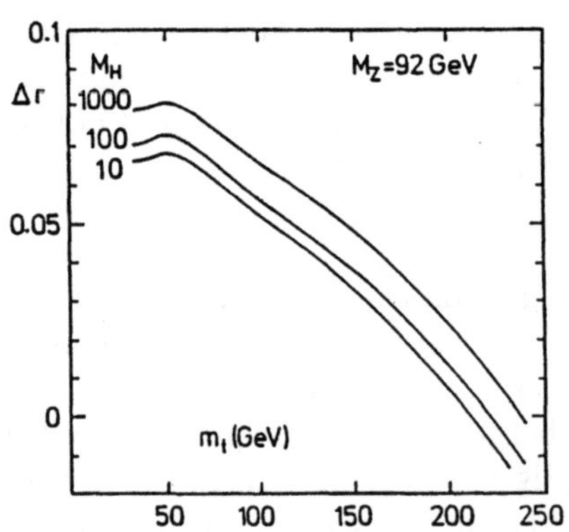

Fig. 4.5. The electroweak correction Δr as function of the top mass M_t for various Higgs masses M_H [8]

For large top masses the correction becomes large since the leading term to Δr is given by

$$(\Delta r)_{\text{top}} \approx -(\alpha/4\pi)(3\cos^2\theta_W/4\sin^2\theta_W)\cdot(M_t/M_W)^2 \tag{4.11}$$

hence a quadratic dependence on M_t. Since this contribution is negative it tends to cancel the other corrections and for $M_t \approx 230$ GeV the weak radiative correction would go to zero.

The correction Δr is relatively large but this fact can be readily understood physically [7]. In Eq. (4.10) the parameters α and G_F are specified from experiments done at very low energies whereas M_W and $\sin^2\theta_W$ involve a scale of $O(100\text{ GeV})$. One can expect that the correction should be small if the coupling constants are taken also at high energies. As can be shown G_F does not run whereas $\alpha(M_W^2) \approx 1/128$ and hence one can estimate

$$1 - \Delta r \approx \frac{\alpha}{\alpha(M_W^2)}\cdot\frac{G_F(M_W^2)}{G_F} \approx \frac{128}{137} = 0.934$$

which is very close to the properly calculated result.

Defining $\sin^2\theta_W$ by (4.1) requires a change of the relation (4.2) but also (4.3), the coupling constants for the neutral currents, have to be modified due to radiative corrections. This can be done by the replacement

$$J_3 - \sin^2\theta_W J_{\text{em}} \rightarrow J_3 - k\cdot\sin^2\theta_W J_{\text{em}} \tag{4.12}$$

where the correction k is relatively large since it contains leading $\alpha\log(M_Z/M_f)$ and $\alpha\,(M_t/M_Z)^2$-terms [8]. Correspondingly one has to modify also (4.3b) which now reads [7, 8]

$$\frac{\sigma(\nu_\mu e - \nu_\mu e)}{\sigma(\bar\nu_\mu e - \bar\nu_\mu e)} = \frac{3 - 12\sin^2\theta_W + 16\sin^4\theta_W + \alpha C[-12 + 32\sin^2\theta_W]}{1 - 4\sin^2\theta_W + 16\sin^4\theta_W + \alpha C[-4 + 32\sin^2\theta_W]} \tag{4.13}$$

where C has to be calculated from higher order graphs.

Now one can adopt two strategies:

i) one considers α, M_Z and $\sin^2\theta_W$ (together with Γ) as independent parameters to fit the line width. In this case it seems [21] that the dependence of the line shape on $\sin^2\theta_W$ is sufficiently strong due to the definition of K in (4.6a). It is the peak cross section which has to be determined very accurately. A detailed analysis showed that with a luminosity error of $\pm 2\%$ an error of $\delta\sin^2\theta_W = \pm 0.003$ could be obtained which is somewhat better than what has been achieved in neutrino scattering experiments ($\approx \pm 0.005$). A high precision experiment has to be performed, however, to determine $\sin^2\theta_W$ from the line shape but it might be possible. Usually it is thought that the weak angle would have to be determined from the asymmetries (Sect. 4.4).

ii) an alternative approach which is occasionally called [21] the modified-on-mass-shell procedure consists in replacing the quantity K in (4.6a) by

$$K = (G_F/\sqrt{2})\cdot(M_Z^2/8\pi\alpha) = 0.37239\,(M_Z/91\,\text{GeV})^2.$$

In this way the dependence of the line shape on $\sin^2\theta_W$ is weak and only G_F and M_Z (together with Γ) are free parameters for the line fit. The weak angle can then be calculated by the formula

$$\sin^2\theta_W = \frac{1}{2}\left[1 - \sqrt{1 - \frac{4A}{(1-\Delta r)M_Z^2}}\right] \tag{4.14}$$

with $A = \pi\alpha/\sqrt{2}G_F = (37.281\,\text{GeV})^2$. Since Δr is a function of M_Z and M_W and hence also of $\sin^2\theta_W$, Eq. (4.14) has to be solved by iteration. Of course, M_t and M_H are at present unknown parameters entering Δr. As has been shown above the Higgs mass does not enter in a significant way whereas the top mass does. Tables of radiative corrections for different masses have been published [8, 15].

Another important correction must be mentioned. The interference between the photon- and Z-propagator has the effect that the width of the Z is energy dependent. It can be shown [8, 21] that in very good approximation one can put

$$\Gamma(s) = (s/M_Z^2)\cdot\Gamma \tag{4.15}$$

where Γ is independent of energy. As a consequence of (4.15) the resonance denominator $D^2 = (s - M_Z^2)^2 + M_Z^2\Gamma^2$ becomes

$$D^2 = (s - M_Z^2)^2 + (s\Gamma/M_Z)^2\ . \tag{4.16}$$

This replacement causes a shift of the resonance peak of about 35 MeV to lower energies and therefore has a non-negligible effect.

A comparison of the MOMS and OMS procedure has been made by a group at Zeuthen and Dubna [24]. In Table 4.2 the errors resulting for M_Z and Γ are shown for the two methods, broken down according to the source of origin. The statistical error was calculated assuming a luminosity of $10^{31}\,\text{cm}^{-2}\text{s}^{-1}$ and 100% detection efficiency. A scan covering 13 energies in steps of 2 GeV was foreseen running for 10 days for each energy (in brackets errors for 2 days' runs). The error in the absolute luminosity calibration ΔL_{abs} was asumed to be 2%, in the relative luminosities $\Delta L_{rel} = 2\%/10\,\text{GeV}$, a beam spread of $\Delta\sqrt{s}_{rel} = 55\,\text{MeV}(\sqrt{s}/100\,\text{GeV})$ and an error of the absolute energy calibration $\Delta\sqrt{s}_{abs} \approx 10$ to 30 MeV. One notices

Table 4.2. Expected errors for the Z mass and width for two different kinds of approaches: MOMS (modified-on-mass-shell) and OMS (on-mass-shell) [12]

	ΔN_{stat}	ΔL_{rel}	ΔL_{abs}	$\Delta\sqrt{s}_{rel}$	$\Delta\sqrt{s}_{abs}$
MOMS					
ΔM_Z	13 (29)	7	1	15	$\Delta\sqrt{s}$
$\Delta\Gamma_Z$	16 (37)	5	2	5	0
OMS					
ΔM_Z	14 (32)	7	1	15	$\Delta\sqrt{s}$
$\Delta\Gamma_Z$	18 (41)	5	2	5	0
$\Delta\sin^2\theta_W$	0.0015 (0.0033)	0.0005	0.0030	0.0005	0

that the errors of M_Z and Γ are similar for the two methods. The reason, of course, is that $\sin^2\theta_W$ is derived essentially from the peak cross section which on the other hand has little efect on M_Z and Γ. These authors have also investigated the best strategy as to how many energy points should be included in the scan. The results will be discussed in Sect. 4.3.

4.2.2.2 Introduction of an Effective Weak Angle

The fact that the on-mass-shell definition of $\sin^2\theta_W$ leads to large radiative corrections and in particular to a sizeable influence of the top mass is used sometimes as an argument for introducing a different definition of $\sin^2\theta_W$ which leads to smaller corrections and which are less dependent on M_t. This procedure is less accurate but should suffice for the first round of LEP experiments [8, 25].

In (4.12) one puts $k = 1$ for one particular process involving neutral currents expecting that the radiative corrections for other reactions with neutral currents would also be small. By doing this one has to introduce an effective weak angle which will be denoted by a bar. The relation between the weak angle defined in this way with the previously defined one turns out to be

$$\overline{\sin^2\theta_W} = \sin^2\theta_W + \cos^2\theta_W\,\delta\rho \tag{4.17}$$

where $\delta\rho$ is a correction. Also a new radiative correction $\overline{\Delta r}$ is introduced by

$$\overline{\sin^2\theta_W}\,\overline{\cos^2\theta_W} = (A/M_Z^2)/(1-\overline{\Delta r}) \tag{4.18}$$

instead of (4.10). The advantage is that $\overline{\sin^2\theta_W}$ and $\overline{\Delta r}$ depend little on the top mass (Fig. 4.6).

Using this effective weak angle one can derive an "improved Born approximation" which allows to fit the Z-line in an almost model independent way, i.e. without a knowledge of the top and Higgs mass. For the cross section one finds [8, 15]

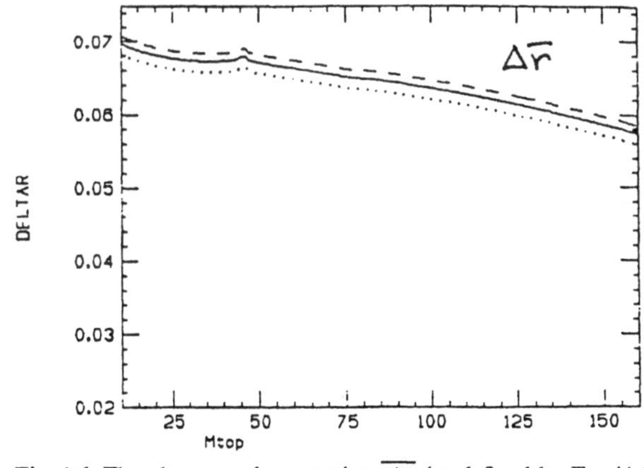

Fig. 4.6. The electroweak correction $\overline{\Delta r}$ (as defined by Eq. (4.18)) as function of the top mass M_t. The influence of various Higgs masses is indicated by the dashed (1000 GeV) and pointed lines (10 GeV) [15]

$$\sigma(e^+e^- \to \mu^+\mu^-) \cong 12\pi\{\Gamma_e \Gamma_\mu / D^2\} \left[\frac{s}{M_Z^2} + R_f \frac{s - M_Z^2}{M_Z^2} + \frac{\Gamma}{M_Z} J_f + \ldots\right] \quad (4.19)$$

where

$$R_f \cong \frac{2Q_e Q_f V_e V_f}{(V_e^2 + A_e^2)(V_f^2 + A_f^2)} \overline{\sin^2\theta_W}\,\overline{\cos^2\theta_W} + \ldots$$

For the fit one has the 3 parameters M_Z, Γ and R_f and the normalisation of the cross section which is essentially determined by $\Gamma_e \Gamma_\mu (1 + \Gamma/M \cdot J_f)$. Tables for the various quantities have been prepared for different M_t and M_H [15].

Expression (4.19) can be folded analytically with the QED corrections. The formula is rather lengthy [15] and will not be reproduced here. It should be accurate to about 1 to 2% and should be adequate for the first round of LEP experiments. This modified Born approximation is, however, not sufficiently accurate for the asymmetries which will be discussed in Sect. 4.4.

4.2.3 Hadronic Decays and QCD Corrections

Corrections due to the strong interaction are, of course, only relevant for Z-decays into hadrons. They can simply be taken into account by multiplying the colour factor c_f by a correction factor P^{ecc} which in lowest order is

$$c_f \to c_f[1 + \alpha_s(M_Z^2)/\pi + \ldots] \approx c_f(1.04 \pm 0.01)$$

Recently not only second order but also third order graphs have been calculated [26] with the result (N_f number of flavours)

$$1 + \alpha_s(M_Z^2)/\pi + (1.98 - 0.115 N_f)[\alpha_s(M_Z^2)/\pi]^2 + 64.835[\alpha_s(M_Z^2)/\pi]^3 \,. \quad (4.20)$$

The last coefficient is surprisingly large but nevertheless the second and third order terms are (with $\alpha_s(M_Z^2) = 0.12 \pm 0.02$) not significant for the accuracies aimed at here.

4.2.4 Summary of the Radiative Corrections

As has been explained radiative corrections, both QED and electroweak, are important for the interpretation of experimental data even for experiments with moderate luminosity. Recently perfect agreement between different groups of theoreticians could be achieved in the methods and also the results for the radiative corrections. A summary of the results obtained by different working groups has been presented by Altarelli [15]. A number of tables for the radiative corrections for different parameters and energies were given. Here we shall summarize these results in the form of "rules of thumb" which relate the peak energy with the Z mass and its width with the points at half-maximum $\sqrt{s_+}$ and $\sqrt{s_-}$:

for $e^+e^- \to \mu^+\mu^-$

$\sqrt{s_{peak}} = M_Z + 130\,\text{MeV} - 35\,\text{MeV} = M_Z + 95\,\text{MeV}$
$\Gamma = (\sqrt{s_+} - \sqrt{s_-})/1.16$

where the first term comes from QED and the second from $\Gamma(s)$ (see (4.15)).

for $e^+e^- \to$ hadrons

$\sqrt{s_{peak}} = M_Z + 96\,\text{MeV}$
$\Gamma = (\sqrt{s_+} - \sqrt{s_-})/1.15$

with

$$(\sqrt{s_+} - \sqrt{s_-})_{hadrons} = (\sqrt{s_+} - \sqrt{s_-})_{\mu\mu} - 35\,\text{MeV}.$$

From precision measurements it will also be possible to determine $\sin^2\theta_W$ with an acuracy comparable to that obtained from the asymmetries.

The uncertainties of Δr are ± 0.0007 from the experimental uncertainty of R_{ee} characterizing the contributions of the light quarks.

For top masses between 75 GeV $< M_t <$ 200 GeV Δr would vary between 0.06 and 0 (and $\overline{\Delta r}$ between 0.07 and 0.05).

The influence of the Higgs is much weaker and is about $\Delta r \approx \pm 0.011$.

If the top and Higgs masses were known Δr could be calculated with a precision of about ± 0.0013. This value should set the standard for experimental accuracies one should try to obtain.

4.3 Strategy to Determine the Z-Parameters from the Line Shape

The first aim of LEP experiments will be the determination of M_Z and Γ. (How the number of neutrino families can be deduced from Γ will be discussed later.) The next aim is to determine $\sin^2\theta_W$ which, together with α and G_F, allows to determine Δr. This then can be compared with the theoretical predictions of the SM. Apart from checking the SM from such a comparison one might get information on the masses of the top and even of the Higgs. If one of these particles would be found more rigorous test could be performed.

Let us discuss what experimental accuracy would be needed for M_Z and $\sin^2\theta_W$ to carry out a meaningful test and whether it will be possible to achieve these accuracies at LEP.

For the sensitivity of Δr on M_Z and $\sin^2\theta_W$ one easily finds from (4.12)

$$\delta\Delta r = 2(1 - \Delta r)(\delta M_Z/M_Z) \quad \text{and}$$
$$\delta\Delta r = (8/3)\,\delta(\sin^2\theta_W) \tag{4.21}$$

From these relations one derives the needed experimental accuracies provided the experimental and theoretical errors of Δr are required to be both $\approx \pm 0.0013$:

$$\delta M_Z = \pm 50\,\text{MeV}$$
$$\delta(\sin^2\theta_W) = \pm 0.004. \tag{4.22}$$

It was mentioned above that statistical errors smaller than (4.22) can be achieved and one has now to consider systematic experimental errors. These have been studied by several authors [27, 13]:

- background: for Z – muon decays even after only mild cuts the background should be negligible. The trigger efficiency should be close to 100%.
- luminosity: absolute measurement not needed for M_Z and Γ
  relative scale error: linear in E_{beam} affects M_Z, not Γ
  estimate $\approx 2\%, \delta M_Z = \pm 10\,\text{MeV}$
  quadratic in E_{beam} affects Γ, not M_Z
  estimate $\approx 1\%,\ \delta\Gamma = \pm 10\,\text{MeV}$.
- beam energy
  from the field strength of dipole magnets $\Delta B/B \approx 3 \cdot 10^{-4}$ (due to temperature changes, etc.)
  resulting $\delta M_Z = \pm 30\,\text{MeV}$
  from the transverse polarization $\Delta B/B \approx 10^{-4}$ and short term instabilities of about same size
  resulting $\delta M_Z = \pm 10\,\text{MeV}$
- energy spread of beam
  $\Delta E/E \approx 0.7 \cdot 10^{-3}, \Delta E \approx 64\,\text{MeV}$
  no effect on mass and width.

First we discuss a low statistics experiment with a total luminosity of about 30 pb^{-1}.

A relatively crude scan will be performed with, let us say, 12 energy points between 80 and 104 GeV. With about 2 pb^{-1} per point, i.e. a total luminosity of 24 pb^{-1} one would have about 4500 pairs of which about 2000 in the peak. This could be done in 1989/90. The expected errors are:

Table 4.3

	δM_Z (MeV)	$\delta\Gamma$ (MeV)
Statistics	15	40
rel. Luminosity	10	5
abs. Energy (Field)	30	5
(Polaris.)	(10)	
QED corr.	15	
Total	≈ 50	≈ 45

With transverse polarization to monitor the energy one could reduce the error of M_Z to about 20 MeV and with appreciably more statistics the error of Γ could also be brought down to about 20 MeV (Fig. 4.7). Even without these later improvements the errors are sufficiently small to make a crude test of the SM and to count the number of neutrino families.

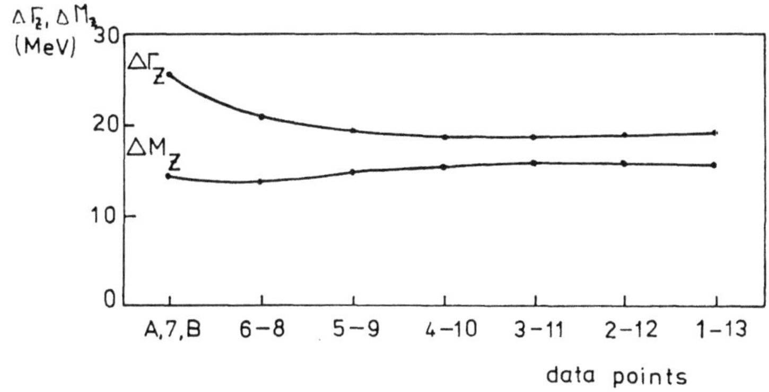

Fig. 4.7. The errors of the mass and total width of the Z for different numbers of data points (scanned energies). Point 7 is at the peak, (6 and 8), (5, 6 and 8, 9), etc. are symmetrically distributed on both sides of the peak. The points A and B are taken at $M_Z \pm \Gamma/2$ [12, 21]

A high statistics experiment (200 pb^{-1} or more) would allow an independent determination of $\sin^2\theta_W$ as was explained in Sect. 4.2.2. In this case because of statistics it would be more favourable to stay close to the Z-peak (see e.g. [25]) and scan only 3 to 5 points. The statistical errors of M_Z could then be reduced to less than 10 MeV. From a determination of Δr limits on the top ($M_t > 150\,\text{GeV}$) and the Higgs mass ($M_H \approx 800$ to 1000 GeV) may be obtained.

All these considerations do not take into account the acceptances and trigger inefficiencies of real detectors. Indeed cuts will in some cases increase the higher order corrections. It might also happen that due to bremsstrahlung emission in the initial state the topology of events might change (e.g. of hadronic events) which leads to efficiencies which could change with energy. A considerable amount of work has been done by the LEP groups but since these results apply in most cases only to a particular detector we shall not consider them here (see e.g. [28, 29]).

4.4 Asymmetries at the Z^0 Peak

4.4.1 Definition of Asymmetries

Further test of the SM model are possible if one measures asymmetries at the Z^0 peak (for summaries see [8, 30, 31, 32]). One can distinguish the following kinds of asymmetries:

a) Forward–Backward Asymmetry (or Charge Asymmetry). One can determine the forward-backward asymmetry A_{FB} if one measures the number $N(F)$ of fermions emitted under an angle Θ with respect to the direction of the incident electron (Fig. 4.8) and the number of the same fermions $N(B)$ emitted under the angle $(180° - \Theta)$ with the definition

$$A_{FB}(e^+e^- \to f\bar{f}) = \frac{N(F) - N(B)}{N(F) + N(B)} \quad \text{at } \Theta \tag{4.23a}$$

The charge asymmetry can be defined by

$$A_{ch}(e^+e^- \to f\bar{f}) = \frac{N(f) - N(\bar{f})}{N(f) + N(\bar{f})} \quad \text{at } \Theta \tag{4.23b}$$

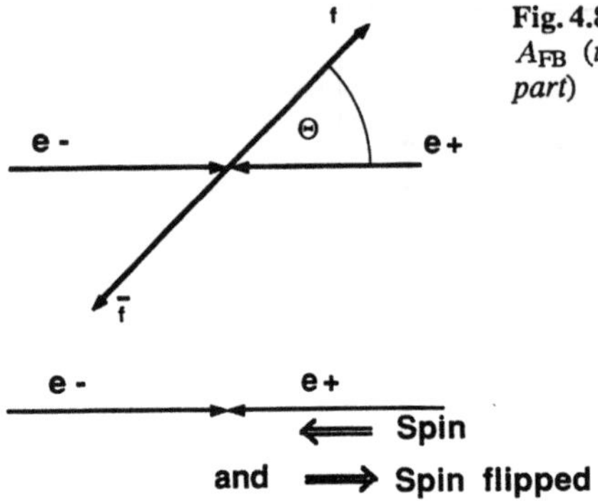

Fig. 4.8. Definition of the forward-backward (charge) asymmetry A_{FB} (*upper part*) and the polarization asymmetry A_{pol} (*lower part*)

where one measures the number $N(\mathrm{f})$ of fermions and the number $N(\bar{\mathrm{f}})$ of antifermions emitted under the same angle Θ with respect to the direction of the incident electron (Fig. 4.8).

For two-body reactions such as $e^+e^- \to f\bar{f}$ the two asymmetries A_{FB} and A_{ch} are identical. Because of small rates one very often integrates over the hemisphere $0 < \Theta < \pi/2$.

b) Asymmetry with Final State Polarization. If the fermion f is in the final state produced with left or right helicity (L, R) one can define

$$A^{\mathrm{pol}}(e^+e^- \to f\bar{f}) = \frac{N(f_L) - N(f_R)}{N(f_L) + N(f_R)} \quad \text{at } \Theta \tag{4.24}$$

where one has to measure the helicity of the fermion. The only practical case seems to be the reaction $e^+e^- \to \tau^+\tau^-$ where the tau helicity can be determined from its decay products.

This asymmetry gives the same information as the left-right asymmetry provided universality for the coupling constants holds.

c) Left-right Asymmetry with Polarized Beam. If the incident electron beam is 100% longitudinally polarized (Fig. 4.8) and if the polarization is flipped to provide electrons with either left (L) or right (R) helicity one can measure the asymmetry

$$A_{LR}(e^+e^- \to f\bar{f}) = \frac{N(e_L) - N(e_R)}{N(e_L) + N(e_R)} \quad \text{at } \Theta\,. \tag{4.25}$$

Again one can integrate over the hemisphere.

d) Polarized Forward-Backward Asymmetry. Sometimes an asymmetry is considered where one combines the forward-backward with the left-right asymmetry and which is defined by

$$A_{LR}^{\mathrm{pol}}(e^+e^- \to f\bar{f}) = \frac{N(e_{L,F}) - N(e_{L,B}) - [N(e_{R,F}) - N(e_{R,B})]}{N(e_{L,F}) + N(e_{L,B}) + N(e_{R,F}) + N(e_{R,B})} \tag{4.26}$$

at Θ and with a 100% longitudinally polarized electron beam (for formulae with partially polarized beams see [32]).

e) Transverse Polarization Asymmetry. The polarization asymmetries discussed so far require a longitudinal beam polarization which is difficult to provide (Sect. 2). On the other hand a transverse polarization perpendicular to the beam plane is naturally obtained by the emission of synchrotron radition. Hence one can define a transverse polarization asymmetry which corresponds to an azimuthal asymmetry by

$$A_T = N(\phi)/N(\phi + 180°) \tag{4.27}$$

where ϕ is the azimuthal angle around the beam direction counted from the direction of the polarization.

As will be shown A_T measures essentially the same information as the unpolarized leptonic forward-backward asymmetry if the leptonic couplings are universal. Hence it is normally considered as of little interest. For the general formulae for partially polarized beams see [31, 32].

For a particular experiment it will be necessary to integrate over a specific range of angles Θ. Therefore we give the differential cross section for the process $e^+e^- \to f\bar{f}$ as function of the c.o.m. energy $\sqrt{s}$ and angle Θ

$$\frac{d\sigma}{d\cos\Theta} = \frac{2\pi\alpha^2}{s}[(1+\cos^2\Theta)F_1(s) + 2\cos\Theta F_2(s)] \quad \text{with} \tag{4.28}$$

$$F_1 = 1 + \frac{2s(s-M_Z^2)}{D^2}\cdot\frac{V_e^2}{16K} + \frac{s^2}{D^2}\cdot\frac{(V_e^2+A_e^2)^2}{(16K)^2} \quad \text{and}$$

$$F_2 = \frac{2s(s-M_Z^2)}{D^2}\cdot\frac{A_e^2}{16K} + \frac{4s^2}{D^2}\cdot\frac{(V_e^2A_e^2)^2}{(16K)^2}\,.$$

The quantities D and K are defined in (4.6) and (4.6a) (see also Sect. 4.2.2.2).

To get the energy dependences alone one can integrate over the angle and obtains e.g.

$$A_{FB} = \frac{3}{2}\cdot\frac{A_e^2B_1 + 2V_e^2A_e^2B_2}{1+2V_e^2B_1+(V_e^2+A_e^2)^2B_2} \quad \text{with} \tag{4.29}$$

$$B_1 = gM_Z^2 s(s-M_Z^2)/D^2$$
$$B_2 = g^2M_Z^4 s^2/D^2$$

where $g = K/M_Z^2 \approx 4.5\times10^{-5}$.

At the resonance one has $s = M_Z^2,\ B_1 = 0,\ B_2 = (KM_Z/\Gamma)^2$.

At the Z^0 peak the Z diagram dominates and one can neglect the photon diagram since its contribution is of order $(\Gamma/M_Z)^2 \approx 7\times10^{-4}$. Integrating over Θ one obtains for the different asymmetries :

$$\begin{aligned} A_{FB} &\approx (3/4)A^e A^f \\ A^{pol} &\approx 2A^f \\ A_{LR} &\approx 2A^e \\ A_{FB}^{pol} &\approx (3/4)A^f \end{aligned} \tag{4.30}$$

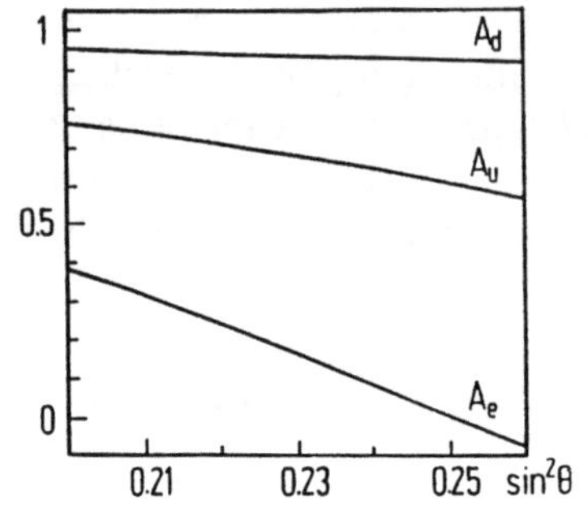

Fig. 4.9. The quantity A^{f} (see Eq. (4.30)) for quarks (u, d) and leptons (electrons) [8]

where $A^{\mathrm{e}} = [2V_{\mathrm{e}}A_{\mathrm{e}}/(V_{\mathrm{e}}^2 + A_{\mathrm{e}}^2)]$ and $A^{\mathrm{f}} = [2V_{\mathrm{f}}A_{\mathrm{f}}/(V_{\mathrm{f}}^2 + A_{\mathrm{f}}^2)]$ (e = electron, f = fermion in final state). In the SM one has $A_{\mathrm{e}} = A_\mu = A_\tau = 1$ and $V_{\mathrm{e}} = V_\mu = V_\tau = \nu = 1 - 4\sin^2\theta_{\mathrm{W}}$. The dependence of the A^{f} for different fermions on $\sin^2\theta_{\mathrm{W}}$ is shown in Fig. 4.9 [8]. As one can see A^{lepton} depends much more on $\sin^2\theta_{\mathrm{W}}$ than A^{quark}.

Finally it may be mentioned that $A_{\mathrm{T}} \approx -1 + V_{\mathrm{e}}^2/A_{\mathrm{e}}^2$ which gives essentially the same information as A_{FB}.

With these relations the approximate $\sin^2\theta_{\mathrm{W}}$ - dependence of the asymmetries is obtained and shown in Table 4.4.

Table 4.4

$A_{\mathrm{FB}} \approx 3\nu^2$	$\delta A_{\mathrm{FB}} \approx 24\nu\ \delta\sin^2\theta_{\mathrm{W}}$
$A^{\mathrm{pol}} \approx 2\nu$	$\delta A^{\mathrm{pol}} \approx 8\ \delta\sin^2\theta_{\mathrm{W}}$
$A_{\mathrm{LR}} \approx 2P_{\mathrm{e}}\nu$	$\delta A_{\mathrm{LR}} \approx 8P_{\mathrm{e}}\ \delta\sin^2\theta_{\mathrm{W}}$
$A_{\mathrm{FB}}^{\mathrm{pol}} \approx (3/2)P_{\mathrm{e}}\nu$	$\delta A_{\mathrm{FB}}^{\mathrm{pol}} \approx 6P_{\mathrm{e}}\ \delta\sin^2\theta_{\mathrm{W}}$

with $\nu = 1 - 4\sin^2\theta_{\mathrm{W}}$ and P_{e} longitudinal electron polarization.

To provide a feeling on how large these asymmetries are their values are given [33] for $\sin^2\theta_{\mathrm{W}} = 0.23$ (in lowest order without radiative corrections) in Table 4.5.

Table 4.5.

f	A_{FB}	A^{pol}
μ, τ	0.019	0.119
u, c	0.080	0.505
d, s, b	0.112	0.702

As one notices the asymmetries for leptons are quite small (and with the new value $M_{\mathrm{Z}} = 91$ GeV even smaller) whereas those for quarks are much bigger. Flavour

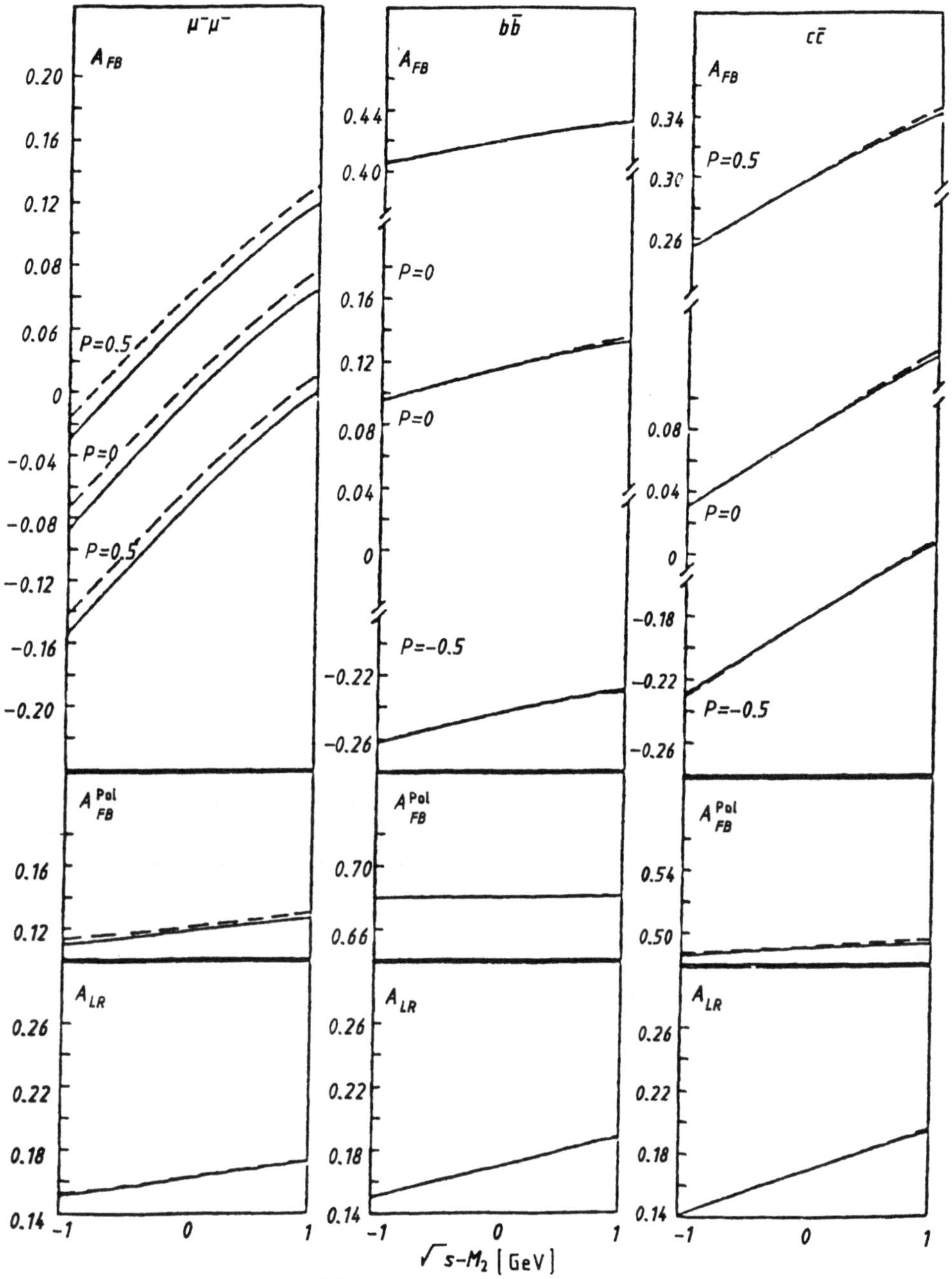

Fig. 4.10. The asymmetries A_{FB}, $A_{\mathrm{FB}}^{\mathrm{pol}}$, A_{LR} (= A_{pol}) as function of the c.o.m. energy around the Z-peak [32]

tagging could therefore become quite important and it will be discussed later. One further sees that A^{pol} is much larger than A_{FB} and hence less statistics is required, but to measure the helicity in the final state is not so easy. The value for A_{LR} is equal to that for A^{pol} for 100% polarized electrons. It is proportional to the electron polarization and in order to profit from its larger value it is necessary to measure P_e with an error of less than 10%. An advantage of A_{LR} is that any channel can be used since it suffices to measure the counting rate when flipping the electron spin.

The energy dependence of the various asymmetries is shown in Fig. 4.10. One sees that A_{FB} for leptons has a very steep dependence whereas the other asymmetries, in particular those of the heavy quarks change much less. This is important for the radiative corrections as will be discussed in the following sections.

One can make the following general remarks concerning the difficulties to measure these asymmetries:

- A_{FB} is proportional to ν^2 and therefore small. The error depends explicitly on ν and hence on $\sin^2\theta_W$ and is large. If one wants to extract $\sin^2\theta_W$ from a measurement of A_{FB} one finds $\delta\sin^2\theta_W \approx (2 \text{ to } 1) \times \delta A_{FB}$ depending one the precise value of $\sin^2\theta_W$ (assumed to vary from 0.22 to 0.24) and hence the error is large. These circumstances compensate to a large extent the relative ease with which A_{FB} can be measured. An additional difficulty arises from the fact that because of the steep energy dependence QED corrections are large (up to a factor 2) and have to be taken into account carefully (see next section).
  In principle A_{FB} and A_{ch} give the same information. From an experimental point of view, however, they are different. A_{ch} involves the difference between acceptances for particles with opposite charge, whereas A_{FB} requires a comparison of acceptances in different hemisperes.
- A^{pol} is more difficult to measure but more sensitive to $\sin^2\theta_W$, since $\delta\sin^2\theta_W \approx (1/8)\,\delta A^{pol}$ and independent of $\sin^2\theta_W$.
- A_{LR} gives in principle the same information (except that initial and final state are interchanged) as A^{pol} but requires longitudinally polarized beams which are not easy to get, and the error of P_e enters in the error of $\sin^2\theta_W$. A measurement of A_{LR} and A^{pol} in the same experiment is interesting since it would allow a simultaneous test of universality of the coupling constants.

4.4.2 A_{FB} and A_{LR} for Leptons in the Final State

The forward-backward and left-right asymmetries show a very strong dependence on $\sin^2\theta_W$ as is shown in Fig. 4.11 and hence it is thought that a good value of $\sin^2\theta_W$ could be obtained from its measurement. However, A_{FB} also changes rapidly with energy (Fig. 4.10) and therefore is very sensitive to radiative corrections. In particular QED corrections are important as Fig. 4.12 shows. The change $\delta A_{FB}^{QED} \approx 0.02$ due to QED corrections is much larger than the expected experimental accuracy of about 0.005 and therefore QED corrections have to be taken into account in a very precise way. Fortunately the calculations of different groups agree now with a difference of only about 0.03% which is fully sufficient compared to the expected experimental errors [15]. On the other hand they have to be folded with the experimental acceptance which might reduce the accuracy considerably.

The weak corrections and therefore A_{FB}, A_{LR}, etc are, of course, again functions of $\sin^2\theta_W$, M_{top} and M_H and one can consider similar strategies as to the determination of the main parameters as in Sect. 4.3. The parameters deduced from the asymmetries should agree with those derived from the line shape and thus a test of

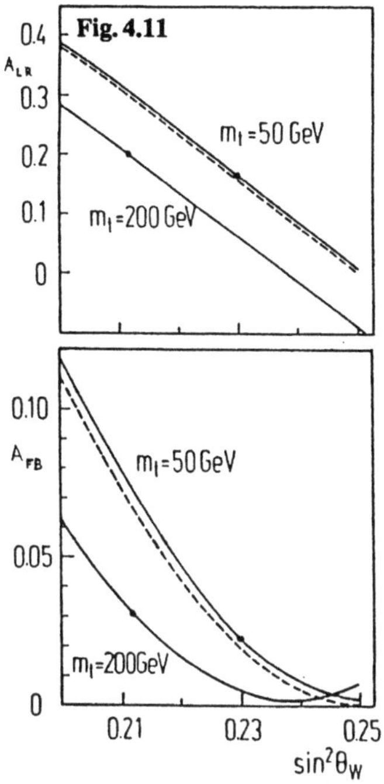

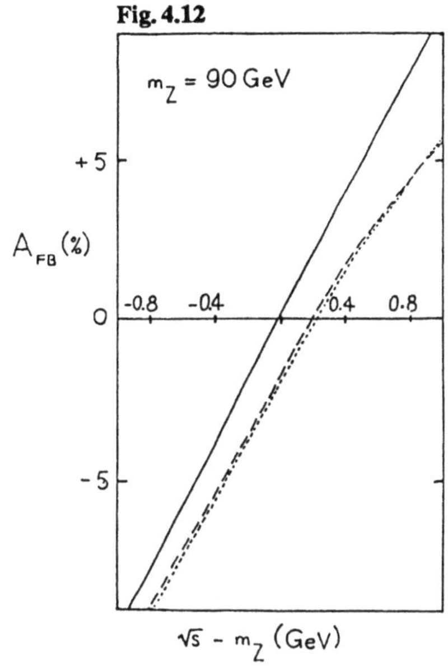

Fig. 4.11. The asymmetries A_{FB} and A_{LR} as a function of the weak angle at the Z-peak (92 GeV) in lowest order (*full lines*) and with weak corrections included (*dashed line*). The different curves correspond to different top masses. A Higgs mass of 100 GeV was assumed [8]

Fig. 4.12. First-order radiative corrections to the forward-backward asymmetry A_{FB} near the Z-peak. Solid line: Born approximation; short-dashed: pure QED corrections; long-dashed line: first-order weak corrections included (parameters $\sin^2\theta_W = 0.2435$, $\Gamma = 2.26$ GeV) (Ref. Altarelli et al. in [1])

the SM is obtained. It can be noted that A_{LR} is relatively sensitive to M_H (more than A_{FB}) but unfortunately the dependence on M_{top} is even stronger.

The weak corrections have recently been reviewed [15] and contrary to the past an excellent agreement between different groups has been achieved with differences being reduced to less than 0.2%. This is important since in contrast to the Z^0 line shape exact formulae are mandatory for the asymmetries due to their small values. Results are shown for A_{FB} in Fig. 4.13.

The left-right asymmetry A_{LR} and A^{pol} are practically independent of the fermion final state species and lend themself therefore to an inclusive measurement. They are practically indepent of QED corrections since they depend only on P_e or the helicity in the final state and therefore the correction cancels when the helicity (or electron spin for A_{LR}) is flipped. On the other hand their dependence on M_t and even on M_H are relatively strong (Figs. 4.11 and 4.13).

The experimental accuracy that can be achieved for A_{FB} has been discussed in detail by Baillon et al. [34]. With a luminosity of $10^{31}\,\mathrm{cm}^{-2}\mathrm{s}^{-1}$ and a run of 200 days (giving about 200,000 muon pairs) a statistical accuracy of about 0.5% for $\sin^2\theta_W$ is estimated. Can a similar systematic accuracy be obtained? One can derive from Eq. (4.29) the following relation

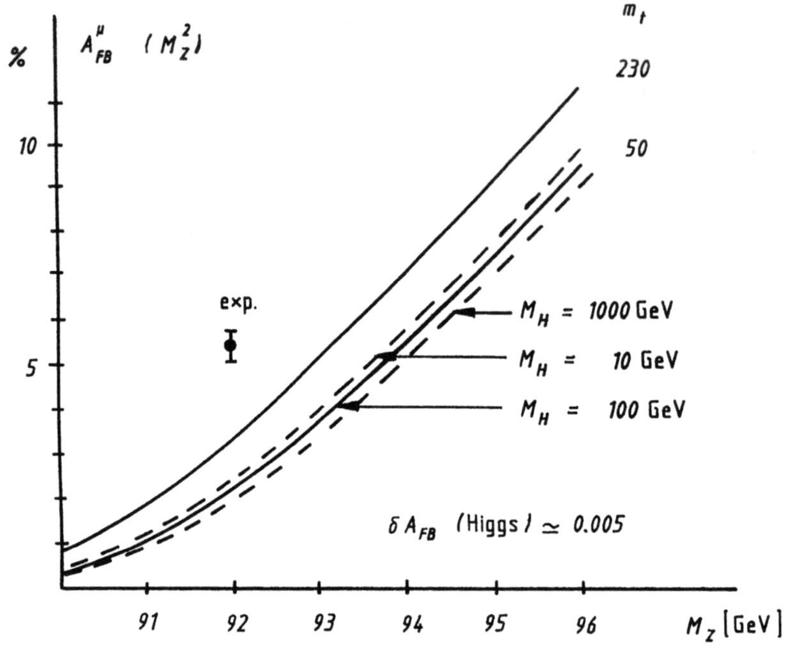

Fig. 4.13. The forward-backward asymmetry $A_{\rm FB}$ as function of the Z-mass including the weak corrections. The pure QED corrections are subtracted. The different curves show the influence of the top and Higgs mass [15]

$$\frac{\delta\nu}{\nu} = \frac{1}{2}\frac{\delta A_{\rm FB}}{A_{\rm FB}} + \frac{1}{4}\frac{\delta\sigma_{\rm tot}}{\sigma_{\rm tot}} + \frac{1}{2}\frac{\delta\Gamma}{\Gamma} - \frac{1}{4}\frac{M_{\rm Z}}{\Gamma}\frac{1}{\nu^2}\frac{1}{\sqrt{B_2}}\frac{\delta(s - M_{\rm Z}^2)}{s} \tag{4.31}$$

in which the last term is the most relevant one because of the factor $1/\nu^2$ which is of order 100. The absolute energy calibration is not so important since $s - M_{\rm Z}^2$ enters. With a relative energy accuracy of 2×10^{-4} one can estimate that the error of $\sin^2\theta_{\rm W}$ will be less than the statistical error.

This leaves as the main systematic error the different efficiencies for oppositely charged particles. To achieve $\delta A_{\rm ch} = 0.002$ the error of the difference of the acceptances has to be smaller than 0.4%. At PETRA and PEP errors of less than 1% could be obtained but there the background due to Bhabha scattering, tau pairs etc. was larger and therefore one might hope that the required accuracy can be achieved.

In conclusion it seems that by taking data for about one year $A_{\rm FB}$ could be measured with an accuracy of about ± 0.005 yielding an error of $\sin^2\theta_{\rm W}$ of about ± 0.001 depending on its precise value. The main errors are statistical and uncertainties of the acceptance. A very careful elimination of QED effects will also be essential. With a Z^0 mass of about 91 MeV $A_{\rm FB}$ will be quite small. It might therefore be important to measure its energy dependence in the neighbourhood of the resonance peak. This would not decrease the statistics remarkably but would give much more confidence in the measured value.

4.4.3 Asymmetries for Heavy Quarks

The asymmetries for quark decays are quite interesting since

- they can be larger than those for leptons,
- they are much less energy dependent (Fig. 4.10) and hence less sensitive to radiative corrections than the asymmetries for leptons,
- they are very sensitive to $\sin^2\theta_W$,
- they are more sensitive to some new physics even at the Born level, e.g. for additional Z.

Of course, QCD corrections have to be applied and one finds [35, 36]

$$\delta A_{FB}^{QCD}/A_{FB} \approx (\alpha_s/\pi)(1/3 + k) \tag{4.32}$$

where $0 < k < 1$. For $\alpha_s(M_Z)/\pi \approx 0.045$ the estimate $k \approx 0.2$ is obtained. Therefore one expects that $\delta A_{FB}^{QCD}/A_{FB} < 0.015$ which is acceptable.

The electroweak corrections have to be integrated into a strategy depending on which parameters are to be determined in a similar way as discussed in Sect. 4.3.

The essential question is, of course, whether individual flavours can be separated ("tagged"). A strategy has been developed for this purpose by Drees et al. [36] which consists of the following steps:

- define a signature for a given flavour, e.g. for
  b pairs: lepton-kaon pair of same charge signs in one jet and with an angle $> 5°$ between lepton and kaon (to suppress c pairs)
  c pairs: lepton-kaon pair of same charge signs in one jet and with an angle $< 5°$ between lepton and kaon (to suppress b pairs)
  s pairs: leading K on both sides
  u pairs: leading proton
- estimating the efficiency of the procedure and its uncertainty
- estimating the purity and its inaccuracy.

To apply these methods a careful comparison between different fragmentation models and the experimental data for quark fragmentation will be essential.

To check the feasibility of flavour tagging with a real detector Monte Carlo simulations were performed [36] for DELPHI. Besides the signature mentioned above ("sequential cuts") other methods were used. The results are shown in Table 4.6. The systematic experimental errors were also investigated by these authors. They also consider a combination of different asymmetries in order to obtain quantities which are more sensitive to new physics and depend less on corrections.

A summary of the errors of A_{FB} expected for different final states is given in Table 4.7. One notices that a measurement of A_{FB} for quarks is quite competitive with that for leptons.

For the b-quark the $b\bar{b}$ mixing must be taken into account [37, 15] by

$$A_{measured} = A_{FB}(b)(1 - 2\chi) \tag{4.33}$$

where χ is the mixing parameter. With

Table 4.6. Errors of the forward-backward asymmetry A_{FB} for muons and different quark pairs [32]

Measurement	Value of the asymmetry	Error	Precision in $\sin^2\theta_W$
$A_{FB}^{\mu\mu}$	0.02	0.0035	0.0017
$A_{FB}^{s\bar{s}}$	0.12	0.007	0.0012
$A_{FB}^{c\bar{c}}$	0.08	0.007	0.0015
$A_{FB}^{b\bar{b}}$	0.12	0.005	0.0009
$A_{FB}^{u\bar{u}}$	0.08	0.01	0.0020

Detail for $A_{FB}^{\mu\mu}$:

Statistical error	$\pm$ 0.002
Systematic error due to uncertainty on M_Z	$\pm$ 0.0015
Systematic error due to QED correction	$\pm$ 0.002
Systematic error due to apparatus effects	$\pm$ 0.001
	$\pm$ 0.0035

Table 4.7. Simulations for flavour tagging for the DELPHI detector [36]

quark		efficiency (%)	background (%)								
					u,c		d,s		$\bar{u},\bar{c}$		$\bar{d},\bar{s},\bar{b}$
b	(sequential cuts)	6.3	16.1	=	1.2	+	2.6	+	5.5	+	6.8
	(Fisher analysis)	11.2	16.1	=	1.2	+	1.4	+	7.8	+	6.5
					u		d,s,b		$\bar{u},\bar{c}$		$\bar{d},\bar{s},\bar{b}$
c	(sequential cuts)	2.1	35.4	=	0.4	+	24.9	+	0.9	+	9.3
	(D^* analysis)	1.8	32.2	=	1.8	+	24.2	+	2.2	+	4.0
					u,c		d,b		$\bar{u},\bar{c}$		$\bar{d},\bar{s},\bar{b}$
s	($K^{+/-}$ analysis)	3.4	45.0	=	20.8	+	10.6	+	7.3	+	6.3
	(K^{*0} analysis)	2.4	46.1	=	15.4	+	7.8	+	6.6	+	16.3
					c		d,s,b		$\bar{u},\bar{c}$		$\bar{d},\bar{s},\bar{b}$
u		1.4	30.5	=	6.6	+	16.0	+	1.4	+	6.6

$$\delta A_{FB}(b) = \frac{\delta A_{meas}}{1-2\chi} + A_{meas} \cdot \frac{2\chi}{(1-2\chi)^2} \cdot \frac{\delta\chi}{\chi} \tag{4.34}$$

and an expected $\chi = 0.15 \pm 0.02$ and $\delta A_{meas} \approx 0.005$ one finds that the two errors contribute about equal amounts to the total error of about $\delta A_{FB}(b) \approx 0.01$.

The forward-backward asymmetry for b-jets in Z-decays has recently been studied in detail by Djouadi et al. [37].

4.4.4 Polarization Asymmetry for Tau Decays

The only case where it seems possible to measure the polarization in the final state seems to be the process $e^+e^- \to \tau^+\tau^-$ which has a branching ratio of about 3%. As will be explained the polarization of the τ can be determined from its decay products [38, 39]. Because of the short lifetime of the τ it travels only a few millimeters and hence its decay products can be easily detected. To measure the polarization of muons in the final state seems difficult since, because of their long lifetime, most of them leave the detector. The only way to detect the polarization of electrons in the final state would be to measure the circular polarization of the emitted bremsstrahlung which is completely hopeless.

The τ decays into the following channels with the branching ratios indicated:

τ	$\rightarrow$	$\mu\,\nu\,\bar{\nu}$	$18.5 \pm 1.1\ \%$
		$\mathrm{e}\,\nu\,\bar{\nu}$	$16.5 \pm 0.5\ \%$
		$\pi\,\nu$	$10.3 \pm 1.2\ \%$
		$\rho\,\nu$	$22.1 \pm 2.4\ \%$

In the detector tau pairs appear as low-charged multiplicity events and since the τ leptons are quite energetic the topology of the events are two almost back-to-back jets, each containing 1 or 3 charged particles. If one discards events with 2 electrons (Bhabha scattering) and 2 muons it is expected that the background is very small and the detection efficiency should still be around 90%.

The polarization of the τ is given by [38]

$$P_\tau(\cos\theta) \approx A^\tau + (P_\mathrm{e} + A^\mathrm{e})\frac{2\cos\theta}{1+\cos^2\theta} \tag{4.35}$$

where P_e is the longitudinal electron polarization and θ is the angle between the τ^- and the e^- directions (Fig. 4.14). The combinations of coupling constants A^τ and A^e have been defined in (4.30).

A longitudinal polarization would help to investigate the universality of coupling constants. If on the other hand this is taken for granted ($A^\tau = A^\mathrm{e}$) and with no longitudinal beam polarization one obtains

$$P_\tau(\cos\theta) \approx A^\mathrm{e}\frac{(1+\cos\theta)^2}{1+\cos^2\theta}\ . \tag{4.35a}$$

The dependence of the tau polarization on the weak angle is shown in Fig. 4.15.

This polarization can be measured by using different decay channels. In choosing one channel one must not only consider its branching ratio for statistical reasons but also its sensitivity for the measurement of the polarization.

The two-body decays $\tau \rightarrow \pi\,\nu$ and $\rho\,\nu$ are particularly favourable for the following reasons. We shall first discuss the case of pion decay since it is particularly

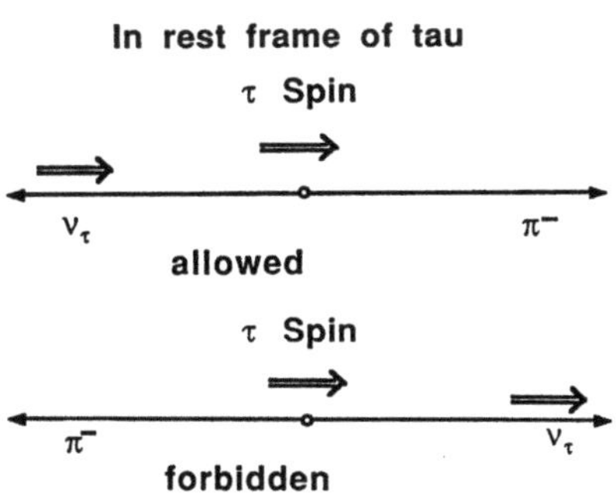

Fig. 4.14. The spin correlation in tau decays

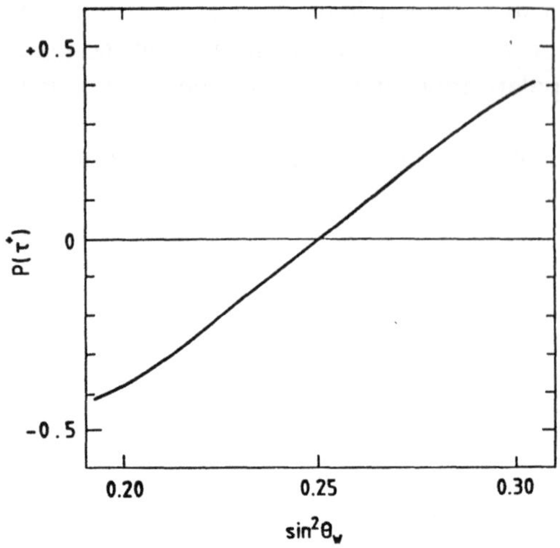

Fig. 4.15. Dependence of tau polarization on $\sin^2\theta_W$

simple. The pion and the neutrino are emitted in opposite directions because of momentum conservation. The pion has spin 0 and since the neutrino has negative helicity (spin opposite its momentum) the pion is preferentially emitted in the direction of the τ spin because of angular momentum conservation (Fig. 4.14). The argument for the ρ decay is more involved since the ρ has spin 1 and therefore more spin combinations are possible. Nevertheless also in this case the meson is emitted preferentially in the direction of the τ spin.

Because of these parity violating decays the decay products of the τ are emitted asymmetrically with respect to its spin direction in its restframe which, however, is also reflected in the labframe.

For instance the pion spectrum from the $\tau \rightarrow \pi\,\nu$ decay is given by [37]

$$\frac{dN(\tau)}{dx_\pi} = 1 - P_\tau(2x_\pi - 1) \tag{4.36}$$

where $x_\pi = E_\pi/E_{\text{beam}}$ is the ratio of the pion energy to the beam energy. This spectrum is shown in Fig. 4.16 for different values of $\sin^2\theta_W$ and one observes that it depends quite sensitively on this parameter.

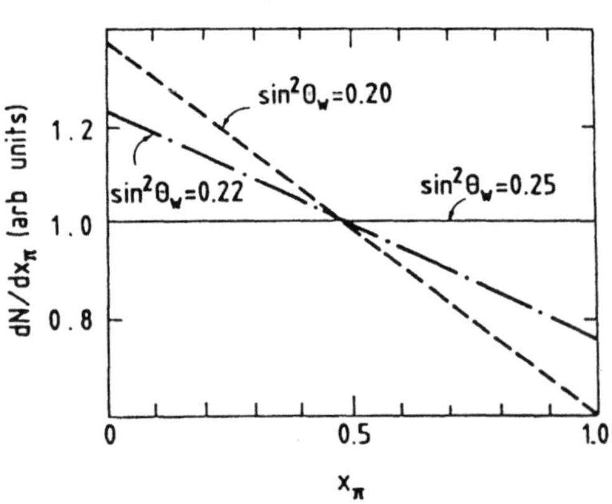

Fig. 4.16. The pion (normalized) momentum spectrum of tau decays for two values of the weak angle

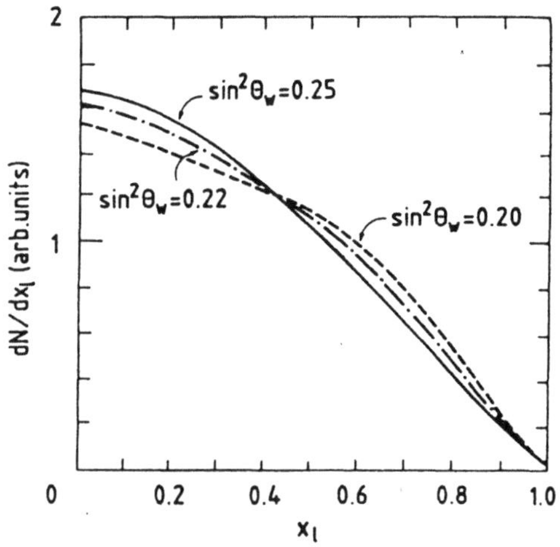

Fig. 4.17. The lepton (normalized) momentum spectrum of tau decays

A similar behaviour is expected for the $\tau \to \rho\,\nu$ decay, although with a reduced sensitivity due to the more complicated spin additions. The ρ spectrum is expected to be given by

$$\frac{dN(\tau)}{dx_\rho} = 1 - 0.46 P_\tau (2x_\rho - 1). \tag{4.37}$$

Finally one could determine the τ polarization from its pure lepton decays which have the advantage of a large branching ratio. On the other hand their sensitivity is low due to the 3-body decays. The lepton spectrum for a pure leptonic decay is given by

$$\frac{dN(\tau)}{dx_\ell} = a(x_\ell) - P_\tau b(x_\ell) \tag{4.38}$$

with

$$a(x_\ell) = \tfrac{5}{3} - 3x_\ell^2 + \tfrac{4}{3}x_\ell^3$$
$$b(x_\ell) = \tfrac{1}{3} - 3x_\ell^2 + \tfrac{8}{3}x_\ell^3.$$

The spectra for different values of $\sin^2\theta_W$ are shown in Fig. 4.17.

The potentials of the different decay modes to determine $\sin^2\theta_W$ are compared in the following table where for the statistical error a total luminosity of $100\,\mathrm{pb}^{-1}$ and for the acceptance an average LEP detector has been assumed [37]. The best results can be expected from the $\pi\,\nu$ mode, the dominant factor being the analysing power. Simulations have been performed for the ALEPH detector [37, 38] with the result that for $100\,\mathrm{pb}^{-1}$ an error $\delta \sin^2\theta_W = \pm 0.0013 \pm 0.0008$ can be obtained with the assumption of universality. This makes the measurement of the polarization in the final state quite attractive.

Table 4.8

Decay mode	Nr. of decays	Visible decays	Background	$\delta \sin^2 \theta_W$
$\mu\nu\bar{\nu}$	30,000	20,000	None	6×10^{-3}
$e\nu\bar{\nu}$			$\rho\nu$	$> 6 \times 10^{-3}$
$\pi\nu$	13,000	5,000	$\rho\nu$ easy to remove	2×10^{-3}
$\rho\nu$	33,000	?	$\pi\nu$, $e\nu\bar{\nu}$, $K^*\nu$ difficult to remove	$> 10^{-2}$

4.5 Neutrino Counting

One of the fundamental questions concerns the number of flavour families. There are two methods to determine the number of neutrino generations on or near the Z-peak.

4.5.1 Neutrino Counting from the Z-Width

If there are more then three kinds of neutrinos there seems to be no reason why they should not be also massless or, at least, very light. Hence each additional generation of neutrinos would give an extra contribution to the width of the Z. Assuming that there are no other hidden decays one may infer the number of neutrino families from the total width which can be written

$$\Gamma = 3\Gamma_{\text{lepton}} + \sum \Gamma_{\text{quarks}} + \Gamma_{\text{inv}} \tag{4.39}$$

where $\Gamma_{\text{inv}} = N_\nu \cdot \Gamma(Z \to \nu\bar{\nu})$ and N_ν denotes the number of neutrino generations. Since the decay widths of the leptons and quarks are quite well calculable in the SM (see Sect. 4.1.1) one can infer from the measured total width Γ the number of neutrino generations. Since $\Gamma(Z \to \nu\bar{\nu}) \approx 165$ MeV one needs a precision on the width better than this to obtain a meaningful result for N_ν.

It turns out that even with only 10,000 Z-events one can get a statistical precision of about 1/3 neutrino generation. To investigate the importance of systematic errors one can express Γ_{inv} in the following way [40]

$$\Gamma_{\text{inv}} = \Gamma_\mu \left\{ \frac{\sqrt{12\pi}}{M_Z} \sqrt{\frac{0.7\, L\, \epsilon}{N}} - 3 - \frac{N_H\, \epsilon_H}{N_\mu\, \epsilon_\mu} \right\} \tag{4.40}$$

where N, N_H and N_μ are the number of total, hadronic and muon events, respectively, ϵ, ϵ_H and ϵ_μ are the corresponding detector efficiencies and L is the integrated luminosity. The factor 0.7 is the radiative correction which has to be applied to the peak cross section [15]. The influence of the different errors has been studied for the DELPHI detector [40] and some results are shown in Fig. 4.18. Here the errors are shown as a function of the number of Z-events. A systematic error of the luminosity of 3% has been assumed. As can be seen the error of the muon efficiency is negligible compared to the other systematic uncertainties. The most important error

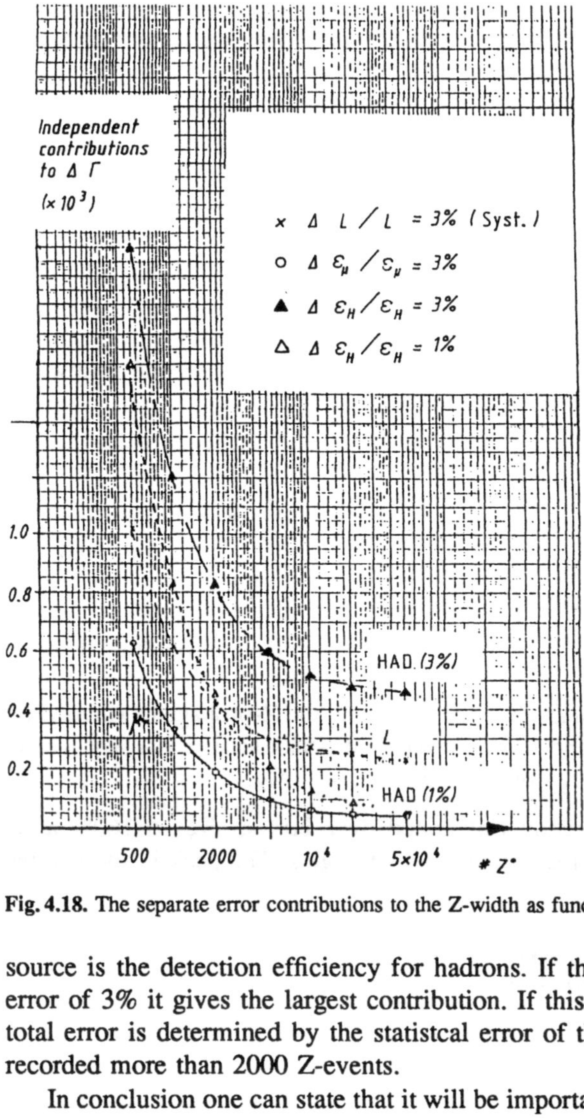

Fig. 4.18. The separate error contributions to the Z-width as function of the number of Z-events [40]

source is the detection efficiency for hadrons. If the efficiency for hadrons has an error of 3% it gives the largest contribution. If this can be reduced to 1% then the total error is determined by the statistcal error of the luminosity provided one has recorded more than 2000 Z-events.

In conclusion one can state that it will be important to understand the systematics of the apparatus. If this is the case even a moderate sample of events (about 10,000 Z) will allow to determine the number of neutrino generations with an accuracy of about 0.5. In a more ambitious later experiment with about 1 pb^{-1} luminosity and systematic errors reduced to the 1% level one could obtain N_ν with an error of about 0.2 from a measurement of the Z-width.

4.5.2 Radiative Neutrino Counting

A complementary and more direct method to count neutrinos or look for neutral particles with masses less than $M_Z/2$ is the observation of the reaction $e^+e^- \rightarrow \nu\bar{\nu}\gamma$ where the photon comes mainly from initial state bremsstrahlung (see Fig. 4.2a) [41]. The signature is a single photon recoiling against a system with the mass of the Z^0. The cross section is given by

$$\frac{d\sigma}{dx\,dy} = \frac{\alpha G_F^2 s(1-x)}{3\pi^2 x(1-y^2)} \left[(1-\frac{1}{2}x)^2 + \frac{1}{4}x^2y^2\right] F(x) \qquad \text{with} \tag{4.41}$$

$$F(x) = 1 + \frac{2M_Z^2[M_Z^2 - s(1-x)](V_e - A_e) + 2M_Z^4 N_\nu (V_e^2 + A_e^2)}{[M_Z^2 - s(1-x)]^2 + M_Z^2 \Gamma^2}$$

and A_e and V_e as defined in Eq. (4.3). The parameters x and y are related to the photon energy E_γ and its angle Θ_γ relative to the incoming electron direction

$$x = 2E_\gamma/\sqrt{s} \qquad \text{and} \qquad y = \cos\Theta_\gamma.$$

The general behaviour of this cross section can be characterized by the following remarks:

- The photon spectrum decreases with $1/x$, the typical dependence of bremsstrahlung processes
- In F(x) the "1" corresponds to the photon diagram, the term proportional to N_ν, the number of neutrino generations, comes from the Z-diagram and the intermediate term is due to the interference between the photon and the Z diagrams.
- Near the Z-peak the energy dependence is approximately proportional to $\sim (\alpha G_F^2/3\pi^2) \cdot s$
- For energies close to the Z-mass the term involving the Z-propagator (involving therefore N_ν) is enhanced and this enhancement peaks at $x = 1 - M_Z^2$ which corresponds to a photon energy of about $E_\gamma \approx \sqrt{s} - M_Z$.

Fig. 4.19 shows the cross section (integrated over $20° < \Theta_\gamma < 160°$) as a function of the photon energy for different c.o.m. energies. The peak in the photon

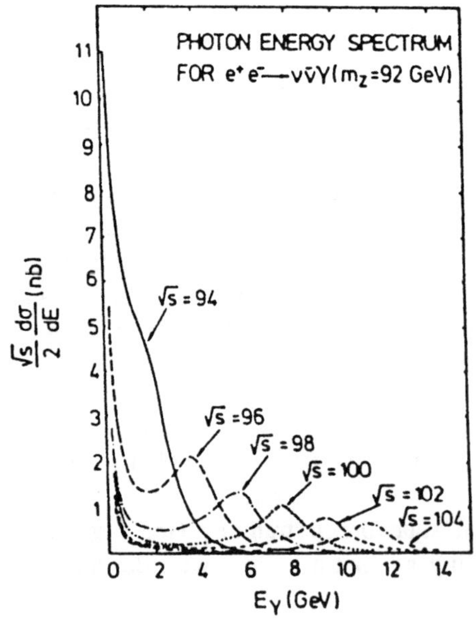

Fig. 4.19. The photon spectrum for the reaction $e^+e^- \rightarrow \gamma$+nothing for 3 neutrino families [41]

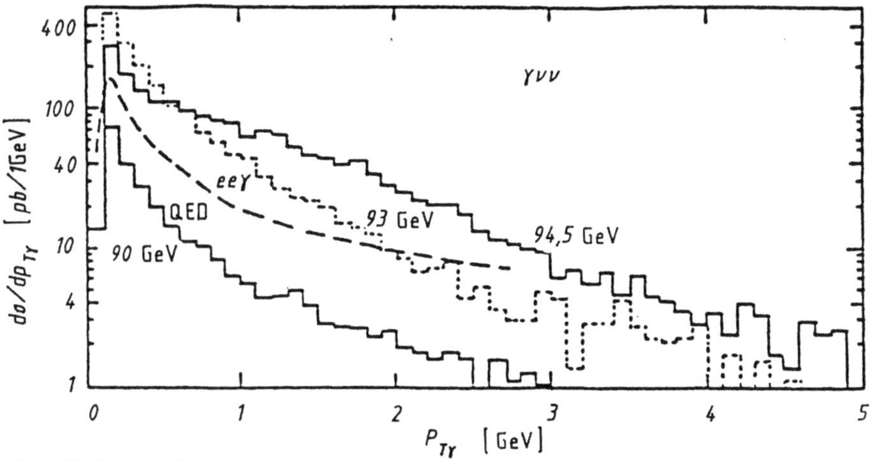

Fig. 4.20. The distribution of transverse photon momentum for single photons from $e^+e^- \to \gamma$+nothing with c.o.m. energies on the Z-peak (*dotted line* for 93 GeV), below and above the Z-mass (solid histograms) as expected for L 3. The QED background is indicated by the dashed line. [43]

spectrum shifts as expected with increasing c.o.m. energy. The height of the peak is proportional to N_ν (assumed to be 3 in the figure). The figure shows also that at energies close to the Z-peak the cross section is higher but the separation from the Z-resonance is bad. In this case the Bhabha background is also more serious as has been shown in simulations for DELPHI [42]. The background can partially be suppressed by an angular cut. A good resolution for the photons will be essential for a reasonable signal-to-background ratio as has been studied [43] for L 3 with its BGO spectrometer which has an excellent resolution (Fig. 4.20).

Somewhat further away from the Z the situation is cleaner but the cross section gets smaller.

If a fourth neutrino generation exists the peak cross section will go up by about 30% which does not require a high statistics run to be detected. However, QED corrections have to be taken carefully into account [16].

Whether such a measurement will be feasible will depend on whether events with a photon with 4 to 6 GeV recoiling against nothing can be isolated. The cross section is big enough so that an integrated luminosity of 1 pb^{-1} would provide sufficient statistics to be sensitive to about $\Delta N_\nu \approx 0.3$ and about the same systematic uncertainty is expected. This relatively low luminosity would make it possible to run parasitically while the line shape is precisely measured. In this way additional data on the number of neutrino generations could be obtained at no extra cost.

5. QCD Tests

Electron-positron annihilation provides many possibilities to test the theory of strong interactions, in particular Quantum Chromodynamics (QCD). Of course, the strong interaction becomes effective only in the final state and in most cases theory has been applied on the perturbation level only. Nevertheless QCD tests have reached a precision level which is impressive and in many cases comparable to those obtained from hadron interactions. Areas of interest are total hadronic cross section, quark and gluon jets, determination of the effective coupling constant, fragmentation models and maybe a first test of the non-abelian character of QCD could be done at LEP. It will not be possible to review here all aspects of QCD problems, e.g. $\gamma\gamma$ physics will not be dealt with (for a full review see Ali [44]).

5.1 Cross Section, Thrust, Oblateness and Sphericity

One of the first quantities which is usually measured at e^+e^- machines is the total hadronic cross section which is usually expressed in terms of the ratio

$$R = \sigma(e^+e^- \to \gamma,\, Z \to \text{hadrons})/\sigma(e^+e^- \to \gamma,\, Z \to \mu^+\mu^-) \tag{5.1}$$

where the point like cross section for the muon channel has been given in (4.19).

In the quark-parton model in zeroth order one has $R_0 = 3\sum Q_i^2$ where Q_i are the charges of the quarks. Threshold factors have been neglected since the masses of the known quarks are small compared to the Z-energy and we do not consider the top quark here since it might be beyond the range of LEP.

In first order perturbation theory one finds

$$R_1 = R_0(1 + \alpha_s(Q^2)/\pi) \tag{5.2}$$

with

$$\alpha_s(Q^2) = \frac{12\pi}{(33 - 2N_f)\ln(Q^2/\Lambda^2)}$$

As was mentioned already in Sect. 4.2.3 this gives a correction of about 4% with $\alpha_s(M_Z^2) \approx 0.12$. There it had also been shown that higher order corrections are negligible.

Since the top mass seems to be well above the Z-mass a measurement of R would provide a very good and almost model independent way [45] to determine the coupling constant $\alpha_s(M_Z^2)$. With an error of the cross section and the Z-width of about 1% one obtains $\alpha_s(M_Z^2)$ with an error of about 20%.

The asymmetries involving hadrons in the final state have already been discussed in Sect. 4.4.3, so we shall not consider them further here.

To investigate final hadronic states and to characterize the topology of events ral quantities are used. The first two are used by experiments which measure momenta by magnetic deflection whereas the third is more useful for experiments with calorimetry. Their definitions will be recalled in the following:

Thrust

$$T = \max \frac{\sum |\boldsymbol{p}_i \cdot \hat{\mathbf{e}}|}{\sum |\boldsymbol{p}_i|} \tag{5.3}$$

where $\boldsymbol{p}_i$ are the momenta of the hadrons and the maximum has to be found by choosing the right unit vector $\hat{e}$. The extreme values are

$$T = \begin{cases} 1 & \text{for back-to-back events} \\ 1/2 & \text{for spherical events.} \end{cases}$$

Sphericity

$$S = \frac{3}{2} \sum (\boldsymbol{p}_{i\perp}{}^2)_{\min} / 2 \sum \boldsymbol{p}_i{}^2 \tag{5.4}$$

where $\boldsymbol{p}_{i\perp}$ are the transverse momenta of the hadrons relative to an axis which has to be chosen in such a way that $\sum \boldsymbol{p}_{i\perp}{}^2$ becomes a minimum. The extreme values are

$$S = \begin{cases} 0 & \text{for back-to-back events} \\ 1 & \text{for spherical events.} \end{cases}$$

Oblateness

This quantity is derived from the energy flow carried by hadrons, i.e. E_i is considered as a vector. First two quantities are defined wich are similar to thrust but use energies (measured e.g., by calorimeters) instead of momenta

$$F_{\text{thrust}} = \max \frac{\sum |\boldsymbol{E}_i \cdot \hat{e}_1|}{\sum E_i}$$

where $\hat{e}_1$ is chosen such as to maximize this quantity;

$$F_{\text{major}} = \max \frac{\sum |\boldsymbol{E}_i \cdot \hat{e}_2|}{\sum E_i}$$

where $\hat{e}_2$ is chosen in a plane perpendicular to $\hat{e}_1$ such as to maximize this quantity;

$$F_{\text{minor}} = \max \frac{\sum |\boldsymbol{E}_i \cdot \hat{e}_3|}{\sum E_i}$$

can then be calculated with $\hat{e}_3$ perpendicular to both $\hat{e}_1$ and $\hat{e}_2$.

From these quantities one derives finally the oblatenes

$$O = F_{\text{major}} - F_{\text{minor}} \tag{5.5}$$

The extreme values are

$$O = \begin{cases} 0 & \text{for spherical events} \\ 1 & \text{for back-to-back events.} \end{cases}$$

5.2 Fragmentation Models

The hadron production in e^+e^- collisions cannot be fully described by perturbative QCD but has to be complemented by fragmentation models. As a consequence the hadronisation schemes are phenomenological, containing a number of parameters

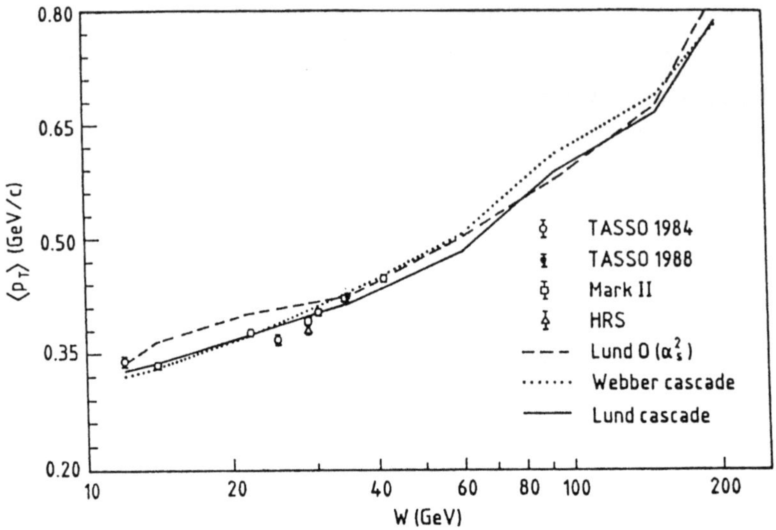

Fig. 5.1. The average transverse momentum as function of the c.o.m. energy [46]

which have to be tuned to describe the experimental data as best as possible. This is regrettable from a theoretical point of view but the phenomenology must itself be a manifestation of QCD from which one might eventually deduce a rigorous description of hadronisation. Fragmentation models provide an intermediate step between perturbative QCD calculations and the experimentally measured final hadronic states.

These models are important for another reason. To understand the response of complex particle detectors simulations are indispensible. Any such simulations are based on event generators implementing fragmentation models. When analyzing real events it is necessary to estimate contaminations from hadronic events which again requires a good knowledge of fragmentation. The same is true for unfolding experimental raw data to correct for effects of detector acceptances.

Since the cross section for hadron production is large the first quantities which are usually measured at e^+e^- machines are charge multiplicities n_{ch}, transverse momentum p_T distributions, thrust, sphericity, etc. In the PETRA, PEP and TRISTAN energy range several fragmentation models could be tuned in such a way as to reproduce the experimental data equally well [46]. However, the predictions for some of the observables differ significantly at LEP energies and therefore a good test of different models will be possible.

If we consider at first average quantities, we take from Fig. 5.1 that the average transverse momentum $< p_T >$ relative to the sphericity axis increases rapidly with the c.o.m. energy but the predictions of various models agree. On the other hand the evolution of the average charge multiplicity $< n_{ch} >$ (Fig. 5.2) and of $< x_p >$ (where $x_p = 2p/W$) (Fig. 5.3) differs considerably for different models. Looking at differential distributions it is found that for instance the distributions for charge multiplicity n_{ch} (Fig. 5.4), aplanarity A (Fig. 5.5) and thrust T (Fig. 5.6) differ for different models. whereas for other variables like p_T and x_p the differences are smaller (Fig. 5.7), providing a less sensitive test of the models. Another interesting

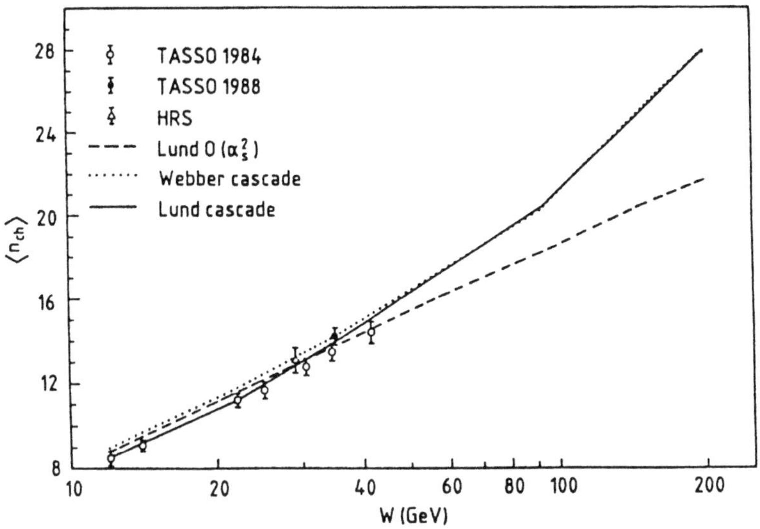

Fig. 5.2. The average multiplicity as function of the c.o.m. energy [46]

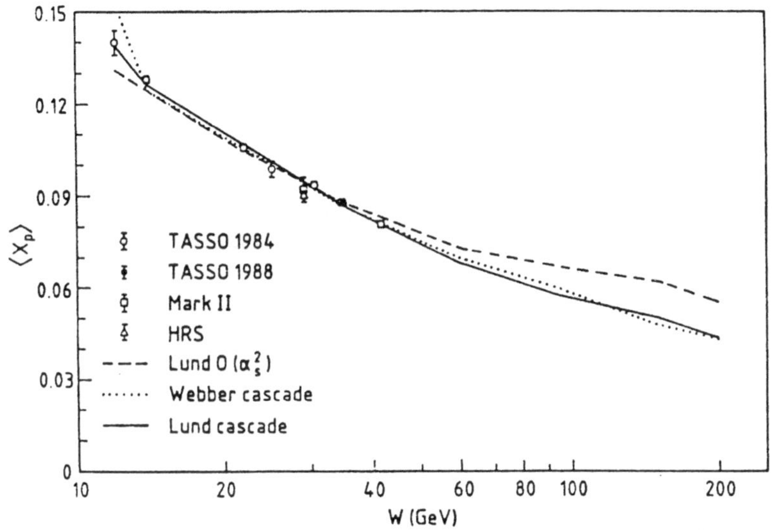

Fig. 5.3. The average normalized particle momentum $x_p = 2p/W$ (p momentum, W c.o.m. energy) as a function of W [46]

question is the evolution of events with a different number of jets. The expected ratios are shown in Fig. 5.8.

5.3 Test of the Non-Abelian Character of QCD

One of the essential features of QCD is the gluon self-coupling due to its non-abelian gauge theory character. There are some indications that this feature of QCD is consistent with experimental data (scaling violations in ν N scattering, 3-jet events

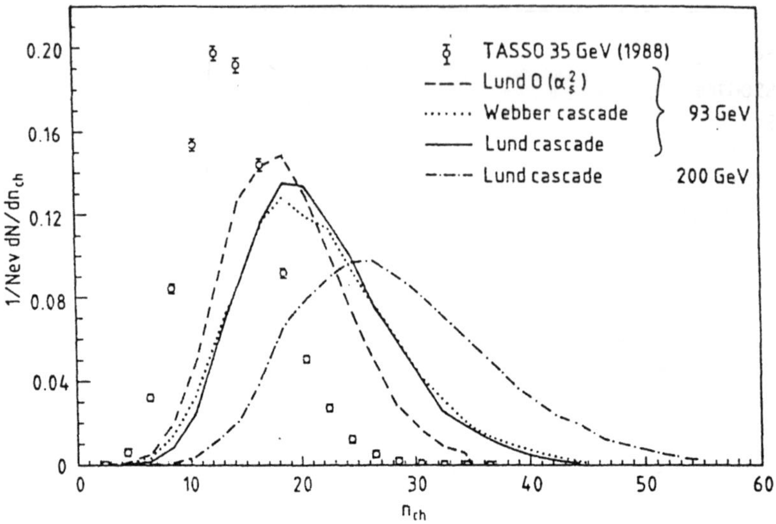

Fig. 5.4. The charge distribution as calculated by different models and for two different c.o.m. energies [46]

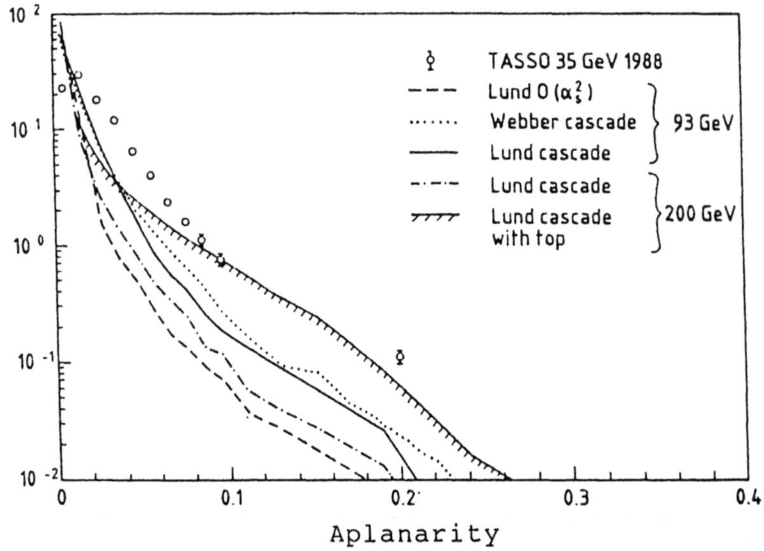

Fig. 5.5. The aplanarity distribution as calculated by different models and for two different c.o.m. energies [46]

in e^+e^-, high E_T-jets in pp) but it is desirable to test the gluon self-coupling in a quantitative way. This may be achieved at LEP (for detailed reviews see Ali [47] and Rudolph [48]).

In e^+e^- annihilation into hadrons, the 3-gluon vertex enters only in second or higher order diagrams. One possibility is to investigate events with 4 jets which can be produced in the reactions $e^+e^- \rightarrow q\bar{q}gg$ or $q\bar{q}q\bar{q}$. The contributing QCD diagrams are shown in Fig. 5.9 where one notices that only the last diagram contains

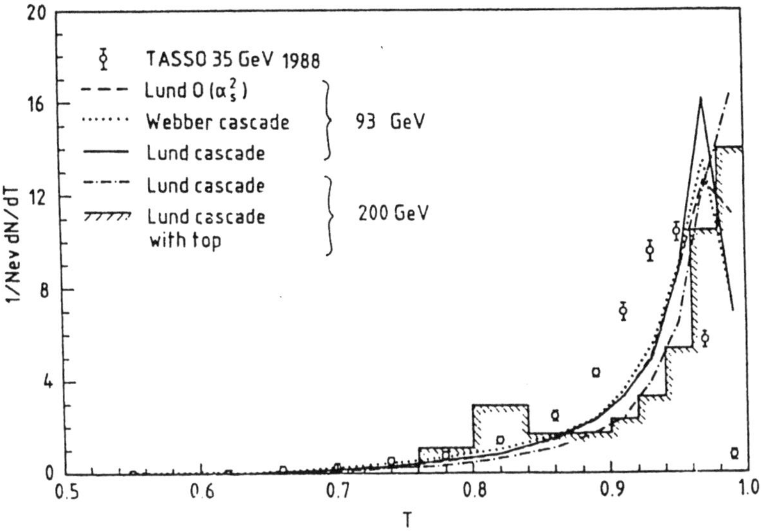

Fig. 5.6. The thrust distribution as calculated by different models and for two different c.o.m. energies [46]

the 3-gluon vertex. The cross section for $q\bar{q}q\bar{q}$ is only about 1/10 of the $q\bar{q}gg$ cross section and can be neglected in most cases.

To test the theory the QCD expectations are compared with a theory for the strong interaction which has the characteristics of an abelian theory like QED. This leads to a "QED" theory where gluons carry no colour but quarks still do. This means that the diagram with the 3-gluon vertex is switched off (Fig. 5.9). In order that QCD and "QED" yield the same 3-jet rate the effective couplings have to be related by $\alpha_{\text{abelian}} = (4/3)\alpha_{\text{non-abelian}}$. To be consistent the 4-quark cross section has also to multiplied by a factor of 8 which makes the $q\bar{q}q\bar{q}$ and $q\bar{q}gg$ rates comparable in the "QED" model. Because of the switching-off of the 3-gluon diagram and the enhancement of the 4-quark channel one expects observable differences.

To select 4-parton events kinematical cuts have to be applied to select the four jets and also to avoid theoretical divergencies. Usually the Sterman-Weinberg cut-off procedure is used since it matches both the experimental and theoretical requirements (see e.g. [49]).

Using these variables the following conditions are imposed: In a 4-jet event all jet energies E_i must satisfy $E_i > \varepsilon\sqrt{s}/2$ and the angles between two jets $\Theta_{ij} > \delta$ where the parameters ε and δ have to be chosen in such a way as to obtain a good jet resolution. Their normal ranges are $\varepsilon = 0.1$ to 0.15 and $\delta = 25°$ to 35°. With such values one expects of the order of 100,000 4-jet events for 10^6 Z-events.

Three methods have been suggested to test the non-abelian character of QCD. The first [50] will not be discussed here since it did not lead to differences for the two models.

The second method to distinguish between the two types of theories was proposed by Koerner, Schierholz and Willrodt [51]. They suggest to study the azimuthal correlation where the azimuthal angle ϕ is defined by

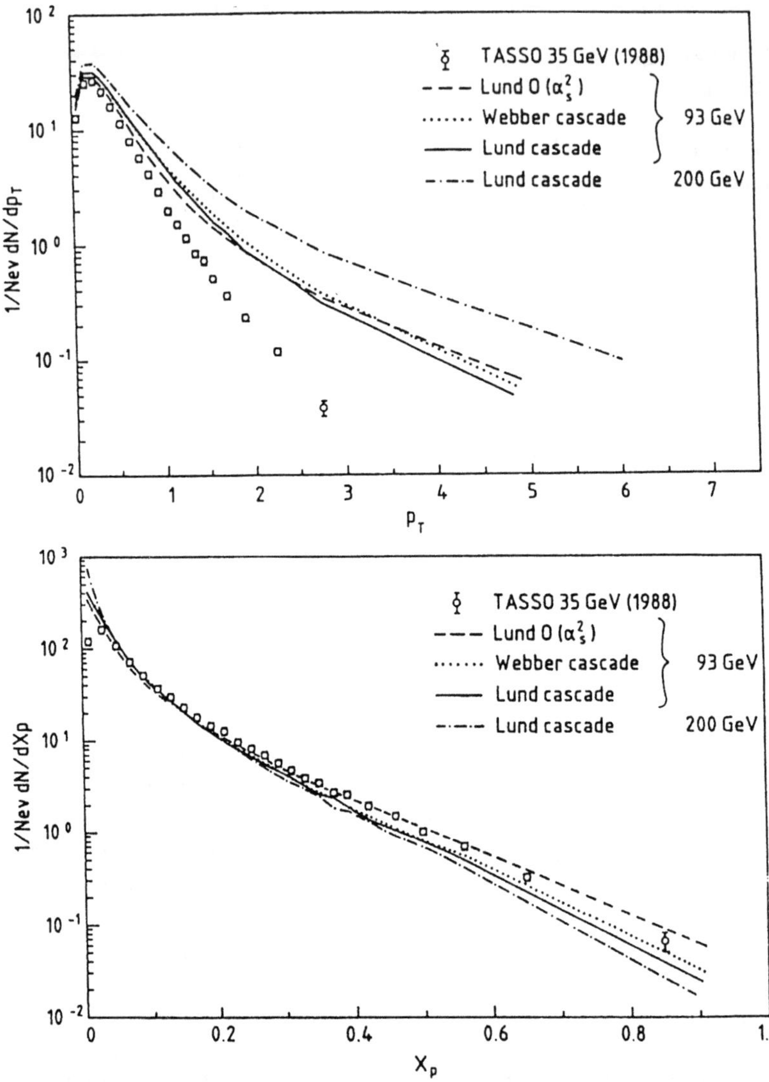

Fig. 5.7. The distributions of transverse momentum p_T and normalized momentum x_p as calculated by different models and for two different c.o.m. energies [46]

$$\cos\phi = \hat{\mathbf{e}}_1 \cdot \hat{\mathbf{e}}_2 \tag{5.6}$$

and the normals $\hat{\mathbf{e}}_1$ and $\hat{\mathbf{e}}_2$ (unitvectors) are defined for the configuration with each hemisphere containing two jets

$$\begin{aligned} \hat{\mathbf{e}}_1 &= (\boldsymbol{p}_1 \times \boldsymbol{p}_2)/(|\boldsymbol{p}_1|\,|\boldsymbol{p}_2|) \\ \hat{\mathbf{e}}_2 &= (\boldsymbol{p}_3 \times \boldsymbol{p}_4)/(|\boldsymbol{p}_3|\,|\boldsymbol{p}_4|) \end{aligned} \tag{5.7}$$

where $\boldsymbol{p}_1,\ \boldsymbol{p}_2,\ \boldsymbol{p}_3,\ \boldsymbol{p}_4$ are the momenta of the 4 partons (jets) in an ordered arrangement with $E_1 > E_2,\ E_3 > E_4$.

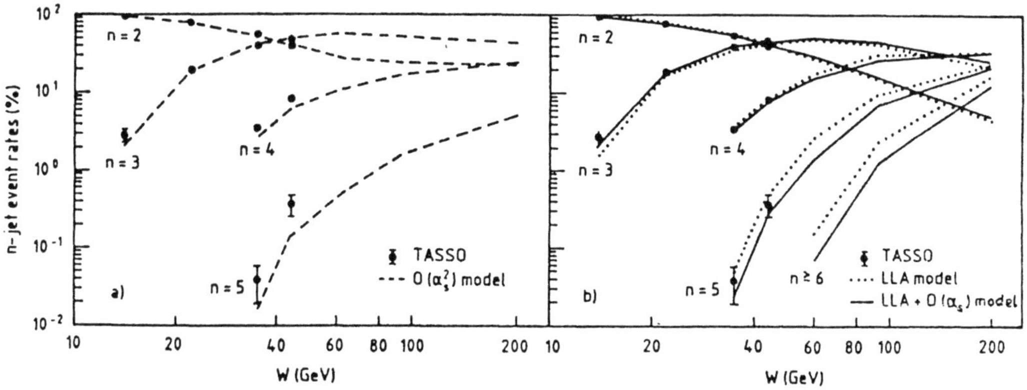

Fig. 5.8. The percentage of n-jet events as function of the c.o.m. energy W

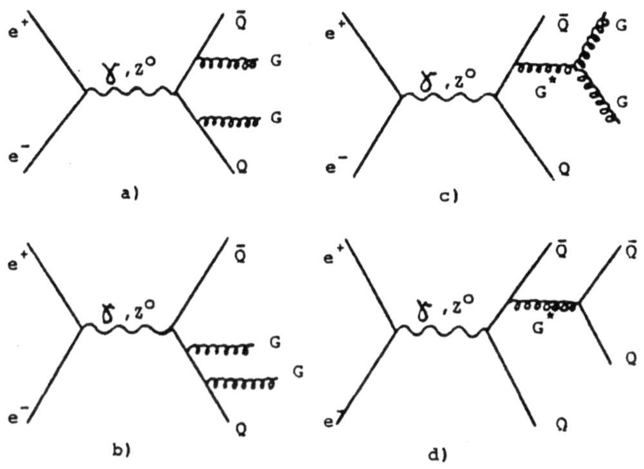

Fig. 5.9. Diagrams contributing to the production of 4-jet events

The normalized distributions in ϕ are shown in Fig. 5.10 for QCD and "QED" for two sets of the parameters ε, δ. The QCD curve shows an enhancement near 180° where the two lower energetic partons, preferentially the gluons, are close together, whereas the "QED" distribution shows less structure. The statistical significance of the test may be demonstrated for 100,000 hadronic events through the average angle $< \phi >$:

		$\varepsilon = 0.10,\ \delta = 25°$	$\varepsilon = 0.15,\ \delta = 35°$
$< \phi >$	QCD	97.0 ± 0.5 °	97.7 ± 0.9 °
	"QED"	91.5 ± 0.6 °	90.3 ± 0.9 °

Already with this number of events a clear distinction seems possible from a statistical point of view. The size of the effect does not depend much on the choice of ε and δ.

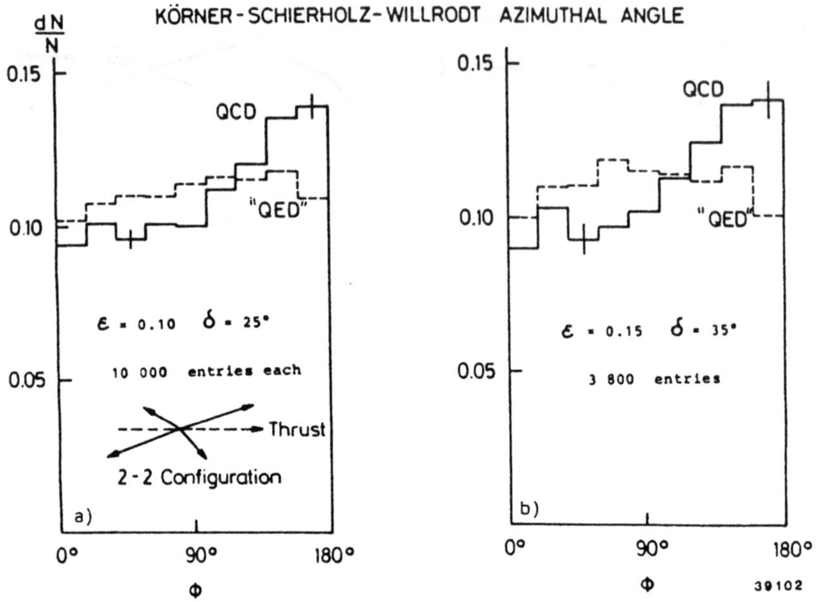

Fig. 5.10. Distribution in the azimuthal angle for two different cuts for the jet definition (a and b). The full curve is calculated from QCD, whereas the broken curve is derived from a "QED"-like abelian QCD

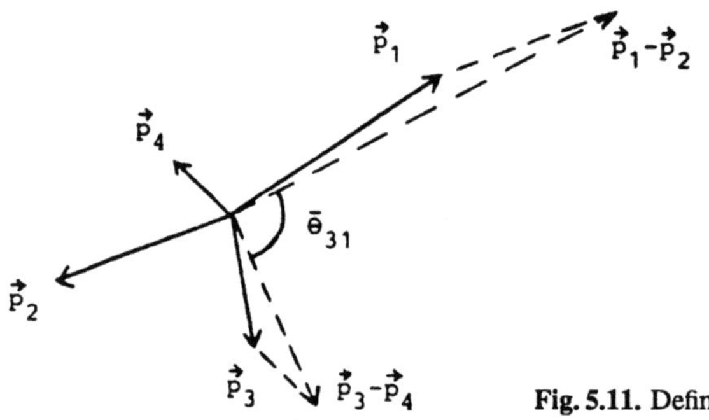

Fig. 5.11. Definition of the polar angle θ_{13} in 4-jet events

Another distribution, proposed by Nachtmann and Reiter [52], exploits the different helicity structure of the $g^* \to g\,g$ and $g^* \to q\bar{q}$ vertices. For this purpose the partons have to be ordered according to their energy $E_1 > E_2 > E_3 > E_4$ and an angle θ_{13} is defined in the following way (see Fig. 5.11)

$$\cos\theta_{13} = \hat{e}_{12} \cdot \hat{e}_{34} \tag{5.8}$$

with

$$\begin{aligned} \hat{e}_{12} &= \boldsymbol{p}_1/|\boldsymbol{p}_1| - \boldsymbol{p}_2/|\boldsymbol{p}_2| \\ \hat{e}_{34} &= \boldsymbol{p}_3/|\boldsymbol{p}_3| - \boldsymbol{p}_4/|\boldsymbol{p}_4| . \end{aligned} \tag{5.9}$$

Two cuts, additional to the ε, δ cuts, are applied namely: $E_3/E_2 > 1/2$ and $\cos\theta_{13} < -0.7$.

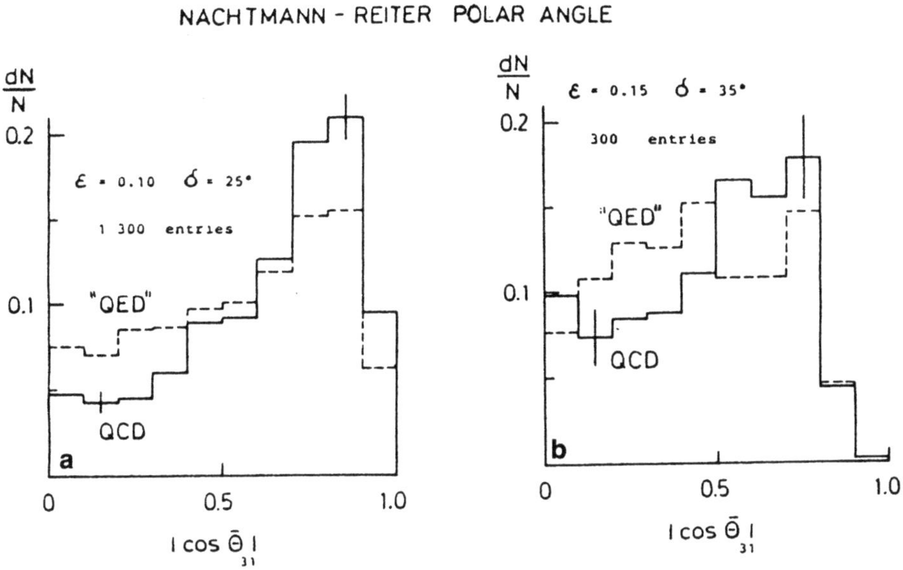

Fig. 5.12. Distribution of the polar angle θ_{13} for two different cuts for the jet definition (a and b). The full curve is calculated from QCD, whereas the broken curve is derived from a "QED"-like abelian QCD

These cuts reduce the 4-jet sample considerably, leaving only about 6% of all 4-jet-events. The normalized distributions are shown in Fig 5.12. The difference between QCD and "QED" is accentuated as compared to the previous method but at the expense of statistics. The significance of the test may be shown again by considering the average $< \cos\theta_{13} >$:

		$\epsilon = 0.10,\ \delta = 25°$	$\epsilon = 0.15,\ \delta = 35°$
$< \cos\theta_{13} >$	QCD	0.627 ± 0.007	0.483 ± 0.014
	"QED"	0.545 ± 0.008	0.445 ± 0.012

Again the statistical significance is satisfactory.

Systematic errors have been studied by Rudolph [48] who considered in particular the influence of fragmentation models. The conclusion is that at LEP energies the uncertainties in the description of fragmentation will not destroy the test but, of course, a good knowledge of the fragmentation mechanism will be important. The fact that the toponium mass now is recognized to be above the Z-mass helps here since the background to 4-jet events from toponium decay is absent.

If toponium would be observable at LEP its decays into 4 jets and γ + 3 jets, respectively, would offer an excellent test of the 3-gluon coupling [47]. Since, however, the top mass seems to be rather high we shall not further discuss this possibility here.

A last method to study the 3-gluon coupling was proposed by Falck et al. [53]. These authors study the production of 5-jet events and find that the $q\bar{q}\,3\,g$ production dominates over the $2\,q\,2\,\bar{q}\,g$ production in analogy to the 4-jet events, where $q\,\bar{q}\,2\,g$ dominates over $2\,q\,2\,\bar{q}$. Distributions for 5-jet events are calculated for various variables for QCD and "QED" and a distinction may be possible.

6. Search for Supersymmetric Particles

The Standard Model (SM) has been very succesful in describing all experimental data for electroweak interactions obtained so far. However, it has many free parameters which are not explained. To find the origins of these parameters one must go beyond the SM and one such extension are supersymmetric theories (SUSY). These theories also try to solve some of the more fundamental problems, like the hierarchy of mass scales or the naturalness problem (corrections to the Higgs mass) [54].

6.1 General Remarks

SUSYs establish a symmetry between constituents of matter (leptons and quarks, i.e. fermions) and particles transmitting the forces (gauge bosons). The SUSY particles differ from the conventional particles by their spin. The SUSY counterparts of fermions (spin 1/2) have spin 0 (squarks, sleptons, sneutrinos) whereas the supersymmetric partners of the gauge bosons (spin 1) have spin 1/2 and are called photino, gluino, wino and zino. The Higgs particles (spin 0) are also given a partner, the shiggs which have spin 1/2. Also a partner for the graviton (spin 2) is required, the gravitino (spin 3/2).

In this way a supersymmetric world is postulated which complements our ordinary world. Sofar none of these supersymmetric particles have been detected and the question whether this SUSY world exists or not is most exciting, and we shall discuss what prospects exist at LEP to find any of these sparticles.

In the limit of an exact supersymmetry the masses of the particles and their supersymmetric partners would be equal. At PETRA, PEP and TRISTAN lower mass limits could be established in the range of 20 to 30 GeV. Obviously sparticles are much heavier than particles and as a consequence supersymmetry must be spontaneously broken.

The couplings of the supersymmetric particles are directly related to those of conventional particles and all one has to do is to make the replacements

$$\underline{\mathrm{f}}\,\mathrm{f}\,\mathrm{g} \rightarrow \underline{\mathrm{f}}\,\tilde{\mathrm{f}}\,\tilde{\mathrm{g}},\ \tilde{\underline{\mathrm{f}}}\,\tilde{\mathrm{f}}\,\mathrm{g},\ \tilde{\underline{\mathrm{f}}}\,\mathrm{f}\,\tilde{\mathrm{g}}$$

where f is a matter fermion (quark or lepton), $\underline{\mathrm{f}}$ is the corresponding antifermion, g a gauge boson and their supersymmetric partners are denoted by tildes.

Similarly one has to replace

$$\underline{\mathrm{f}}\,\mathrm{f}\,\mathrm{H} \rightarrow \underline{\mathrm{f}}\,\tilde{\mathrm{f}}\,\tilde{\mathrm{H}},\ \tilde{\underline{\mathrm{f}}}\,\mathrm{f}\,\tilde{\mathrm{H}}$$

where H is a Higgs and $\tilde{\mathrm{H}}$ its SUSY partner.

It is a general feature of these couplings that they involve always sparticle pairs. Indeed most of the SUSY theories introduce a multiplicatively conserved quantum number R with

$$R = \begin{cases} +1 & \text{for conventional particles} \\ -1 & \text{for supersymmetric partners.} \end{cases}$$

This has three phenomenologically important consequences:

i) sparticles are always produced in pairs, e.g.

$$e^+e^- \to \tilde{\mu}^+\tilde{\mu}^-, \ \tilde{e}^\pm\tilde{\gamma}e^\pm$$

ii) every sparticle decays into one other sparticle, e.g.

$$\tilde{e} \to e\tilde{\gamma}, \ \tilde{W} \to q\underline{q}\tilde{\gamma}$$

iii) the lightest sparticle is stable since it has no allowed decay mode.

If the lightest supersymmetric particle (LSP) were electromagnetically charged or strongly interacting it would have severe cosmological consequences. It would condense along with normal matter into galaxies, stars and planets and form anomalous heavy isotopes. These have not been seen. Therefore the candidates for LSP are sneutrinos (spin 0), photinos or neutral shigges (spin 1/2) or the gravitino (spin 3/2).

The neutral spin-1/2 sparticles (photino, neutral shiggs, zino) can mix and their physical states are called neutralinos. We shall not discuss here the details of this mixing since because of the many free parameters it is rather involved [54]. The physical states are denoted by χ, χ', χ'', χ''' etc.

In a similar way also the charged spin-1/2 particles (wino, charged shiggs) can mix and they are denoted by $\chi^\pm$.

In summary one has to look for

charged sparticles with spin 0	: sleptons, squarks
charged sparticles with spin 1/2	: wino, charged shiggs
neutral sparticles with spin 0	: sneutrinos
neutral sparticles with spin 1/2	: neutralinos
	: (zino, photino, neutral shiggs).

6.2 Experimental Searches for Supersymmetric Particles

Since supersymmetric particles are always produced in pairs the straightforward method is to look for processes with two particles in the final state, like a pair of sleptons or squarks. The recent results from hadron colliders [9] give mass limits for sleptons and squarks which indicate that the production of these particles may not be possible at LEP. An enormous amount of work has been done in discussing the detection of such particles and calculating the relevant cross sections (see [54]) but in view of the new experimental limits we shall not consider the detection of charged particles with spin 0.

Some mass limits for charged particles with spin 1/2 (winos, shiggses) have been established of about 25 GeV. Therefore it is certainly worthwhile to look for such supersymmetric particles at LEP. If the masses of winos or zinos would be lower than those of sleptons and squarks, the process $e^+e^- \to \tilde{W}\tilde{W}$ might be accessible. Winos of such mass would decay via three-body modes $\tilde{W} \to q\bar{q}\tilde{\gamma}$ or $\tilde{W} \to \ell\bar{\nu}\tilde{\gamma}$.

As far as neutral supersymmetric particles are concerned almost nothing is known, both for spin 0 (sneutrinos) and spin 1/2 particles (neutralinos) and this will be one of the very exciting fields of research at LEP.

If both slepton and wino pair production are inaccessible, then the only allowed process could well be $e^+ e^- \to \tilde{z}\tilde{\gamma}$. Indeed for a light photino one expects that $m(\tilde{z}) < M(Z)$ and the zino would then decay according to $\tilde{z} \to q\bar{q}\tilde{\gamma}$ or $\tilde{z} \to \ell\bar{\ell}\tilde{\gamma}$.

In addition to two-body processes one can consider three-body processes in which a sparticle pair is produced in association with a photon, e.g. $e^+e^- \to \tilde{\gamma}\tilde{\gamma}\gamma$. These radiative processes provide a characteristic signature but their cross section is suppressed by an extra power of the fine-structure constant.

In the following we shall discuss these processes individually.

6.2.1 Wino Pair Production

In almost all supersymmetric models with a light photino, the wino, perhaps the only charged sparticle which might be detectable at LEP, is expected to be lighter than the W boson and should be produced at least at LEP 200.

Wino pair production in the process $e^+e^- \to \tilde{W}^+\tilde{W}^-$ takes place via the s-channel exchange of the photon and Z^0 and the t-channel exchange of the sneutrino. Since it is assumed that sleptons would be heavier than the wino, these would decay, as mentioned above, via the three-body modes $\tilde{W} \to q\bar{q}\tilde{\gamma}$ or $\tilde{W} \to \ell\bar{q}\tilde{\gamma}$.

The formulae for the cross section and the decay modes are rather lengthy and contain a number of free parameters and therefore they will not be reproduced here (for details see [54]). The total cross section for wino pair production without any cuts is shown in Fig. 6.1a as a function of the c.o.m. energy for different wino masses under the assumption that the photino mass is 10 GeV. As one can see the cross section is quite high. For the case that both winos decay leptonically (assuming specific mass values for the sleptons, the sneutrino and photino) and applying experimental cuts for the acceptance of electrons (for muons there should be no great difference) one obtains the event rates as shown in Fig. 6.1b. One notices that at the Z-peak the rate is quite large. The cross section is not very sensitive to the photino mass (at $\sqrt{s}$ = 180 GeV the cross section varies only by about 5% if the photino mass changes from 0 to 10 GeV). For a wino mass of 60 GeV, on the other hand, the cross section gets very small (< 0.04 pb) which limits the explorable mass range to about $M(\tilde{W}) < 40\,\text{GeV}$. The characteristic energy and angular distribution of the leptons changes rapidly with $\sqrt{s}$ as is shown in Figs. 6.2 and 6.3. It can be used to distinguish this process from other processes (e.g. the one discussed in the next section) or from background.

The background for leptonic decays should be quite small. The signature for the events is a pair of charged leptons acoplanar with the beams, a substantial amount of missing energy and momentum (carried away by the photinos) and the absence of hadronic activities. The Standard Model background would mainly come from tau pairs decaying leptonically and lepton pairs with a bremsstrahlung photon (or an electron-positron pair) going down the beam pipe.

Tau events can be removed by requiring that each lepton has an energy larger than a preassigned value E_0 and that the opening angle between the leptons should be less than Θ_M given by [54]

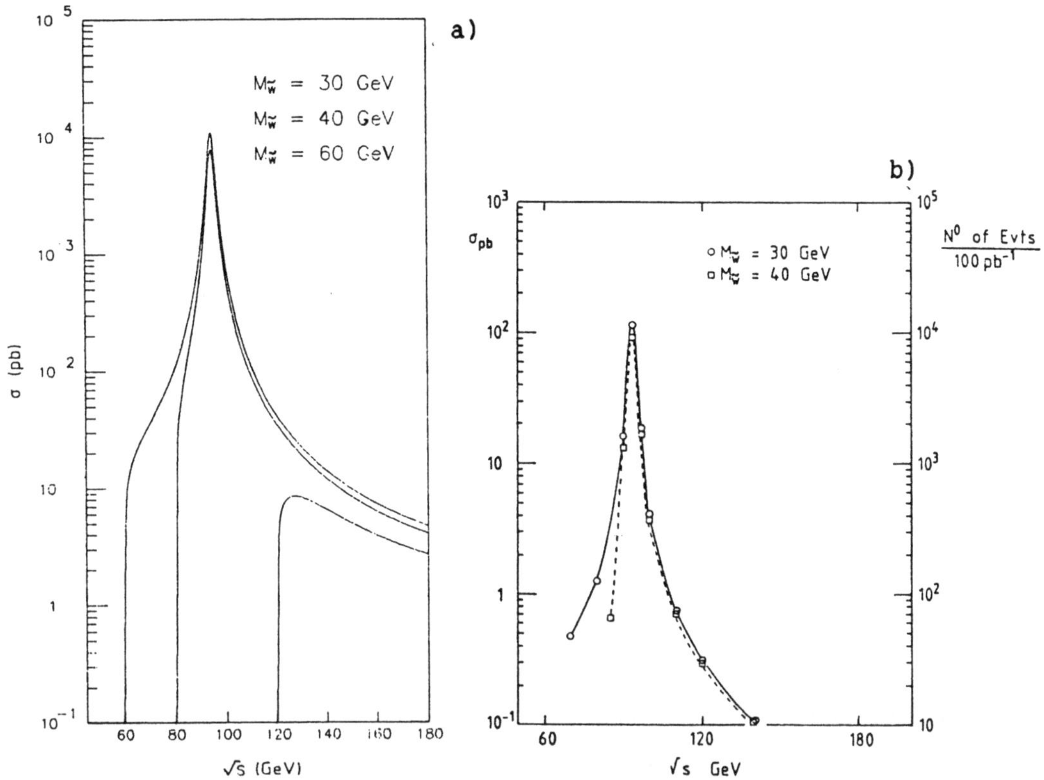

Fig. 6.1. The total cross section for wino pair production as a function of the c.o.m. energy for different wino masses. In Fig. 6.1a the total cross section is shown without any cuts and assuming a photino mass of 10 GeV. In Fig. 6.1b the cross section (times the branching ratio) for $e^+e^- \to \tilde{W}\tilde{W} \to e^+e^-\nu\bar{\nu}\tilde{\gamma}\tilde{\gamma}$ with acceptance and p_T cuts incorporated (with the following masses: photino 6 GeV, selectron 70 GeV, sneutrino 70 GeV)

$$\Theta_M = 180° - 2\arccos\left(\frac{\sqrt{s}E_0 - m_\tau}{\sqrt{s}E_0\beta_\tau}\right) \tag{6.1}$$

where m_τ is the tau mass and the threshold factor is given by $\beta_\tau = (1 - 4m_\tau^2/s)^{1/2}$.

QED processes are associated with small missing p_T and a cut with $\not{p}_T > 0.05\sqrt{s}$ eliminates most of them without substantially reducing the SUSY signal. It should also be noted that in wino pair production the number of $e^+\mu^- + e^-\mu^+$ pairs is almost the same as for $e^+e^- + \mu^+\mu^-$.

If the wino decays hadronically, wino pair production will lead to events with 1 or 2 jets accompanied by missing p_T at a much higher rate than dilepton events. However, they will be more difficult to be separated from background. Hence to look for the rare but clean dilepton events might be more promising than to search for the more abundant jet events.

6.2.2 Associated Zino-Photino Production

The two-body SUSY process with the lowest threshold may be $e^+e^- \to \tilde{z}\tilde{\gamma}$. This is of particular interest for LEP since in a wide class of models one expects $m(\tilde{z}) <$

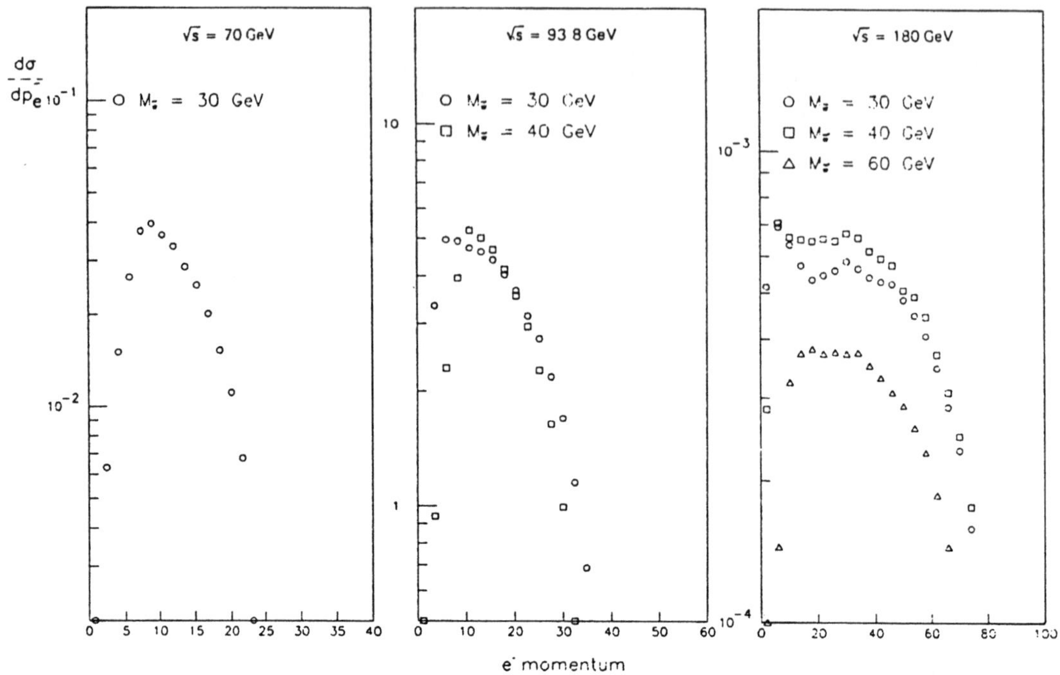

Fig. 6.2. The energy distribution of charged leptons from the three-body decay of the wino for different c.o.m. energies and various wino masses

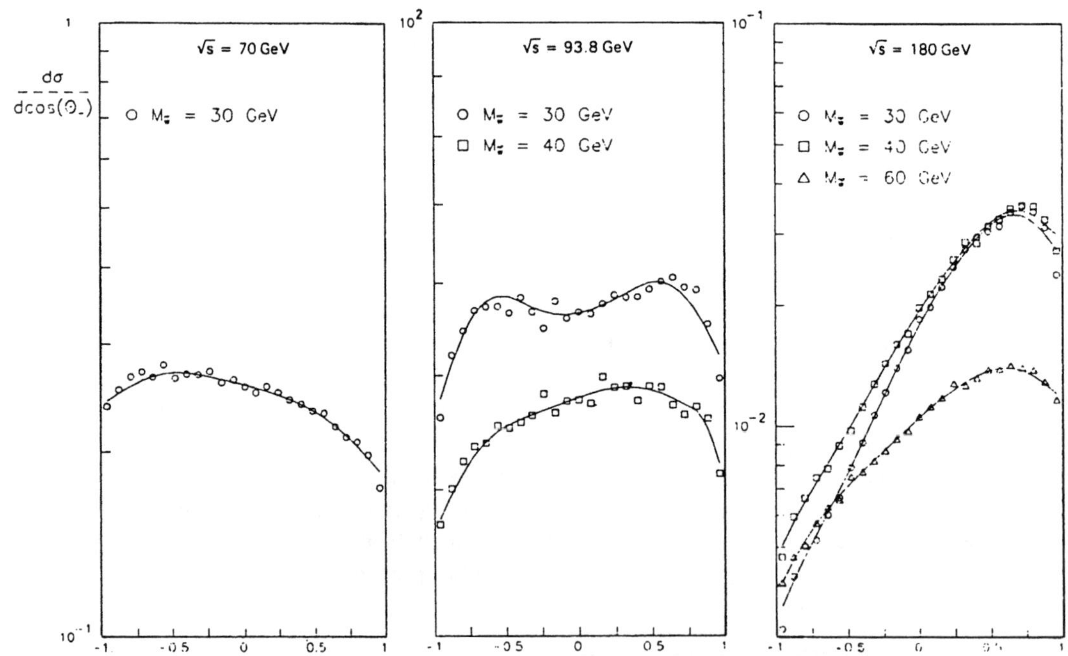

Fig. 6.3. The angular distribution of electrons coming from wino pair production. Here θ is the angle between the incoming electron and the outgoing negative lepton. Acceptance and momentum cuts have been applied

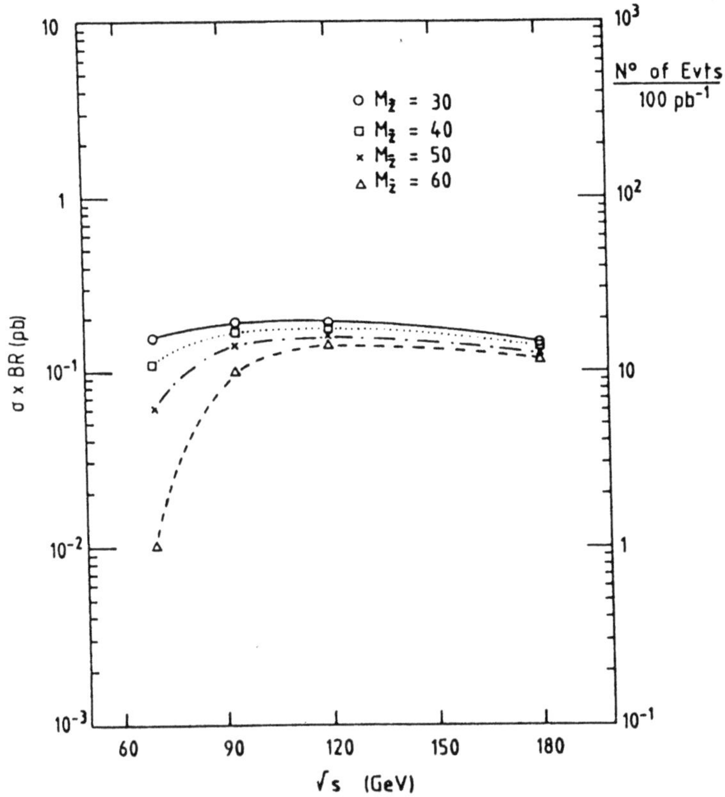

Fig. 6.4. The cross section (times branching ratio) for the associated production of $\tilde{z}\tilde{\gamma}$ followed by the decay $\tilde{z} \to \ell\bar{\ell}\tilde{\gamma}$ as a function of the c.o.m. energy for different zino masses

M(Z^0). In addition, this process should be free of SM background because of its particular signature which is similar to that for the process of Sect. 6.2.1.

The process $e^+e^- \to \tilde{z}\tilde{\gamma}$ proceeds via selectron exchange in the t-channel which has the consequence that it does not peak at the Z-mass and the cross section remains small over most of the energy range in question. The formulae are again rather involved and will not be reproduced here (see [54]).

To get "gold plated" events one should use the decay $\tilde{z} \to \tilde{\gamma}\ell^+\ell^-$. The cross section, including experimental cuts, is shown in Fig. 6.4. Very characteristic is the opening angle of the leptons originating from the $\tilde{z}$ decay. It will be smaller the more the $\tilde{z}$ is produced in motion, with an opening angle of the order of $m(\tilde{z})/\sqrt{s}$ (see Fig. 6.5). At high $\sqrt{s}$ the $\tilde{z}$ may move so fast that both leptons are emitted in the same hemisphere recoiling against nothing.

If hadronic decays of the $\tilde{z}$ are observed the signature is again events with jets and missing p_T. Although 3 to 4 times more abundant than leptonic decays these events will be more difficult to be separated from background.

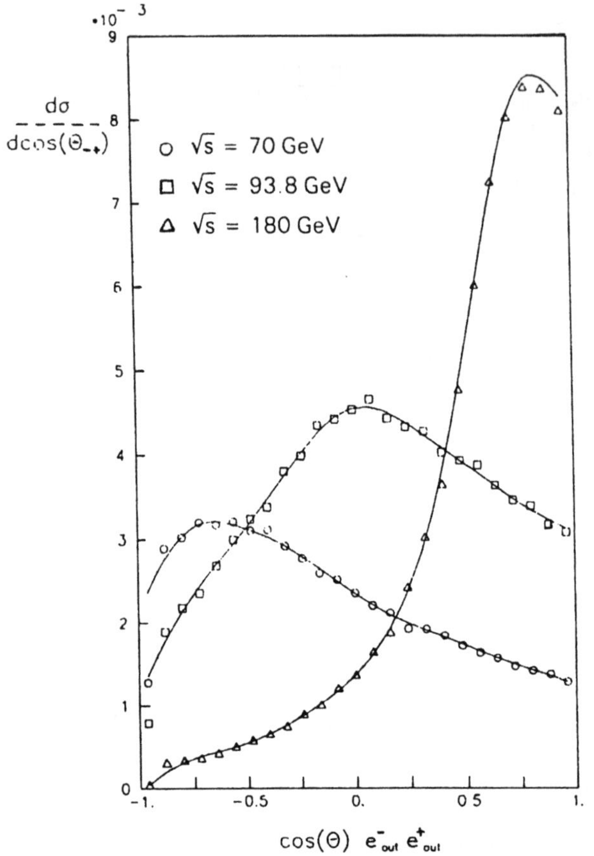

Fig. 6.5. Distribution of the opening angle between the leptons from the zino from the reaction $e^+e^- \rightarrow \tilde{z}\tilde{\gamma}$ for different c.o.m. energies. The zino mass is fixed at 40 GeV. As the energy increases the leptons tend to be emitted more and more in the same hemisphere

6.2.3 Three-Body Final States

a) Radiative Processes. Two SUSY particles which can be searched for even below the Z-peak are the photino and the sneutrino. They can be pair-produced and the events tagged by a bremsstrahlung photon in the final state. This is very similar to the radiative neutrino counting (see Sect. 4.5.2). The two reaction are

$$\begin{aligned} e^+e^- &\rightarrow \gamma + \tilde{\gamma} + \tilde{\gamma} \\ &\rightarrow \gamma + \tilde{\nu} + \tilde{\nu}. \end{aligned} \tag{6.2}$$

The calculation of the cross sections is again rather lengthy and the formulae will not be reproduced (see [54]). The cross section for the second process is shown in Fig. 6.6.

The signature of these two processes, as well as the radiative production of two ordinary neutrinos, is an excess of events with a single high p_T photon with missing momentum and energy and no hadronic activity. Since the signatures are the same for the three processes it will be necessary to measure the energy dependence of

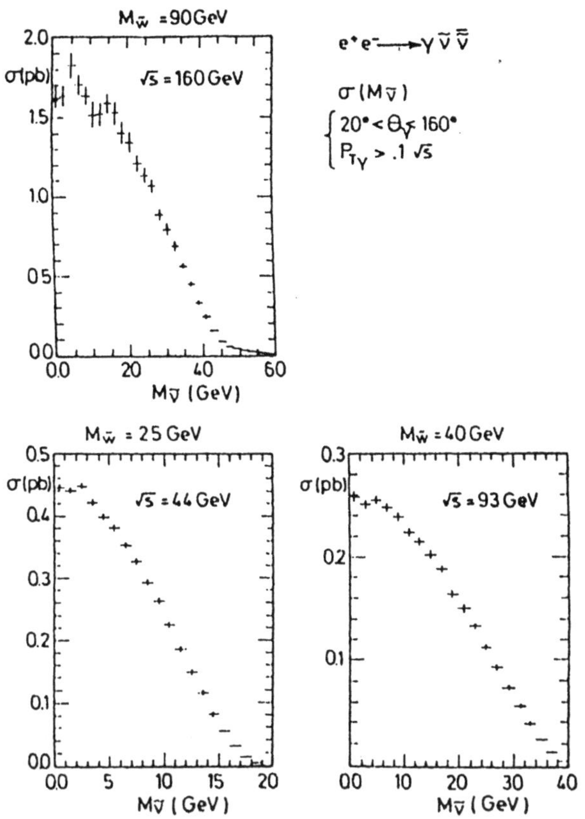

Fig. 6.6. The cross section for the production of a sneutrino pair together with a photon as function of the sneutrino mass for different c.o.m. energies

the cross section to establish the production of SUSY particles. Indeed the energy dependences of the three processes differ remarkably (Fig. 6.7). As one notices the best place to look for sparticles in these radiative processes is below the Z-peak since there the cross section for the production of normal neutrinos is relatively small. The background and its elimination is, of course, similar to that of the radiative production of ordinary neutrinos and the discussion of Sect. 4.5.2 applies here also.

b) Associated production. If the masses of the SUSY particles are larger than the beam energy, one can still search for sparticles via associated production where one light sparticle is produced together with a heavier one. Two reactions of particular interest are

$$e^+ e^- \to e + \tilde{e} + \tilde{\gamma} \tag{6.3a}$$

$$e^+ e^- \to e + \tilde{W} + \tilde{\nu} . \tag{6.3b}$$

Unfortunately the calculation of the cross section is very tedious since more than a dozen diagrams contribute [54], involving both s- and t-channel exchanges. Depending on which exchange dominates the experimental signal is quite different and depends strongly on the c.o.m. energy.

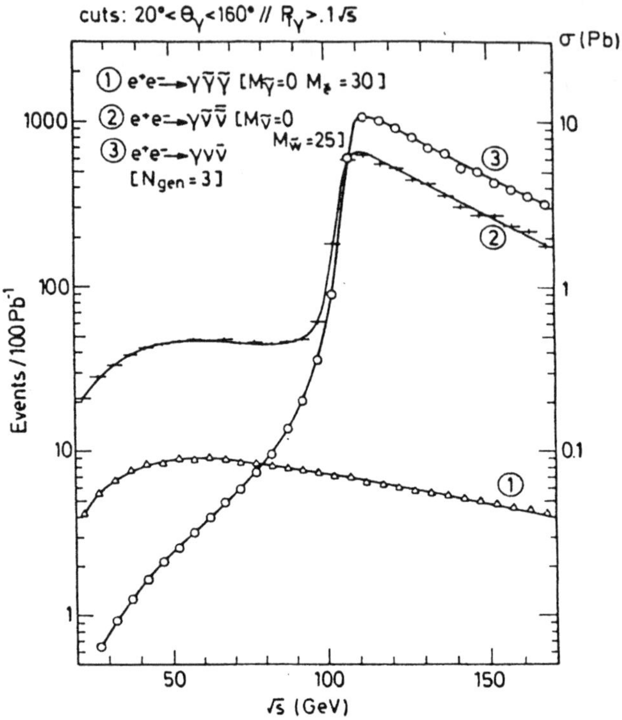

Fig. 6.7. The expected number of single high p_T photon events per 100 pb^{-1} as function of the c.o.m. energy for the pair production of photinos, sneutrinos and 3 types of ordinary neutrinos in association with a photon

Far away from the Z-peak the reactions are dominated by t-channel exchange with the consequence that the final state electron often escapes since it is predominantly produced in forward directions. In that case the signature would be a single high p_T lepton with no other detectable particles.

Near the Z-peak these processes are often dominated by virtual pair production of intermediate states in the s-channel. As a result the rate of lepton pairs acoplanar with the beam (with missing momentum and energy) becomes larger than the events with single leptons.

The reaction (6.3a) is important if the photino mass is much smaller than the selectron mass. In this case the main decay channel of the selectron is $\tilde{e} \rightarrow e\,\tilde{\gamma}$. As discussed above one observes single or dilepton events depending on the c.o.m. energy and the expected rates are shown for different c.o.m. energies in Fig. 6.8. With a total luminosity of about 100 pb^{-1} at the Z-peak one could detect selectrons with masses up to 50 or 60 GeV depending on the photino mass.

The discussion of reaction (6.3b) is more complicated since the wino need not be a pure state but can mix with the charged shigges. We shall assume here that no such mixing takes place (for the general case see [54]). The wino will decay either into $\ell\tilde{\nu}\tilde{\gamma}$ or a jet. Of course, the leptonic decay will again be easier to identify than the more abundant hadronic deacys. Depending on the c.o.m. energy one will find preferentially single or dilepton events, e.g. events with a single μ or an eμ pair.

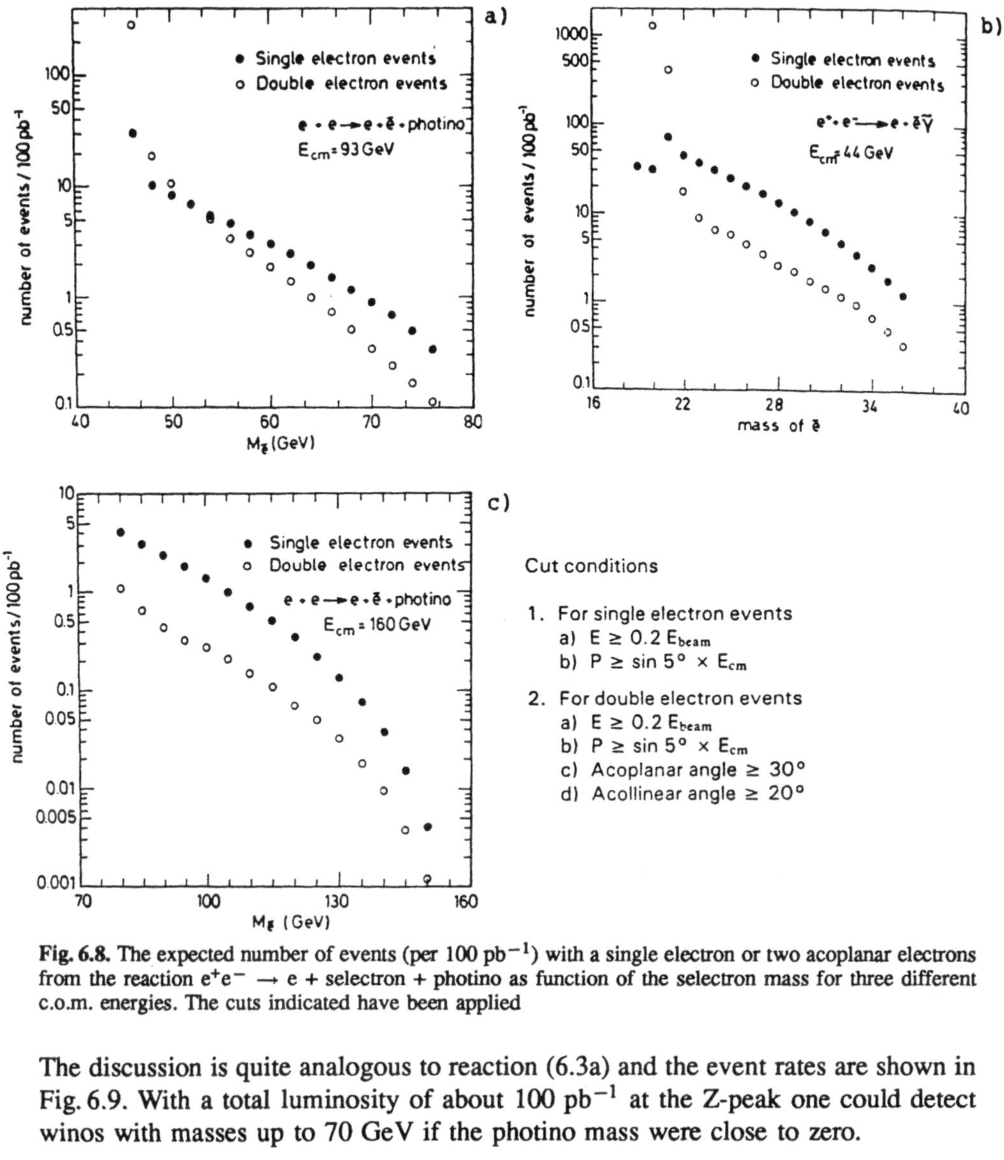

Fig. 6.8. The expected number of events (per 100 pb^{-1}) with a single electron or two acoplanar electrons from the reaction $e^+e^- \rightarrow e$ + selectron + photino as function of the selectron mass for three different c.o.m. energies. The cuts indicated have been applied

The discussion is quite analogous to reaction (6.3a) and the event rates are shown in Fig. 6.9. With a total luminosity of about 100 pb^{-1} at the Z-peak one could detect winos with masses up to 70 GeV if the photino mass were close to zero.

6.3 Conclusions

As far as the search for SUSY particles at LEP is concerned one can summarize the chances in Table 6.1 where the first column indicates existing mass limits and the third column gives these limits as they might be obtained by LEP 200. For LEP 100 they are generally 10 to 20 GeV lower, of course.

In summary one can state that with LEP the mass limits of charged sparticles will be increased by 10 to 30 GeV [55]. New limits for the charged sparticles and the gluino will be obtained in the future at the TEVATRON and they might be extended

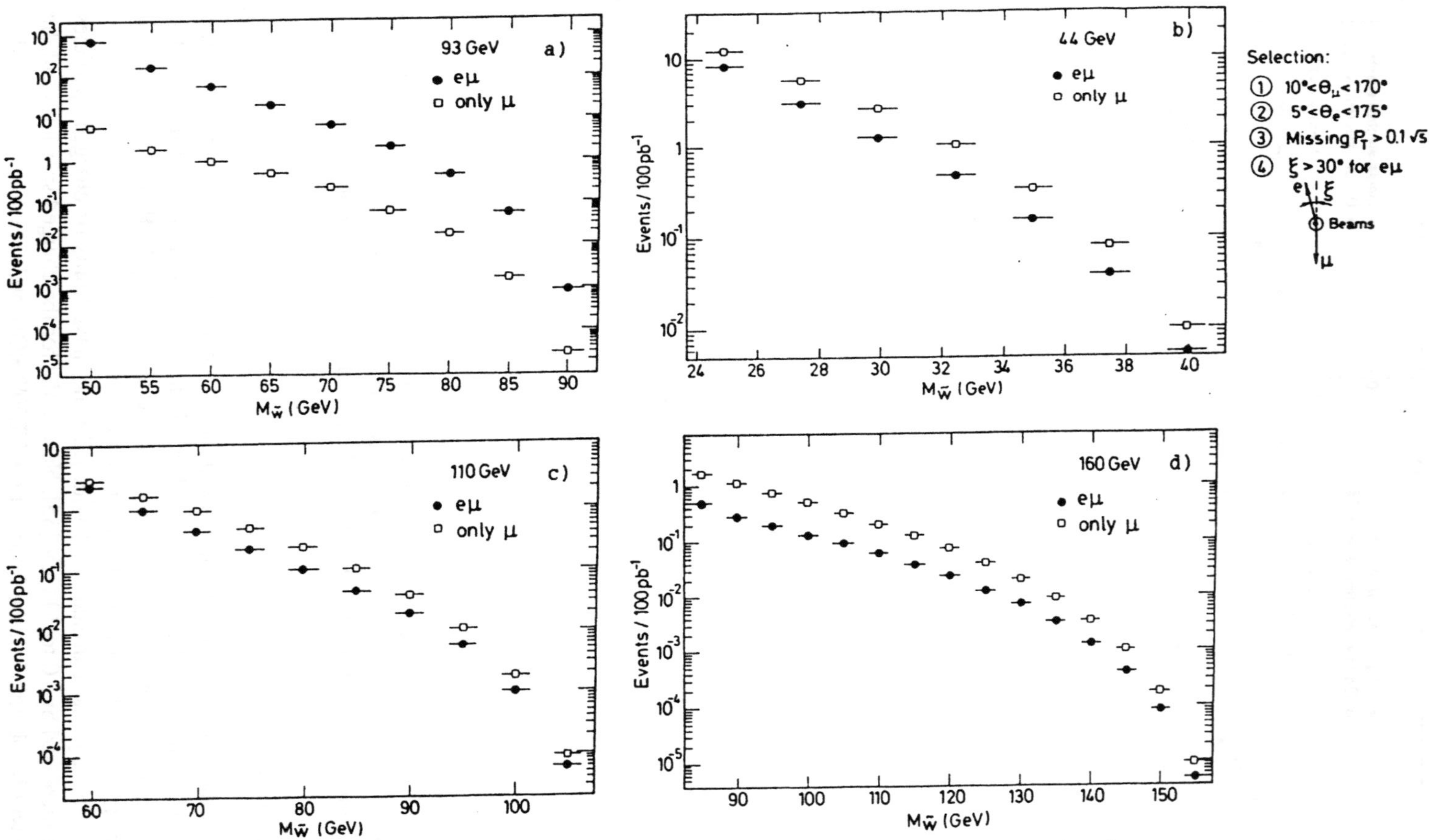

Fig. 6.9. The expected number of events (per 100 pb^{-1}) with a single muon or an acoplanar $e\mu$ pair from the reaction $e^+e^- \rightarrow e$ + wino + photino with the subsequent decay wino $\rightarrow$ muon + sneutrino + photino as function of the wino mass for four different c.o.m. energies. The cuts indicated have been applied

Table 6.1. Mass limits for SUSY particles

Particle	Existing Limit	With LEP 200
gluino	~ 60 GeV	~ 70 GeV
squarks	~ 50 GeV	~ 60 GeV
sleptons	~ 25 to 30 GeV	~ 60 GeV
wino, charged shiggs	~ 25 GeV	~ 70 GeV
sneutrinos	–	~ 50 GeV
zino, photino, neutral shiggs	–	~ 50 GeV

up to about 150 GeV. The most interesting part of the SUSY search at LEP will therefore be the investigation of neutral sparticles where practically nothing is known so far. Of course, it would be most exciting if not only improved mass limits could be given, but if sparticles would be discovered.

7. Search for Higgs Particles

The discovery of the Higgs particle would be a major breakthrough since it would give information on the mechanism of spontaneous symmetry breaking and would provide a clue for understanding particle masses.

Electron-positron colliders offer a unique possibility to detect Higgs-particles since the events are very clean. For hadron colliders it will be much more difficult to find Higgses if not impossible. On the other hand the accessible mass range is more limited for e^+e^- colliders and high luminosities are required. The search for Higgses will be particularly exciting at LEP since almost nothing is known about the Higgs mass. Experimental limits [8] extend from a few hundred keV to theoretical upper estimates [56] of the order of 1 TeV. At LEP 100 it should be possible to explore the mass range up to about 50 GeV and at LEP 200 this could be extended to about 80 GeV. The Higgs search will be a particularly exciting part of the LEP experimental program.

To discuss different experimental possibilities the ordering of the masses of Higgs, top and W are important. A few years ago it was thought that the top mass could be in the range of 30 to 40 GeV and therefore many calculations were performed for such top masses. Recent limits obtained at the CERN proton-antiproton collider and at the TEVATRON at Fermilab indicate, however, that the top mass seems to be heavier than 60 GeV, i.e. toponium would be heavier than the Z^0. Many calculations performed with the lower top mass need to be redone.

7.1 General Properties of the Higgs

We shall briefly summarize here those properties of the Higgs which are relevant for its detection. Exotic theories will not be discussed but we shall limit the discussion to the Standard Model Higgs (SM Higgs) and the Supersymmetric Higgses (for a complete discussion see [57]).

7.1.1 The Standard Model Higgs

The minimal Higgs mechanism of the Standard Model assumes the existence of two complex scalar fields. Three of the four fields are absorbed by the $W^{\pm}$ and Z^0 to give them a mass whereas the fourth field gives rise to a scalar neutral particle, the H^0. Two parameters are left free and hence no prediction can be made about its mass and its coupling constant. The coupling to other particles is proportional to their mass.

7.1.1.1 Higgs Decay into Fermions

For the observation of the Higgs the most important decay mode will be into a fermion-antifermion pair (lepton-antilepton or quark-antiquark pair) and the width is given by [58]

$$\Gamma(\mathrm{H}^0 \to \mathrm{f}\bar{\mathrm{f}}) = \frac{N_c G_F M_f^2}{4\sqrt{2}} M_{\mathrm{H}^0} \beta^3, \tag{7.1}$$

where $\beta = \sqrt{1 - 4M_f^2/M_H^2}$ is the threshold factor, G_F the Fermi coupling constant and the colour factor $N_c = 1$ and 3 for leptons and quarks, respectively.

As is obvious the partial width is largest for the heaviest fermion. The Higgs therefore will decay preferentially into bb quark and $\tau\tau$ lepton pairs (Fig. 7.1). A few examples for Γ for the b quark and hypothetical heavier quarks with mass m_q are given in Table 7.1.

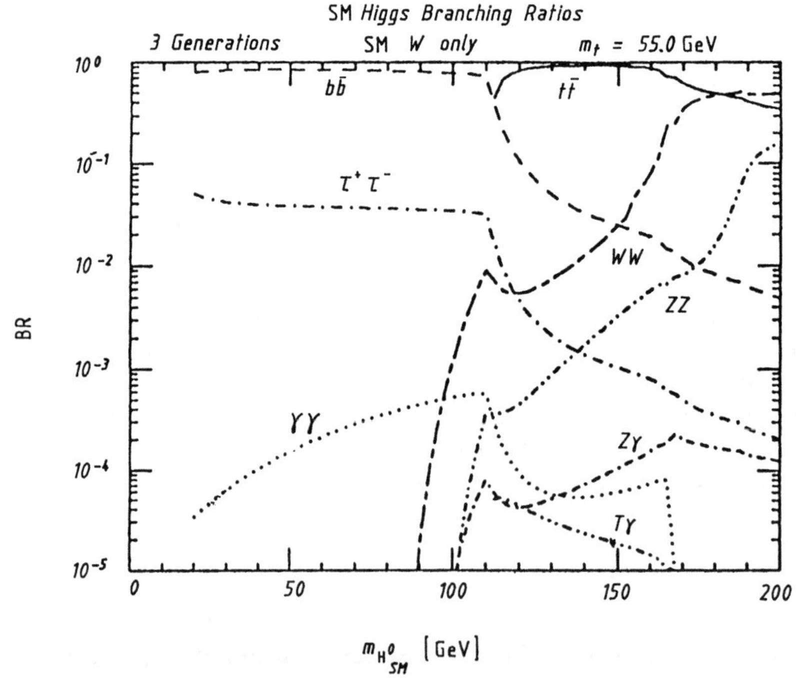

Fig. 7.1. The branching ratios for the decays of a Standard Model Higgs as a function of the Higgs mass. A top mass of 55 GeV was assumed

Table 7.1

M_H (GeV)	20	60	100	140	160
	$\Gamma(H^0 \to f\bar{f})$ (MeV)				
m_q (GeV) = 5	1	3	5	7	8
30	–	–	91	183	226
50	–	–	–	236	375

7.1.1.2 Higgs Decays into Bosons

If the Higgs were sufficiently heavy to decay into WW or ZZ pairs these decays would be the dominant ones (for details see [1], Vol.I, p. 304). However, the production rates of so massive Higgses would be negligible even at LEP 200 and hence we shall not consider these decays.

The decay $H^0 \to \gamma\gamma$ is very rare since it proceeds via fermion or W loops (Fig. 7.1) and would play a role only if the Higgs were so light (less than 200 MeV) that decays into heavier particles are kinematically forbidden (see Sect. 7.4).

Similarly the decay $H^0 \to gg$ goes through quark loops. Depending on the quark masses it can attain a few percent of the decay $H^0 \to q\bar{q}$. However, it will be difficult to identify it and hence it will not be further discussed here.

The total width for the SM Higgs is shown in Fig. 7.2. The essential point is that Γ is always smaller than 1 GeV and since the experimental resolution is larger, a binning of the Higgs mass of about 1 GeV would not lead to any losses of the signal.

7.1.2 The Supersymmetric Higgses

The original Standard Model uses a minimal Higgs mechanism with one doublet. However, there is no reason why the Higgs sector could not be more complicated and indeed more sophisticated models have been proposed to explain CP violation, the generation of neutrino masses and for supersymmetric theories. It is impossible here to discuss the consequences for a Higgs search of all these models and hence only the supersymmetric Higgs mechanism will be considered since the number of free parameters is limited and this model is still relatively simple.

Minimal supersymmetric models require two complex Higgs doublets. Although supersymmetry imposes many symmetry relations five physical Higgs particles are required:

$$H^+ , \quad H^- , \quad H^0 , \quad h_0 , \quad A_0 .$$

The first four particles are scalars whereas the last is a pseudoscalar. In addition to the neutral ones there exist also two charged Higgses. Because of the symmetry constraints the 5 Higgs masses depend only on 2 independent parameters (and the Z^0 mass). Usually the mass of the A_0 and $\tan\beta = v_2/v_1$ (the ratio of two vacuum expectation values) are chosen as free parameters. From the limit on the top mass one can deduce [15] an allowed range for $\tan\beta$ as $1 < \tan\beta < 15$ to 20. The relation for some masses are shown in Fig. 7.3. The most interesting relations for the Higgs search are:

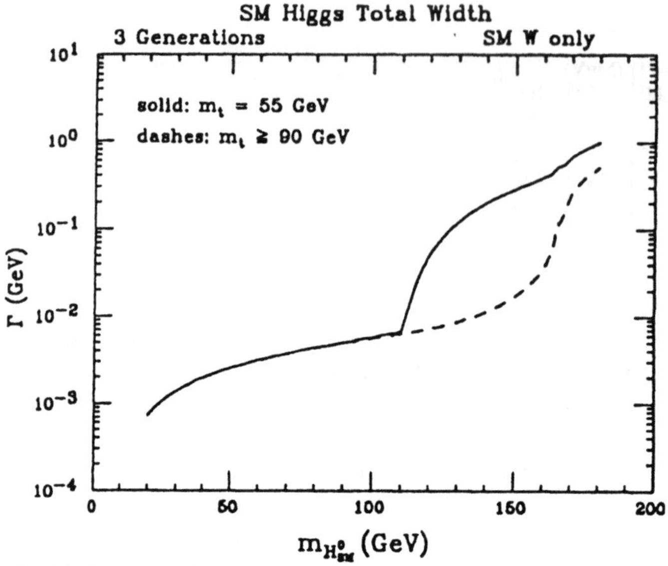

Fig. 7.2. The total width of a Standard Model Higgs as a function of the Higgs mass

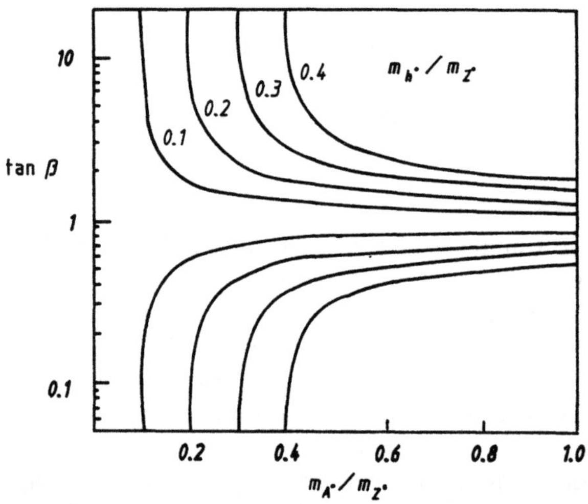

Fig. 7.3. Relation between the masses of the supersymmetric Higgses h_0 and A_0. The free parameter $\tan\beta$ relates different vacuum expectation values

$$
\begin{array}{lll}
M_{h_0} \leq M_{Z^0} & \text{and} & M_{H^\pm} \geq M_{W^\pm} \\
M_{A_0} > M_{h_0} & & M_{H^0} > M_Z \\
M_{h_0} \approx M_{A_0} & & \text{if } \tan\beta \text{ large.}
\end{array}
$$

Hence the h_0 could be rather light and accessible to LEP and also the A_0 could have a mass in the LEP energy range.

Allowed decays are

$$Z_0 \rightarrow h_0\mu^+\mu^- \qquad \text{and} \qquad Z^0 \rightarrow h_0 A_0$$

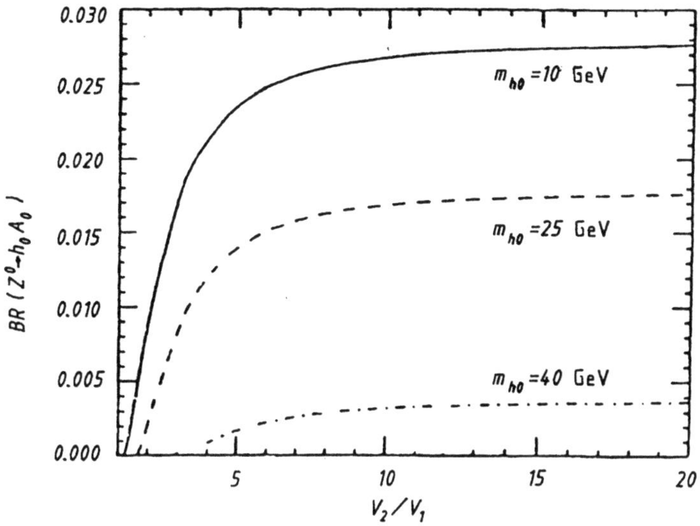

Fig. 7.4. Branching ratios for the decay Z → h_0 A_0 as function of the ratio of the vacuum expectation values for different masses of the h_0

which are complementary in the sense that if the branching ratio for one is big the one for the other is small and vice versa (Fig. 7.4). Branching ratios of the order of 1% are possible.

7.2 Higgs Search at the Z^0 Peak (LEP 100)

We shall now discuss the most promising ways of detecting the various Higgs particles. This implies that one also has to find a production mechanism which at the available energies and luminosities provides a sufficient number of events. At first the situation at LEP 100 and after that the possiblities of LEP 200 will be considered.

7.2.1 Search for the Standard Model Higgs

Let us first consider the Standard Model Higgs. The most important process to produce it is **"Higgs bremsstrahlung"**, where a Z^0 shakes off a Higgs (Fig. 7.5). (Paranthetically it might be mentioned that the process $e^+e^- \rightarrow Z^0 \rightarrow H^0 + H^0$

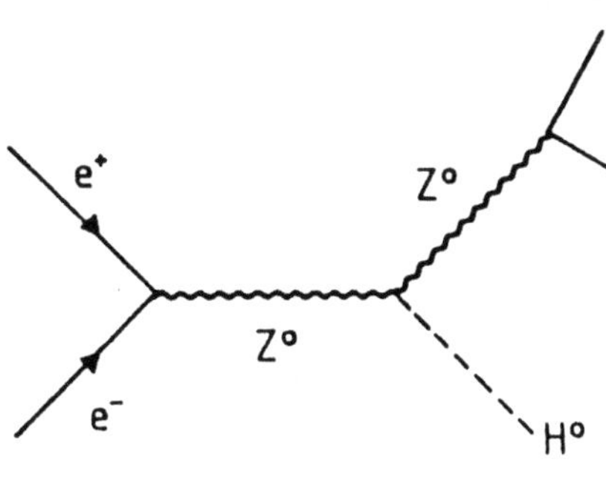

Fig. 7.5. The diagram for the production of a Higss ("Higgs bremsstrahlung"). At the Z-peak the Z is virtual whereas above the Z-peak it can be real

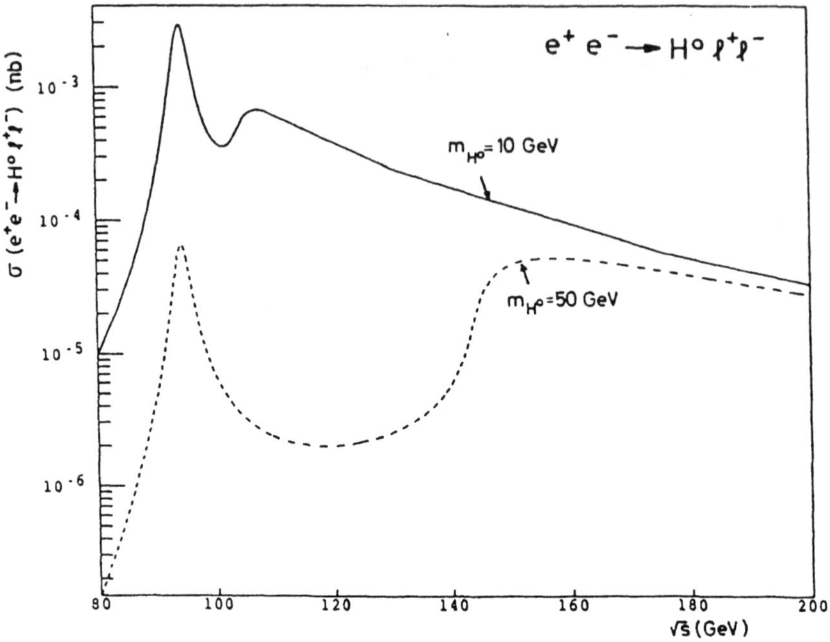

Fig. 7.6. Cross section for $e^+e^- \rightarrow H^0\ell^+\ell^-$ as a function of the c.o.m. energy for two different Higgs masses

is forbidden because of symmetry considerations.) Hence we are dealing with the process:

$$\mathbf{e^+e^- \rightarrow Z^{0*} \rightarrow H^0 + f^+ f^-}$$

where the H^0 will disintegrate preferentially into a bb quark pair.

The cross section at the Z^0 peak depends of course on the Higgs mass and is relatively large (Fig. 7.6). The fermion pair from the Z^{0*} decay can either be a charged lepton pair, a neutrino pair or a quark pair. The branching ratio is largest for the quarks, whereas the branching ratio for the charged leptons (equal for the 3 kinds of charged leptons) is 6 times smaller than for the neutrinos (Figs. 7.1 and 7.7). We shall discuss these 3 cases separately.

i) $e^+e^- \rightarrow Z^0 \rightarrow H^0 + e^+e^-$ or $\mu^+\mu^-$
These decays are most interesting from the experimental point of view since they have a clear signature (two leptons, two jets) and the energy of the leptons can be measured.

From the measured energies E_+ and E_- of the leptons and the angle Θ enclosed by them, one can calculate the invariant mass of the lepton pair, but also the invariant recoil mass. As we shall see the first quantity is useful to suppress backgrounds, whereas the second gives the Higgs mass. Thus the Higgs mass can be determined without observing its decay products.

The mass of the lepton pair, originating from a virtual Z^0 is given by

$$M^2_{Z^*} = 4E_+E_- \sin^2 \Theta/2 \,. \tag{7.2}$$

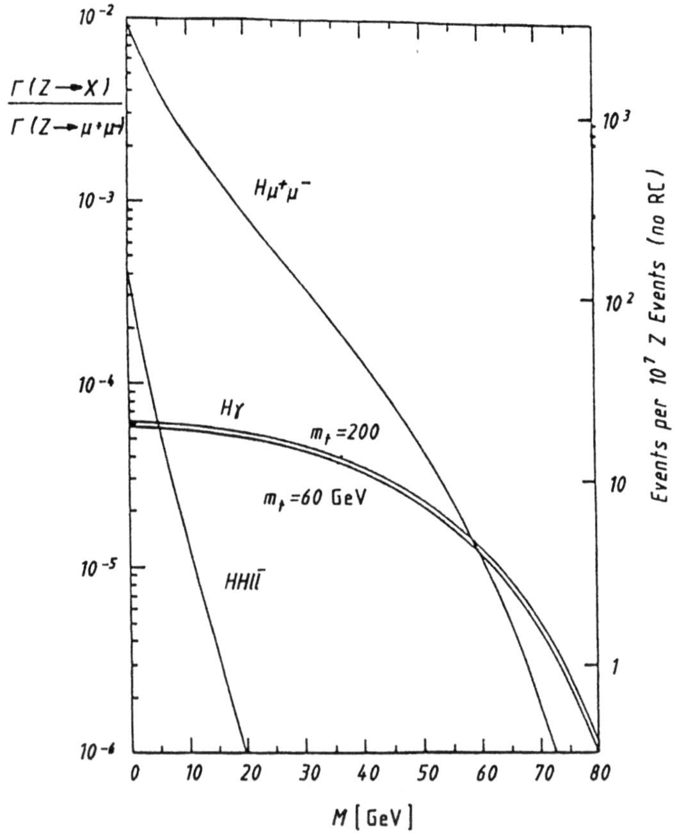

Fig. 7.7. Branching ratio (normalized to muon decay) of the decay of the Z into final states containing at least one Higgs as function of the Higgs mass

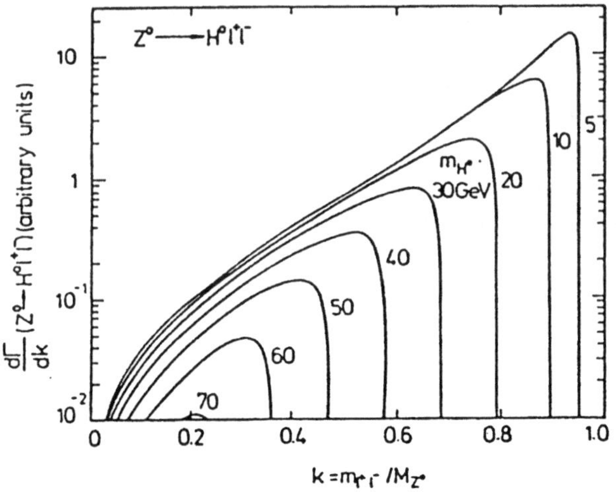

Fig. 7.8. Dilepton mass distribution for the process $Z \to H^0\ell^+\ell^-$ for various Higgs masses

Since the Z is virtual the mass distribution is not sharp but shows a broad distribution, peaking at the highest energy which is kinematically allowed (Fig. 7.8).

The recoil mass is given by

$$M_{\text{recoil}}^2 = (\sqrt{s} - E_+ - E_-)^2 - p_z^2 = s + M_Z^2 - 2\sqrt{s}(E_+ + E_-) \; . \tag{7.3}$$

Measuring errors of s and Θ are negligible and what counts is the precision of E_+ and E_-. One can show that near threshold, i.e. $\sqrt{s} = M_Z + M_H + Q$ with $Q \ll \sqrt{s}$ and $E_+ \cong E_-$, one obtains

$$\delta M_H = (1 + Q/M_H)\sqrt{dE_+^2 + dE_-^2} = (1 + Q/M_H)a\sqrt{E_+ + E_-} \tag{7.4}$$

where the experimental resolution $\delta E/E = a/\sqrt{E/\text{GeV}}$ has been inserted. In order to get a good Higgs mass determination an excellent instrumental energy resolution is necessary. As has been explained in Sect. 3 values of $a \cong 0.1$ are obtained in LEP experiments and hence a mass error of about $\delta M_H \cong 1.5$ GeV or better seems realistic.

The following table gives event numbers for a few examples of Higgs masses.

Table 7.2. $e^+e^- \to Z^0 \to H^0\mu\mu$ events for 10^6 Z^0

Higgs Mass (GeV)	10	20	30	40	50
Number of events	66	26	11	4.2	1.5

These figures show that the possibility for the detection of the Higgs looks rather good, however, high luminosities are needed if the Higgs is heavy.

Background

Some time ago it was thought that the most serious background would come from $Z^0 \to t\bar{t}$ decays and many calculations were made for this case. Since now the mass limit for toponium is above the Z^0 this background seems negligible.

The main background will therefore come from the rare four-fermion decays of the Z^0, like $Z^0 \to \mu^+\mu^-\ q\bar{q}$. The rates for such processes have recently been calculated [15] and some results are shown in Fig. 7.9. As one notices this background can be quite serious for Higgs masses above 40 GeV if no cuts are applied. If one looks, however, at the lepton invariant mass spectrum (Fig. 7.10b) one sees that a cut at around 20 GeV excludes most of the background but suppresses only few of the good events. The result of such a cut at 20 GeV is shown in Fig. 7.10a and the expected event rates without and with cut are given in Table 7.3.

For the L 3 experiment which will have the best energy resolution for leptons (BGO for electrons, precision muon chambers) some calculations have been done which are shown in Fig. 7.11 for luminosities which could be obtained already in 1989 ($L_{\text{int}} = 4\ \text{pb}^{-1}$). If the Higgs is lighter than 20 GeV it should certainly be seen.

ii) $e^+e^- \to Z^0 \to H^0 + \nu\bar{\nu}$

The decay of the virtual Z into a neutrino-antineutrino pair is 6 times more likely than the dacay into charged lepton pairs. As far as rates are concerned this process is therefore more advantageous, but of course, it is more difficult to detect.

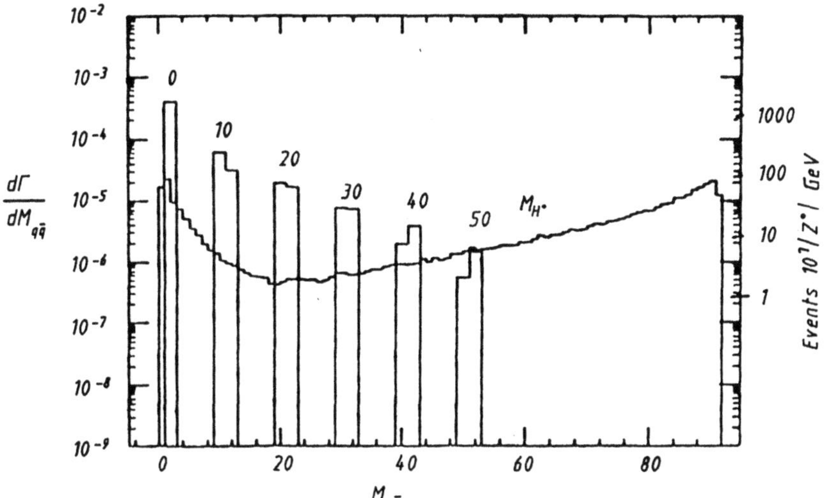

Fig. 7.9. The recoil mass distribution for the decay Z $\rightarrow$ $H^0\mu^+\mu^-$ for various Higgs masses (sharp peaks) and the background from Z $\rightarrow$ $\mu^+\mu^-$ q$\bar{q}$. No cuts were applied

Table 7.3. Number of $Z^0 \rightarrow \mu^+\mu^- + H^0$ for 10^7 Z^0 produced

M_{H^0} GeV	Signal no cut	Signal cut	Background no cut	Background cut
~0	3170	3170	60	60
10	750	749	4.2	3.7
20	291	284	1.8	0.7
30	119	115	2.6	0.27
40	46	42	3.6	0.19
50	15.5	13.2	18	0.19
60	3.5	~3	>18	~0.19

The signature would be 2 jets from the Higgs decay and missing energy. Possible cuts would be: missing energy > 50 GeV, missing mass > 40 GeV and non-collinear jets. Some results from event simulation are shown in Fig. 7.12 with background from multihadron events. A signal might be seen if the Higgs mass is less than about 30 GeV and about 10^6 Z^0 would suffice already to see the Higgs.

iii) $e^+e^- \rightarrow Z^0 \rightarrow H^0 + q\bar{q}$
The decay of the Z^0 into a quark-antiquark pair is most abundant among the H-bremsstrahlung processes. It leads to a final state of 4 jets. However, a separation from ordinary multihadron jet events seems very difficult.

The Process $e^+e^- \rightarrow Z^0 \rightarrow H^0 + \gamma$. So far we have considered only processes which go through the "bremsstrahlung diagram" (Fig. 7.5). The simultaneous production of a Higgs with a photon is an example of a process which goes via W or fermi loops. Although the signature is quite characteristic (mono-energetic photon back-to-back with two jets) this process has hardly a chance to be observed. The reason is that the branching ratio is very small: $BR(Z^0 \rightarrow H^0+\gamma) \sim 10^{-6}$. Therefore the background of ordinary quark-antiquark pairs will be overwhelming.

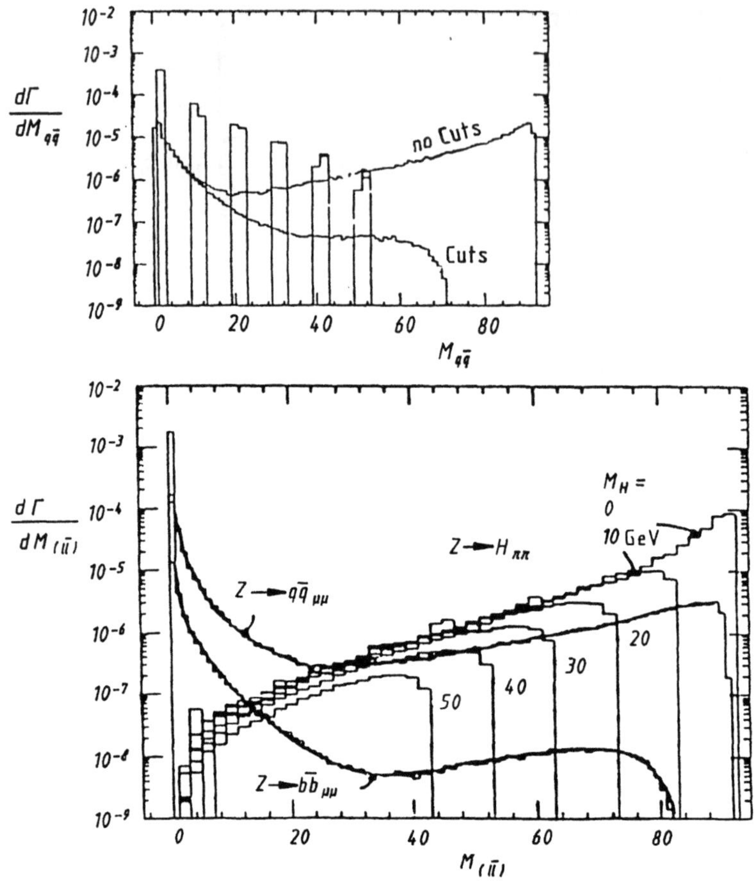

Fig. 7.10. In the lower part the invariant mass distribution of the lepton pairs for the decay $Z \to H^0 \mu^+ \mu^-$ and for the background $Z \to \mu^+ \mu^- q \bar{q}$ is shown. If a cut $M(\mu^+\mu^-) > 20$ GeV is applied one obtains the recoil mass distribution shown in the upper part of the figure

Nevertheless there might be a possibility that this process becomes observable, i.e. if quarks or leptons would have a composite structure. Some model calculations performed by DELPHI show (Fig. 7.13) that if the structure parameter Λ (which determines the average radius $< r > \sim 1/\Lambda$) would be of the order of the Z^0 mass, this process could become observable.

If the toponium mass were in the region of LEP 100 the probability for $Z^0 \to H^0 + \gamma$ would become much bigger since it could go via a toponium intermediate state. Indeed it would have been the most favourable way to look for the Higgs. Since now the toponium seems to be well beyond the Z^0 this case will be discussed in Sect. 7.3.

7.2.2 Search for the SUSY Higgs

As was discussed in Sect. 7.1.2 relations exist between the different masses of SUSY Higgses. The A_0 and the h_0 could well be in reach of LEP 100. For the following we use mainly the results presented in the workshop report by C. Verzegnassi [15].

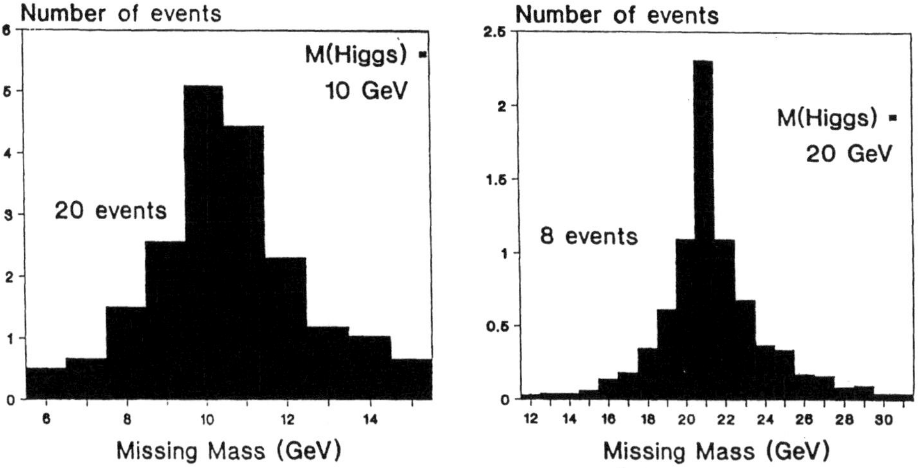

Fig. 7.11. Simulated missing mass distribution for the decay Z $\rightarrow H^0\mu^+\mu^-$ for the L 3 detector for two different Higgs masses

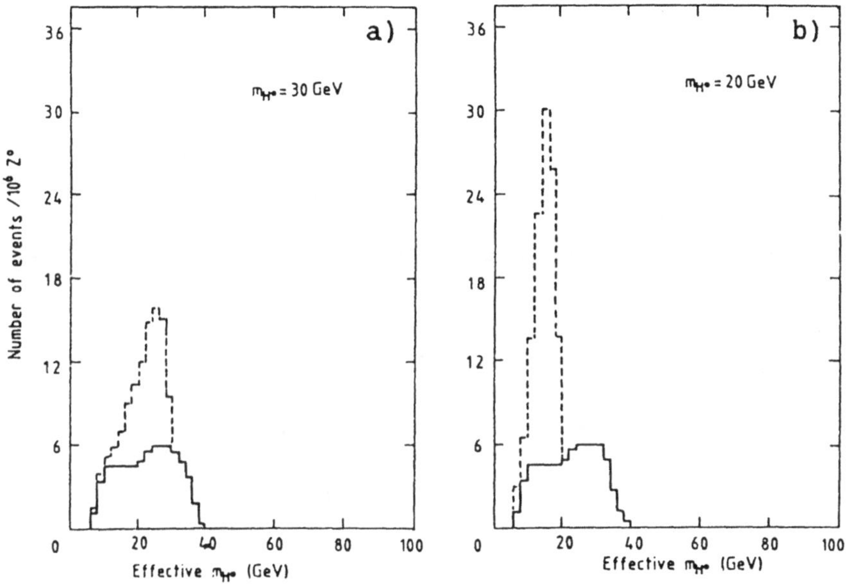

Fig. 7.12. Simulated effective Higgs mass distributions (*solid line* histogram) for the decay Z $\rightarrow \nu\bar{\nu}$ + ($H^0 \rightarrow$ hadrons) and for the multihadronic background (*dashed line* histogram). Monte Carlo results for two Higgs masses are shown

The detection of $e^+e^- \rightarrow Z \rightarrow h_0 + \ell^+\ell^-$ would be easiest if the mass of h_0 were smaller than about 10 GeV. In such a case the branching ratio $BR(Z \rightarrow h_0 + \ell^+\ell^-)$ could be of the order of 0.1% for a large range of A_0 masses (see Fig. 7.14a).

If this processes is not observed it would imply that the masses of A_0 and h_0 are very similar (see Sect. 7.1.2 and Fig. 7.14a) and then $e^+e^- \rightarrow Z \rightarrow A_0 + h_0$ should become observable with branching ratios of the order of $BR(Z^0 \rightarrow A_0 + h_0) \sim 1\%$. The identification of this decay will not be easy, however, since both Higgses will

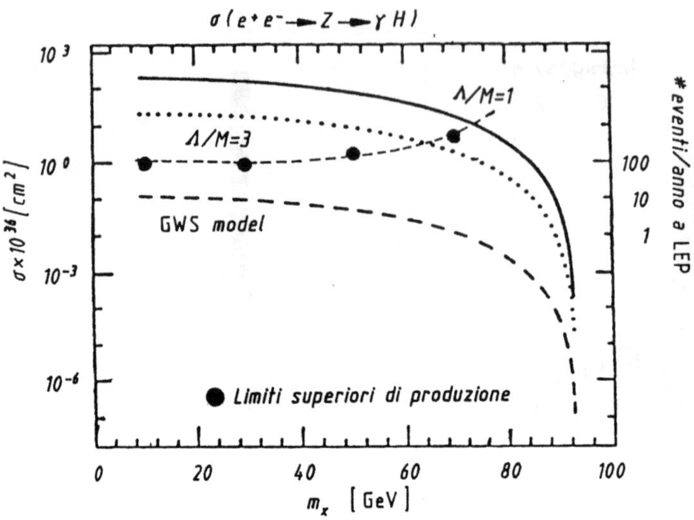

Fig. 7.13. The cross section for the reaction $e^+e^- \to Z \to H^0 + \gamma$ as a function of the Higgs mass. The cross section for a pointlike SM Higgs (*dashed line*) is well below the curve indicating a 2 standard deviation observable effect (dashed line with points). For composite quarks or leptons one obtains the pointed ($\Lambda/M = 3$) and the full line ($\Lambda/M = 1$) where Λ is the cut-off parameter characterizing the composite structure and M is the Z-mass. (DELPHI)

disintegrate into a quark-antiquark pair and the competition from ordinary 4-jet events will be very serious. Maybe the best way to look for this decay channel would be to select cases where one of the two Higgses decays into a $\tau^+\tau^-$ pair. The braching ratios for this case are still large (Fig. 7.14b).

The L 3 experiment can identify leptons within a jet. Event simulation was performed for the decay

$$Z^0 \to H_1 + H_2 \to 4\,b \to 2\,\ell + 4\text{ jets},$$

where H_1 and H_2 stand for any of the neutral Higgses. With a luminosity of 4 pb^{-1} a few thousand events are expected and their invariant mass distributions are shown in Figs. 7.15 and 7.16. As one sees the Higgs peak can be well separated from QCD background.

The search for charged Higgses will be discussed in Sect. 7.5.

7.3 Higgs Search at LEP 200

At center-of-mass energies well above the Z^0 mass the diagram of Fig. 7.5 offers still the most promising way to detect the Higgs. The main difference is that now the Z^0 is not any more virtual but real and hence the invariant mass of the decay fermions will show a clear Z^0 peak instead of being smeared out.

The total cross section for $e^+e^- \to H^0 + Z^0$ is shown in Fig. 7.17 as a function of the beam energy for different Higgs masses. As can be seen above threshold it is a slowly varying function. Assuming that the detection limit will be around 1 pb^{-1} it seems possible to detect Higgses up to about masses of 60 GeV. About 1 event per day is expected (see Table 7.4). However, it is clear that high luminosities (400 pb^{-1} and more) are required in order to find Higgses above 40 GeV.

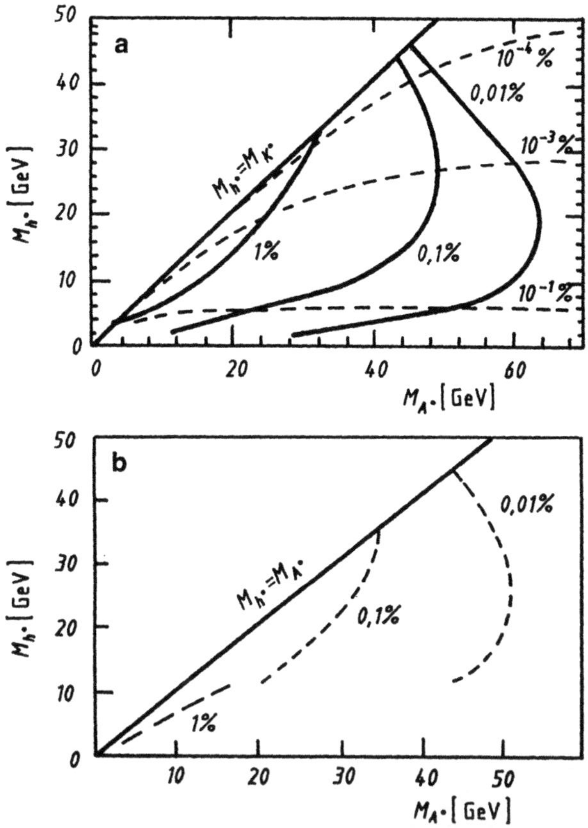

Fig. 7.14. Branching ratios for the Z-decay into the two supersymmetric Higgs particles h_0 and A_0 in the $M(h_0) - M(A_0)$ plane. In Fig. 7.14a the curves indicate constant branching ratios for $Z \to h_0 A_0$ (*dashed curves*) and $Z \to h_0 \mu\mu$ (*full curves*). In Fig. 7.14b the curves give constant branching ratios for $Z \to h_0 \tau\tau$ decays. The parameters used are $M_Z = 93$ GeV, $\Gamma = 3$ GeV and it is assumed that the Higgs mass is large relative to Γ [15]

Table 7.4. Number of events for the process $e^+e^- \to Z + H^0$. Events per day assuming a luminosity of $10^{31}\ cm^{-2}s^{-1}$

Mass of Higgs (GeV)	10	20	30	40	50	60	70	80	90	100
$\sqrt{s}$ (GeV)										
120	12.33	8.55	0.00	0.00	0.00	0.00	0.00	0.00	0.00	0.00
140	5.17	4.72	3.92	2.51	0.00	0.00	0.00	0.00	0.00	0.00
160	2.70	2.58	2.37	2.08	1.66	1.03	0.00	0.00	0.00	0.00
180	1.60	1.56	1.48	1.37	1.23	1.05	0.83	0.51	0.00	0.00
200	1.04	1.02	0.99	0.94	0.87	0.79	0.70	0.59	0.46	0.28
220	0.72	0.71	0.69	0.67	0.63	0.59	0.54	0.49	0.43	0.35
240	0.53	0.52	0.51	0.49	0.47	0.45	0.42	0.39	0.35	0.31
260	0.40	0.39	0.38	0.38	0.36	0.35	0.33	0.31	0.29	0.26

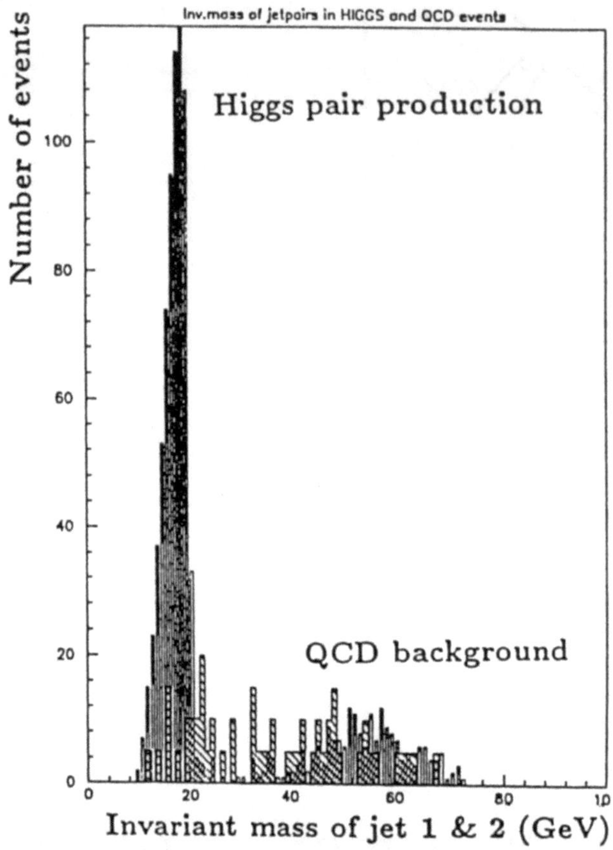

Fig. 7.15. Higgs pair production. Invariant mass distribution of two jets in a 4-jet event. (Events with at least 2 leptons and cut on the invariant mass of the other two jets) (L 3 simulation)

Concerning the backgrounds one does not expect too much difficulties for the Z $\rightarrow \ell^+\ell^-$ decay channels and the observation into neutrino pairs may not be excluded. However, Z^0 decays into quark pairs will again be very difficult to separate from the QCD background.

In order to discuss these processes in more detail two cases will be considered: c.o.m. energies below the W^+W^- threshold (160 GeV) and energies above this threshold (200 GeV). Only SM Higgses and charged Higgses will be dealt with.

7.3.1 Higgs Search Below the WW Threshold (at 160 GeV)

In the event simulations which will be discussed in the following paragraphs it was still assumed that the toponium mass would be in the region of the Z^0 and hence it was thought that it would contribute to the background. Therefore the signal/background relation is expected to be even better than shown in the following simulations.

ALEPH has considered the case $e^+e^- \rightarrow H^0 + (Z^0 \rightarrow e^+e^-)$ for $M_H = 50$ GeV assuming $M_Z = 89$ GeV. Fig. 7.18 shows that after proper cuts in the lepton

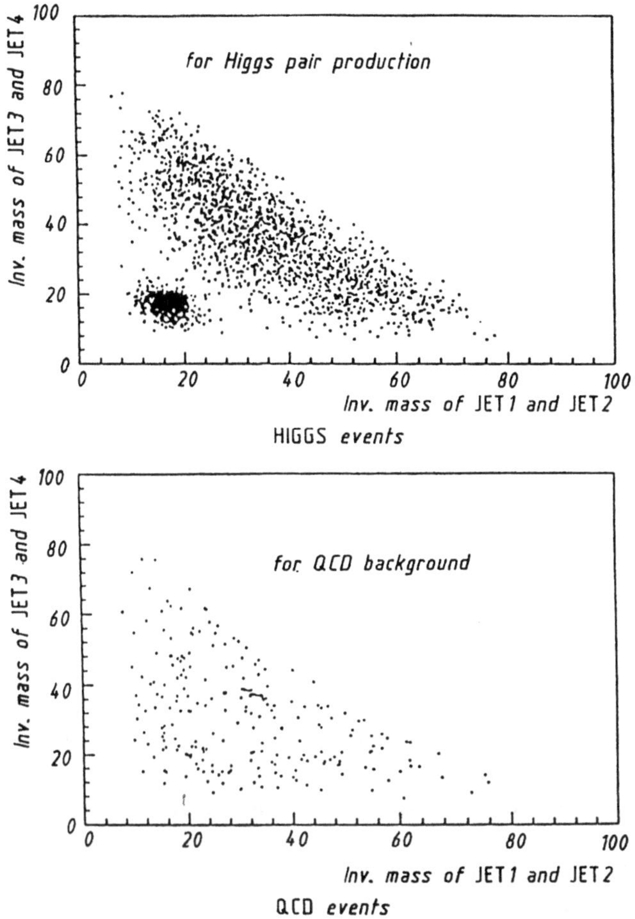

Fig. 7.16. Correlation of the invariant mass of 2 jets with respect to the invariant mass of the other 2 jets in 4-jet events. Events must contain at least 2 leptons. In the upper part events originating from Higgs pair production, in the lower part from QCD background are displayed (L 3 simulation)

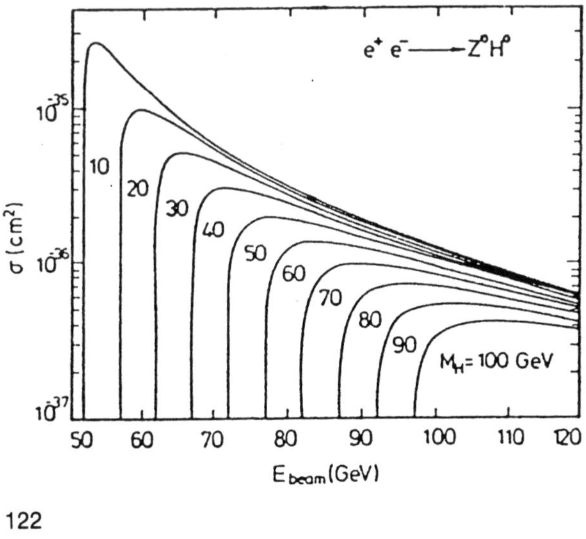

Fig. 7.17. Total cross section for the process $e^+e^- \to Z + H^0$ as a function of the beam energy for different Higgs masses

invariant mass (Fig. 7.18a) a clear Higgs signal (Fig. 7.18b) can be obtained. The situation should be similarly favourable for Z decays into muon pairs.

DELPHI has carried out similar calculations for a c.o.m. energy of 164 GeV and the results are shown in Fig. 7.19. Finally in Fig. 7.20 the results of L 3 are shown.

Summarizing the prospects for the process $e^+e^- \rightarrow H^0 + (Z^0 \rightarrow e^+e^-)$ or $(Z^0 \rightarrow \mu^+\mu^-)$ it seems that the identification of the Higgs against background will not be very difficult, but high integrated luminosities, requiring 1 to 2 years of LEP operation, will be necessary.

The process $e^+e^- \rightarrow H^0 + (Z^0 \rightarrow \nu\bar{\nu})$ has again the advantage that its rate is 6 times higher than the decay into charged leptons but it is much harder to distinguish from background. The signature of the events would be 2 jets (from the H^0 decay) and large missing energy and transverse momentum. The events should also have a relatively large sphericity. ALEPH and DELPHI have performed some calculations and the results are presented in Fig. 7.21 and 7.22. As one notices a Higgs with a mass of 50 GeV could still be detected provided the proper cuts are applied.

Finally it should be mentioned that events for which both the Z^0 and the H^0 decay into 2 jets (4-jet events) have been investigated by ALEPH, and in the invariant mass distribution of two jets they find a Higgs signal over a QCD background of about the same magnitude. Hence this might be a difficult way to identify a Higgs.

7.3.2 Higgs Search Above the WW Threshold (at 200 GeV)

At c.o.m. energies above the WW pair production threshold the reaction $e^+e^- \rightarrow H^0 + Z^0$, with the Higgs decaying into two jets and the Z^0 decaying into either $\ell^+\ell^-$, $\nu\bar{\nu}$ or $q\bar{q}$, is still the most promising way to look for Higgses. With integrated luminosities of about 500 pb^{-1} it should be possible to extend the Higgs search up to masses of about 80 GeV.

For the decay $Z^0 \rightarrow \ell^+\ell^-$ the background includes the following reactions in addition to the normal QCD events:

$$\begin{aligned} e^+e^- &\rightarrow Z^0 (\rightarrow 2 \text{ jets}) + Z^0 (\rightarrow \ell^+\ell^-) \\ &\rightarrow W^+(\rightarrow 2 \text{ jets}) + W^- (\rightarrow e\nu \text{ or } \mu\nu) \\ &\rightarrow t + \bar{t}, \text{ each going to } \ell\nu + \text{jet}. \end{aligned}$$

ALEPH performed some simulations and showed that a detection of a Higgs with a mass of 80 GeV would be difficult but perhaps just possible (Fig. 7.23).

For the decay $Z^0 \rightarrow \nu\bar{\nu}$ the background events would be of similar kind as above but with

$$\begin{aligned} e^+e^- &\rightarrow Z^0 (\rightarrow 2 \text{ jets}) + Z^0 (\rightarrow \nu\bar{\nu}) \\ &\rightarrow W^+ (\rightarrow 2 \text{ jets}) + W^- (\rightarrow \nu\tau) \rightarrow 2 \text{ jets} + \nu\bar{\nu} + \text{hadrons} \\ &\rightarrow \text{QCD events with quarks and gluons (most important background).} \end{aligned}$$

ALEPH has shown that even in this case a Higgs signal for masses up to about 80 GeV might not be completely excluded (Fig. 7.24).

For the decay $Z^0 \rightarrow q\bar{q}$ leading to 4-jet events ALEPH found that the main background comes again from higher QCD effects and some 4-jet events from $e^+e^- \rightarrow W^+W^- \rightarrow$ 4 jets and $e^+e^- \rightarrow Z^0Z^0 \rightarrow$ 4 jets. Nevertheless a Higgs signal could perhaps be seen up to Higgs masses of about 60 GeV (Fig. 7.25).

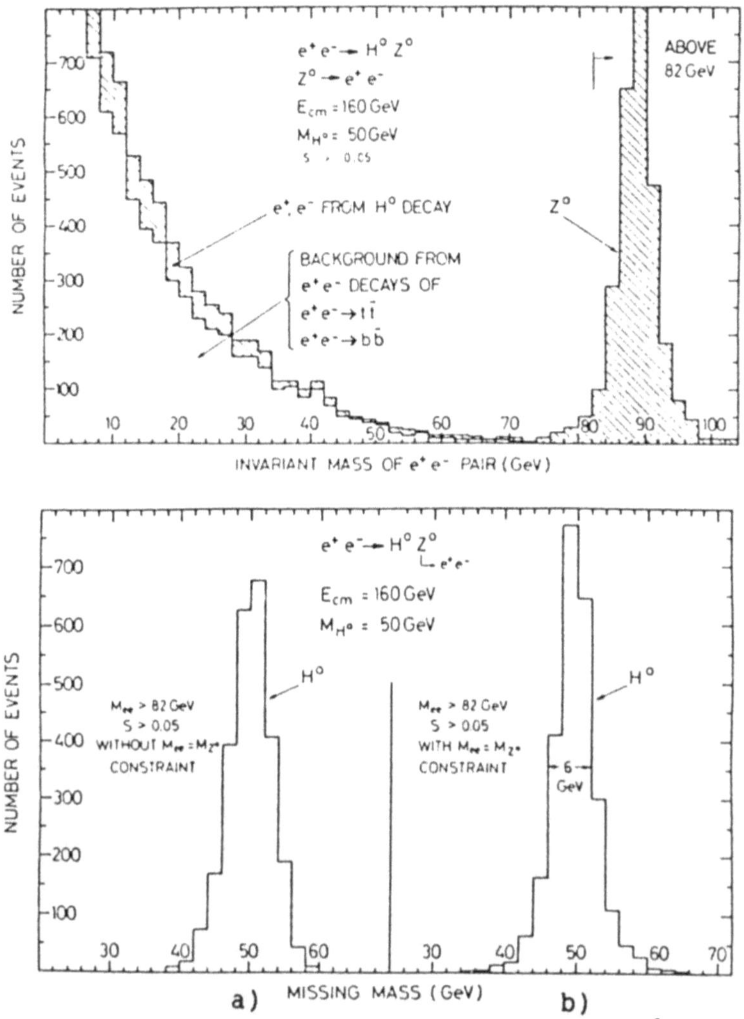

Fig. 7.18. Mass distributions for the process $e^+e^- \to (Z \to e^+e^-) + H^0$ at a c.o.m. energy of 160 GeV and assuming a Higgs mass of 50 GeV. In the upper part of the figure the invariant mass of the electron pair is shown displaying the Z-peak and a broad distribution due to the Higgs decay and backgrounds. The lower part of the figure shows the recoil mass distribution of the electrons if only events with $M(e^+e^-) > 82$ GeV and a sphericity $S > 0.05$ are accepted. a) and b) show the obtained recoil mass spectrum without and with the additional constraint $M(e^+e^-) = M_Z$. (ALEPH simulation)

7.4 Higgs from Toponium Decay

If the mass of toponium T were smaller than the maximum c.o.m. energies available at LEP (e.g. $M_{\text{toponium}} < 200$ GeV) the reaction

$$e^+e^- \to T(t\bar{t}) \to H^0 + \gamma$$

would offer an excellent, maybe the best opportunity to search for the Higgs.

The signature is very clear: a monoenergetic photon with two jets in the opposite direction. The branching ratio could be quite high, $\sim 5\%$ of the toponium decays. Sitting on the toponium peak one could expect a few events per day for $M_H = 40$

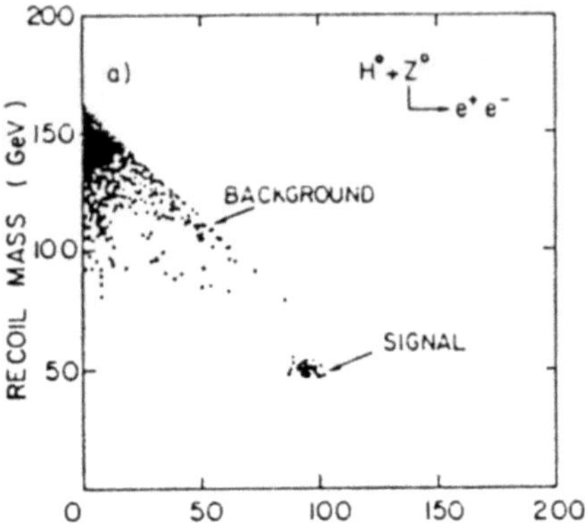

Fig. 7.19. Mass distributions for the process $e^+e^- \to (Z \to e^+e^-) + H^0$ at a c.o.m. energy of 164 GeV and assuming a Higgs mass of 50 GeV. In (a) a scatter plot of recoil mass of the electrons against their invariant mass is shown. The other two parts of the figure show the projections of the scatter plot as a function of the invariant and the recoil mass of the electrons, respectively. (DELPHI simulation)

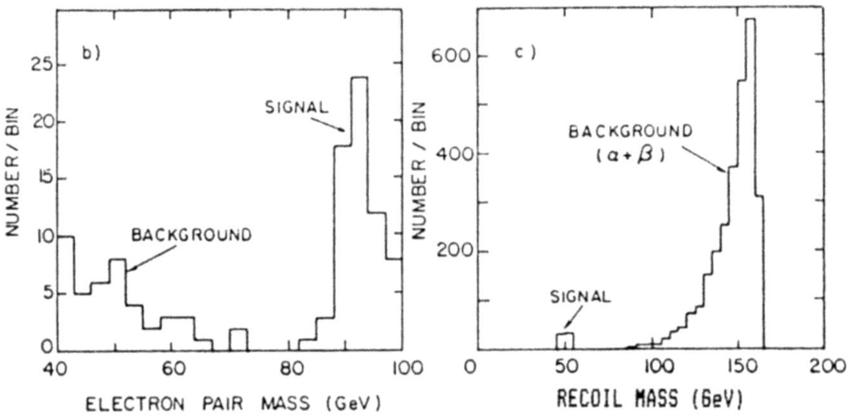

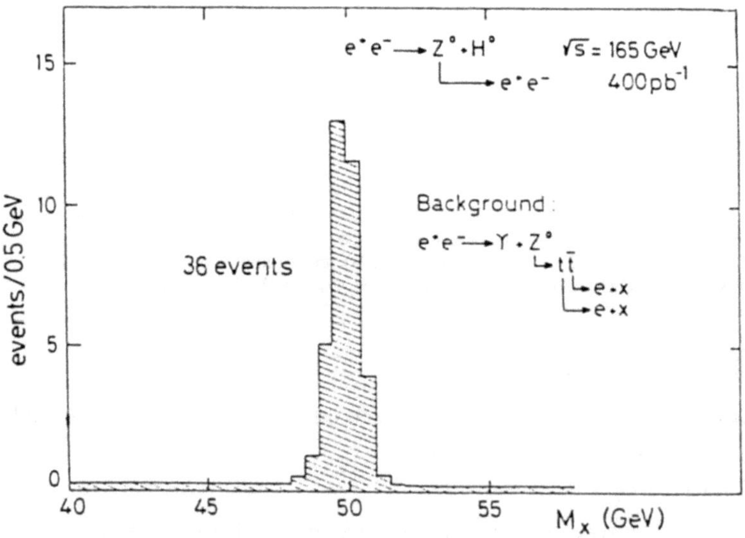

Fig. 7.20. Recoil mass spectrum for the process $e^+e^- \to (Z \to e^+e^-) + H^0$ at a c.o.m. energy of 165 GeV and assuming a Higgs mass of 50 GeV. The number of events for an integrated luminosity of 400 pb^{-1} is shown. (L 3 simulation)

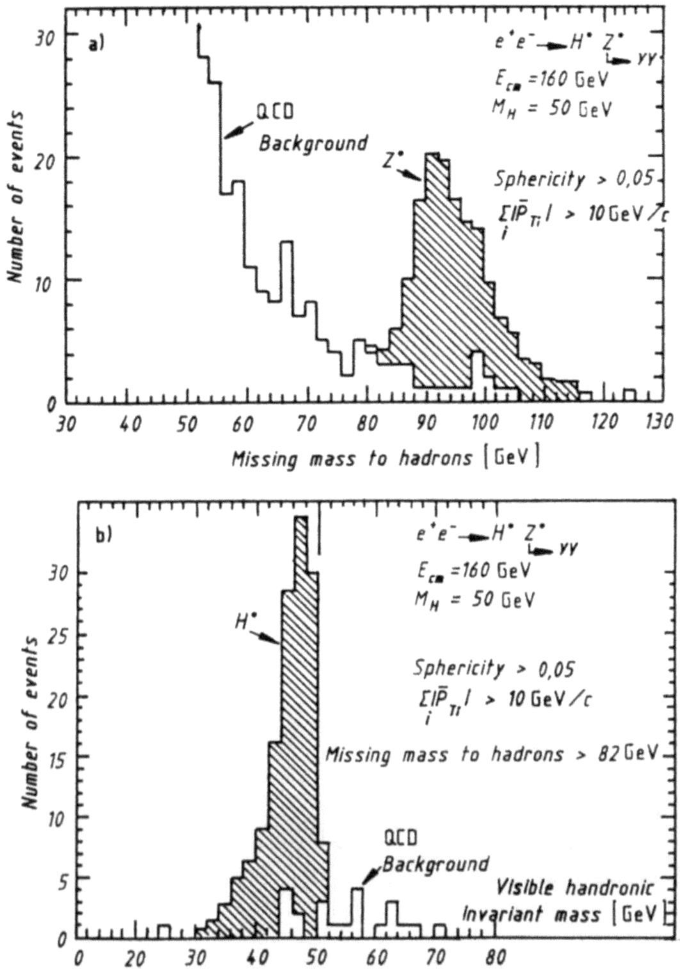

Fig. 7.21. Mass distributions for the process $e^+e^- \to (Z \to \nu\bar{\nu}) + H^0$ at a c.o.m. energy of 160 GeV and assuming a Higgs mass of 50 GeV. In the left part the distribution of the missing mass recoiling against the observed hadrons is shown. The Z-peak is clearly visible. In the right part the distribution of hadronic invariant mass recoiling against the missing masses above 82 GeV is displayed. A clear Higgs peak appears. (ALEPH simulation)

GeV. This has to be compared to a few events per year for the decay $Z^0 \to H^0 + \ell^+\ell^-$, indicating the enormous advantage in rate.

The background would be mostly due to $e^+e^- \to f\bar{f} + \gamma$. Calculations performed by DELPHI for $M_{\text{toponium}} = 70$ GeV are shown in Fig. 7.26. One sees that Higgs masses between 10 and 60 GeV could be measured.

7.5 Search for Charged Higgses

The masses of the charged SUSY Higgses are expected to be higher than the Z^0 (see Sect. 7.1.2). If charged Higgses were found by LEP 100 they could hardly be associated with minimal SUSY Higgses. However, in a model with 2 Higgs

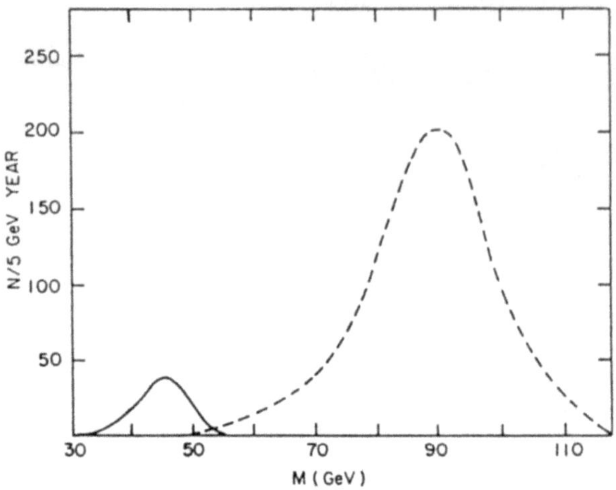

Fig. 7.22. Reconstructed mass for the reaction $e^+e^- \rightarrow (Z \rightarrow \nu\bar{\nu}) + H^0$ at a c.o.m. energy of 164 GeV and assuming a Higgs mass of 50 GeV. (Cuts: visible energy > 35 GeV, p_T > 6 GeV). The full line shows the Higgs peak, the dashed line the Z-background. (DELPHI simulation)

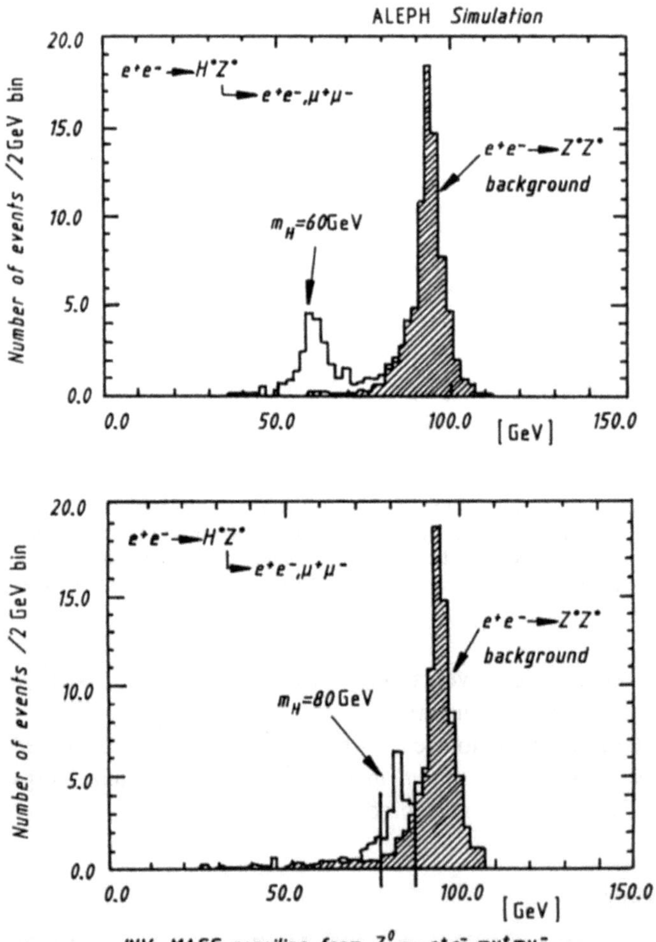

Fig. 7.23. Recoil mass distribution for the process $e^+e^- \rightarrow (Z \rightarrow \ell^+\ell^-) + H^0$ at a c.o.m. energy of 200 GeV for Higgs masses of 60 GeV and 80 GeV. Integrated luminosity 500 pb^{-1} (ALEPH simulation)

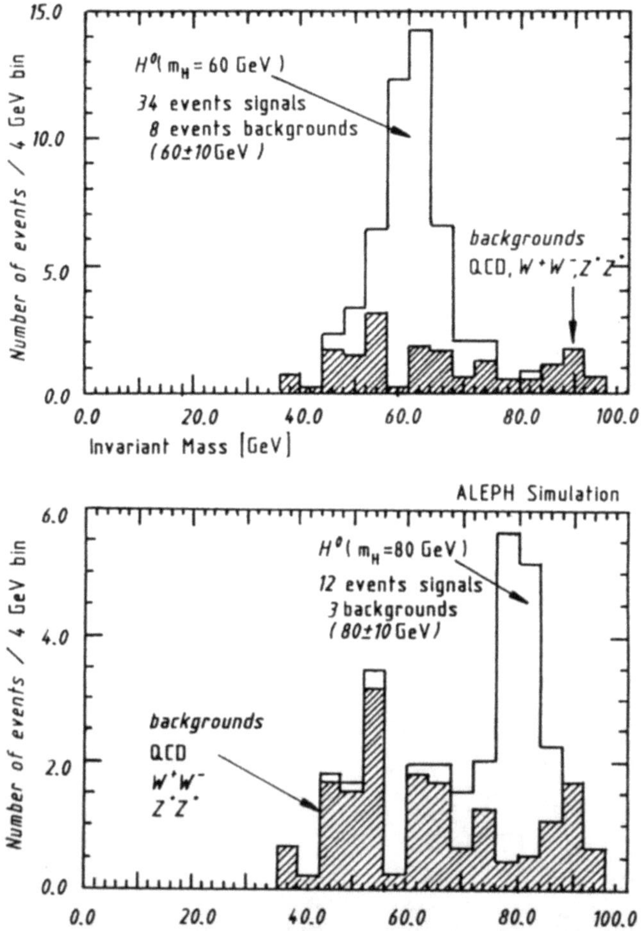

Fig. 7.24. Invariant mass distributions for the process $e^+e^- \to (Z \to \nu\bar{\nu}) + (H^0 \to 2$ jets) at a c.o.m. energy of 200 GeV for Higgs masses of 60 GeV and 80 GeV. Integrated luminosity 500 pb^{-1} (ALEPH simulation)

doublets such light charged Higgs particles could exist. They could be identified by their typical angular distribution (see Sect. 7.3.4) and cross sections are estimated to be reasonable. The present limit (from DESY, TRISTAN) is $M_{H\pm} > 18$ to 20 GeV. This limit could be extended by LEP 100 to about 40 to 45 GeV.

The main production mechanism would be $e^+e^- \to Z^0 \to H^+H^-$. One has to distinguish three cases:

i) $M_{H+} < M_{top}$. If this condition holds the dominating decay of the top quark would be $t \to b + H^+$ and indeed this decay would go so fast that toponium would not exist.
ii) If toponium is found it would conversely mean that $M_{H+} > M_t - M_b$. The main decay of the Higgs would be $H^+ \to t\bar{\nu}$ and $c\bar{s}$.
iii) $M_{H+} > M_t + M_b$. In this case the Higgs would decay predominately according to $H^+ \to t\bar{b}$.

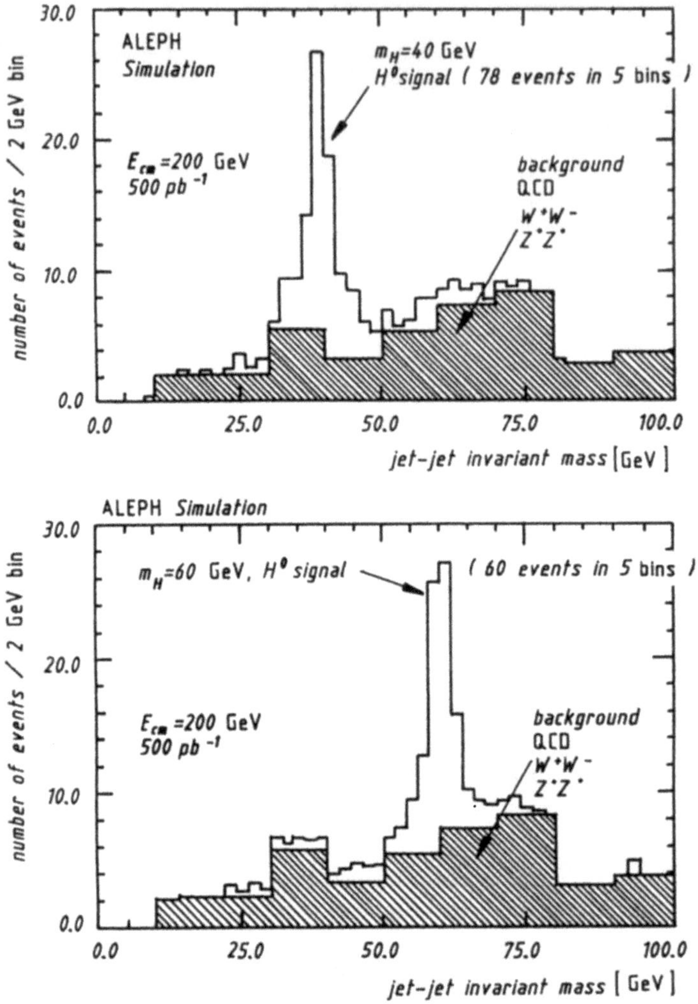

Fig. 7.25. Invariant mass distributions of 2 jets for the process $e^+e^- \to (Z \to 2 \text{ jets}) + (H^0 \to 2 \text{ jets})$ at a c.o.m. energy of 200 GeV for Higgs masses of 60 GeV and 80 GeV. Integrated luminosity 500 pb^{-1} (ALEPH simulation)

In Fig. 7.27 the total cross section for $e^+e^- \to Z^0 \to H^+H^-$ is displayed as a function of beam energy with different Higgs masses as parameter.

Not many simulations have been done, in particular no calculations are known for high top masses. The main background would be QCD proccesses. The results of some simulations by OPAL are shown in Fig. 7.28 (at the Z^0 peak) and in Fig. 7.29 for a c.o.m. energy of 160 GeV. In both cases 2-jet events are looked for with proper cuts to suppress QCD events. It seems that Higgs masses of 30 GeV and 60 GeV, respectively, could be observed.

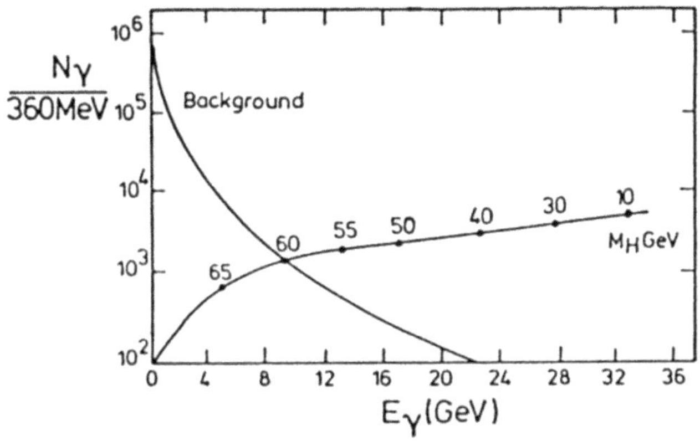

Fig. 7.26. The number of photons with energy E_γ originating from the process $e^+e^- \to$ toponium $\to$ H + γ for different Higgs masses. A toponium mass of 70 GeV was assumed. The estimated background is also shown.

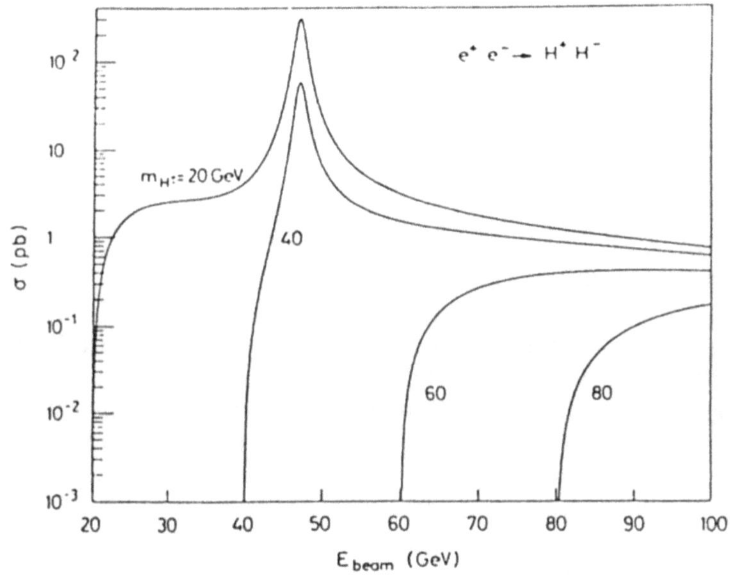

Fig. 7.27. Total cross section (tree level) for $e^+e^- \to H^+H^-$ as a function of beam energy for different Higgs masses

7.6 Light Higgses

Some time ago it was thought that experimental lower limits (mainly from Υ decays) for the Higgs mass could be established with $M_{H^0} > 3.5$ GeV. Subsequently it was shown, however, that the QCD corrections used to derive such a limit were uncertain and hence the possibility of a very light Higgs was revived [59].

To produce Higgs particles in sufficient numbers requires high energies, even if they are very light. The reason, of course, is that the coupling to fermions is proportional to their mass, while the coupling to the weak boson is proportional to

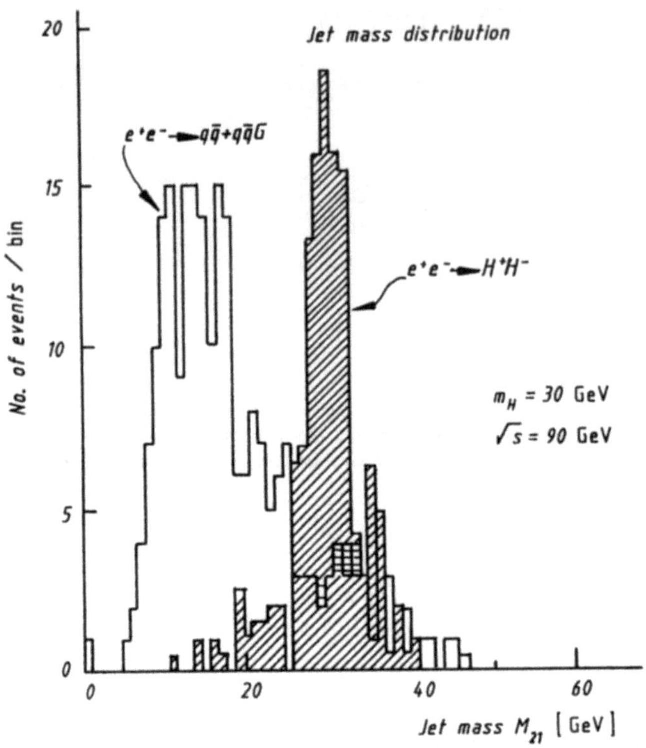

Fig. 7.28. Invariant mass distribution of 2 jets for the process $e^+e^- \to H^+H^-$ at the Z-peak and assuming a charged Higgs mass of 30 GeV. The QCD background is also shown (OPAL simulation)

the square of the boson mass. Hence LEP or SLC are necessary to produce light Higgses. The production mechanism is still Higgs bremsstrahlung (Fig. 7.5).

To detect the Higgs the recoil mass method gets very bad for masses below about 10 GeV since the experimental resolution for lepton energies deteriorates appreciably. On the other hand for masses below 1.5 GeV the Higgs decays into two particles and these 2-body decays provide a very good signature. The decay rates for different mass regions of the Higgs are shown in Table 7.5.

For the search of such a light Higgs one can distinguish the following mass regions (we follow the discussion of Hulth and Shellard [60]):

i) $2\,M_e < M_H < 2\,M_\mu$
In this case the Higgs decays predominantly into an electron pair with some contributions from $H \to \gamma\gamma$ which can reach up to 25% at the upper limit of the region. The partial width for this decay is given by $\Gamma(H \to \gamma\gamma) = 1.2 \times 10^{-12}(M_H/\mathrm{MeV})^3$ in eV. The main feature of a Higgs in this region would be its long lifetime leading to measurable decay length. The average energy of the Higgs particles would be about 8 GeV and the average decay distance is shown in Fig. 7.30. One sees that the Higgs could well traverse a large detector and for 10^6 Z-decays one could observe about 1000 Higgs decays within a volume of 1 m radius for a Higgs mass around 100 MeV.

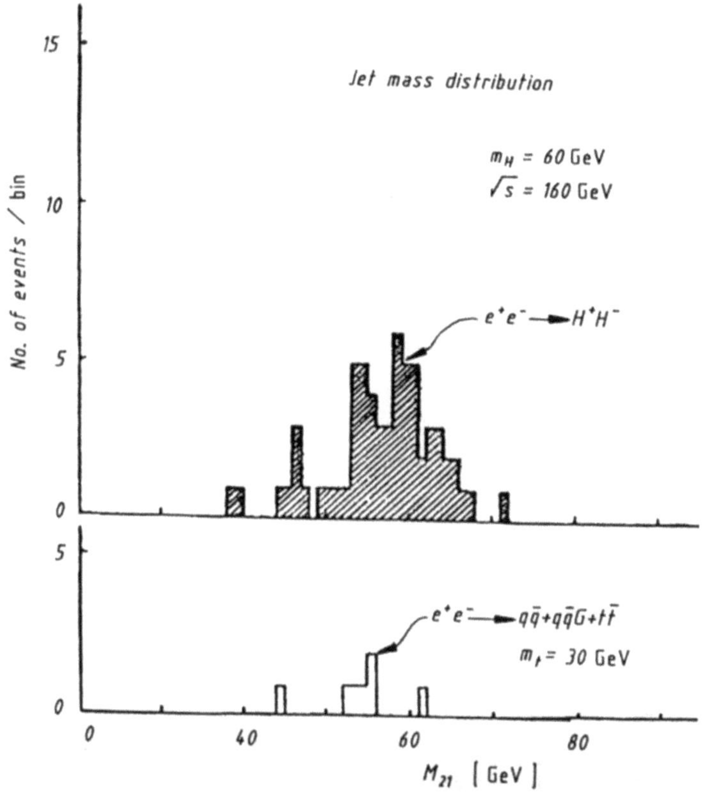

Fig. 7.29. Invariant mass distribution of 2 jets for the process $e^+e^- \rightarrow H^+H^-$ at a c.o.m. energy of 160 GeV and assuming a charged Higgs mass of 60 GeV. The QCD background is also shown (OPAL simulation)

The Higgs decaying into e^+e^- with the Z^{0*} decaying either into neutrinos or into charged leptons resulting in 4 charged particles is a very distinctive signal. Just by counting the number of such events, either prompt or with a vertex in the detector, would make it possible to rule out a Higgs in this region. The Higgs angular distribution in $\cos\theta_H$ is very flat, while the radiative background is divergent in the forward and backward direction, so that a cut could reduce considerably the background.

At the lower limit of this mass region the radiative background could, however, become quite serious and most of the Higgs could decay predominantly outside the detector. This could be overcome by a different technique. For Higgs masses below e.g. 20 MeV the acolinearity of the lepton pair from the Z^{0*} decay is rather large, due to the large average momentum of the light Higgs and a cut in acolinearity would remove most of the background.

ii) $2\,M_\mu < M_H < 2\,M_\pi$

The Higgs decays promptly into a muon pair, giving a distinct signal. For the decay of the Z^{0*} into 2 muons one obtains a 4-muon event and for 10^6 Z^0 one expects 156 of these events. Another clear signal would be provided by events where the Z^{0*} goes into 2 neutrinos and the Higgs to a muon pair. In this case one could observe about 900 events.

Table 7.5. Number of events for the process $e^+e^- \rightarrow (Z \rightarrow \mu^+\mu^-) + H^0$ for 10^6 Z decays

M_{Higgs} (GeV/c^2)	$\gamma\gamma$	e^+e^-	$\leftarrow H_0$
0.040	2	157	
0.080	7	151	
0.120	15	143	
0.160	25	132	
0.200	36	122	

M_{Higgs} (GeV/c^2)	$\mu^+\mu^-$	$\pi\pi$	$K\bar{K}$
0.250	156		
0.400	10	143	
0.600	7	141	
0.900	6	134	
1.200	4	99	27
1.600	3	68	47
2.000	2	55	51
2.400	2	46	50
2.800	2	43	48
3.200	2	38	45

M_{Higgs} (GeV/c^2)	$\pi\pi$	$K\bar{K}$	$\tau^+\tau^-$	$c\bar{c}$	$b\bar{b}$
4.0	16	21	18	19	
6.0	3	4	17	32	
8.0	2	2	13	26	
10.0	1	1	10	22	
15.0	0	0	2	4	15
20.0	0	0	1	2	11
30.0	0	0	0	0	5
40.0	0	0	0	0	2

iii) $2\,M_\pi < M_H < 2M_K$

The Higgs can decay into a muon or a pion pair in this mass region. The branching ratios are difficult to calculate, due to strong interaction parameters which on the other hand could be determined from the Higgs if the Higgs mass were in this region.

The signal to look for would be two charged leptons from the Z^{0*} decay plus two charged tracks of either muons or pions.

iv) $2\,M_K < M_H < 2\,M_\tau$

The dominant decay modes of the Higgs would be a K or a pion pair. Again the decay rates are uncertain because of strong interaction parameters. The K^+K^- mode could be used to reconstruct the Higgs mass.

v) $2\,M_\tau < M_H < 2\,M_b$

If this condition holds the Higgs would decay into charmed or bottom mesons and tau pairs, with ratios depending on the Higgs mass (Table 7.5). Here there is no direct way of measuring the Higgs mass using its decay products and one has to use the recoil mass technique with lepton pairs from the Z^{0*} which, however, is very poor in this mass range. At least 25% of the events would have only 4 charged tracks, due to the Higgs decay into a tau pair. Therefore at least the signal would be clear.

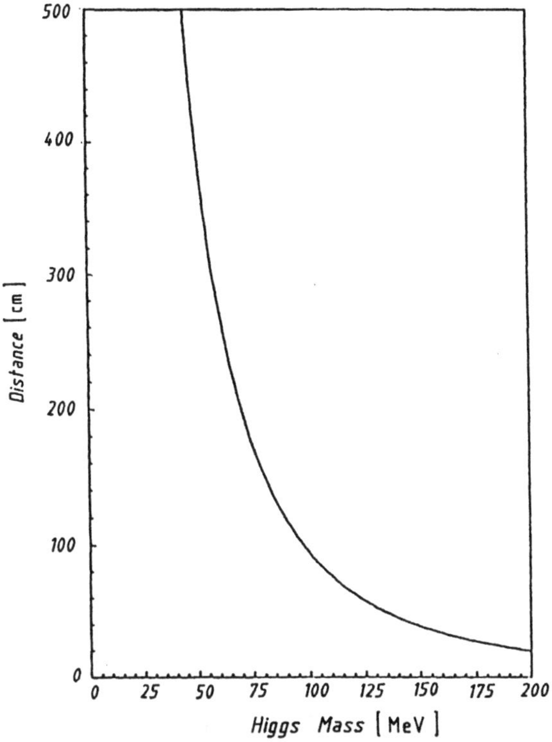

Fig. 7.30. The average decay distance of a light Higgs produced in Z-decays as a function of the Higgs mass

8. Miscellaneous Topics

In this last chapter a few topics will be shortly discussed, not because they are of lower physics interest but either they seem to be out of the reach of LEP energies or their theoretical motivation is not so strong. Of course it will be very important to keep an open eye on unexpected phenomena since they might be more exciting than any theoretically predicted events.

8.1 Top Search

One of the most tantalizing questions concerns the existence of the top quark. Hardly anybody doubts, of course, that it exists since it has to complete the third quark family. The knowledge of its mass would also be important, not only to provide a better understanding of the mass spectrum, but also as input parameter for theories like the Standard Model, in particular for radiative corrections.

Unfortunately there is practically no guidance from theory with respect to its mass. When PETRA and PEP came into operation there were predictions for the top mass in the range 20 to 30 GeV. Later they moved up and have arrived now at masses of about 130 GeV; however, they should be taken with great care since no convincing theory for the understanding of the fermion mass spectrum has been developed.

From the experimental side there is some indirect evidence for the existence of the top from measurements of the forward-backward asymmetry at PETRA and TRISTAN. From an analysis of the radiative corrections of the electroweak interaction (see Sect. 4.2) one can deduce an upper limit of about 180 GeV [15]. The top search at different accelerators has given so far [9] (by August 1989) the following lower mass limits:

	$M_{top} >$
Tristan	30.4 GeV
SLC/MarkII	37.5 GeV
Argus/Doris	50 GeV
UA1/CERN	65 GeV
UA 2/CERN	67 GeV
CDF/Tevatron (prelim.)	77 GeV

These results indicate that it is unlikely to produce top quarks at LEP, certainly not at LEP 100.

This implies that a lot of calculations which have been done for top masses in the range from 40 to 60 GeV are no more useful [7, 61, 62]. The beautiful physics which could have been done if the toponium mass would have been close to the Z-mass (interferences) or the observation of the toponium spectrum will now be difficult or impossible.

Indeed if the top mass were 100 GeV or more, its lifetime would be of the order of 10^{-23} s, too short to allow a top quark to pick up another quark to form a bound state, e.g. toponium or particles with open top.

If toponium T is heavier than a neutral higgs it would decay via $T \rightarrow H^0 + \gamma$ with a width so wide that it might be difficult to see the toponium at all.

Still one should look for the top at the highest energies available at LEP 200 and the easiest way, not knowing what its mass could be, should be to observe threshold effects in the event shape, i.e. measuring quantities like sphericity, thrust or oblateness (see Sect. 5.1) as function of the energy.

8.2 B-Physics at LEP

The production rate of b quarks at the Z-peak is enormous (Fig. 8.1). The cross section for the production of b-pairs amounts to about 20% of the hadronic cross section. Compared to the production rate one hopes to get with dedicated b-factories at low energies (but high luminosities) and which use the Υ_{4S} resonance for b-production, LEP does still quite well since the production cross section at the Z is 5 times higher than at the Υ_{4S} [63]. For 10^6 Z^0 one expects about 10^5 b pairs.

However, b-factories and LEP should rather be considered as complementary ways of studying b-physics. At LEP the energy resolution will not be sufficient to separate B^0 and B^0_s which on the other hand can be done at b-factories.

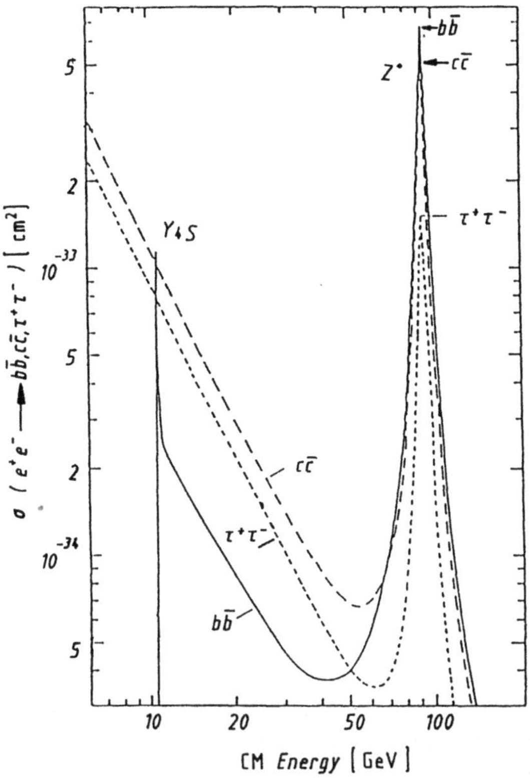

Fig. 8.1. The production cross section of tau and quark pairs as a function of the c.o.m. energy

Alternatively b-quarks are produced at the Z-peak with large energies and hence high velocities. Their travel path is given by $L = c\,E\,\tau/M_{\mathrm{B}} \approx 1.9 \times (E/30\,\mathrm{GeV})\,\mathrm{mm}$ for a lifetime of $\tau \approx 1.2 \times 10^{-12}$ s. Flight paths of several mm are therefore observable. This can be well measured with a micro-vertex-detector for which resolutions of 25 microns are possible. By observing the vertex b-quarks can be tagged. However, even without a vertex detector the tagging of b-quarks seems possible by selecting high p_{T} leptons, which has also the merit of distinguishing between quarks and antiquarks. It seems that about 10% of b-events may be tagged via the primary lepton with a tagging efficiency for c-quarks of only 1% [64].

Two more advantages are provided by nature for b-physics. Initial state bremsstrahlung is suppressed and thus the jet axis can be well defined. Secondly the branching ratio for b is about 30% higher than for c-quarks because of the weak coupling constants and hence this c-quark background is easier to get rid of.

The observation of the forward-backward asymmetry for b-quarks has already been discussed in Sect. 4.4.3 (see also Dornan [64]).

The interesting question is whether it will be possible to observe b-mixing at LEP. The easiest way to study the mixing is to use semi-leptonic decays ($\mathrm{B}^0 \to \ell^+ + \bar{\nu}$ + hadrons) and ($\overline{\mathrm{B}^0} \to \ell^- + \nu$ + hadrons) with a branching ratio of about 11% both for e and μ. Mixing will be characterized by events in which one of the B's was transformed into its antiparticle and hence one obtains events with like-sign leptons [65, 66]. A quantitative way to determine the mixing is provided by the quantity

$$R = \frac{\ell^+\ell^+ + \ell^-\ell^-}{\ell^+\ell^+ + \ell^-\ell^- + \ell^+\ell^-} .$$

For the L 3 detector it has been estimated [67] that during several years of LEP operation at nominal luminosity one could obtain about 60,000 dilepton events with a cut $p_T > 1$ GeV which would determine R to 0.5%.

As mentioned at LEP it will not be possible to separate the B and B_s mesons and as a result one gets an incoherent mixture. It is rather difficult to derive a reliable theoretical value of R for this kind of mixture [68]. Under certain assumptions (maximum mixing, neglecting backgrounds) a reasonable estimate seems to be $R \approx 0.24$. However, it is not exculded that the incoherent mixing can be small and than R may be reduced.

The theoretical interpretation would be much easier if one could measure the mixing only for the B_s^0 state. For this purpose it has been suggested to use specific channels.

One possibility is to look for strangeness correlations by observing the decays $B_s \to \ell^+K^-K^-X$, $\ell^+K^-\Lambda^0X$ i.e. determining the admixture of "wrong" strangeness by measuring

$$R_s = \frac{\ell^-K^-K^-X - \ell^-K^+K^+X}{\ell^-K^-K^-X + \ell^-K^+K^+X}$$

where all three identified particles should be in the same jet.

Applying for the lepton a p_T-cut of $p_T > 1.5$ GeV and for the K a cut $p_K > 0.5$ GeV one finds [68] that R_s decreases from 0.41 to 0.23 for not too complete mixing. Such values might well be measurable but several years of LEP operation will be necessary to obtain the necessary statistics.

It would be very interesting to observe the time dependence of the $B - \bar{B}$ oscillations but this seems not to be feasible because of lack of statistics.

Another interesting problem concerns PC violation in the B-system. Whereas in the K-system PC violation stems mainly from the mixing of states with opposite PC (ε) and the violation through the decay amplitude is small (ε') the situation is the opposite for the B-system.

PC violation would manifest itself in semi-leptonic decays by finding the quantity R_{PC} different from zero, where

$$R_{PC} = \frac{\ell^+\ell^+ - \ell^-\ell^-}{\ell^+\ell^+ + \ell^-\ell^-} .$$

Unfortunately estimates indicate that R_{PC} will be quite small. For B_d decays it would be of order 10^{-3} and for B_s decays 10^{-4}. To measure it one would need about 10^9 B events which seems completely hopeless.

Another possibility to observe PC violation is offered by specific hadronic decays which have the special feature that the final state, being its own CP conjugate, can be reached both from B and $\bar{B}$ decays. The most promising example is the decay into a J particle. Both $B^0 \to J + K_s$ and $\overline{B^0} \to J + K_s$ are allowed. PC violation would manifest itself in the quantity

$$R_{CP} = \frac{(B^0 \to J + K_s) - (\overline{B^0} \to J + K_s)}{(B^0 \to J + K_s) + (\overline{B^0} \to J + K_s)}$$

which would differ from zero in case of PC violation [66]. The theoretical expectations vary between 0.05 and 0.25 and hence this quantity could be reasonably large.

L 3 has investigated this case [67]. The J particle is defined by its dilepton decay and the K_s by the π-π mode. The masses of the J and K_s can be well reconstructed and the opposite jet can be identified to be from a B or a $\bar{B}$ by tagging on the lepton charge. The B mass could be reconstructed in this channel to a few tenth of a GeV allowing a clean identification of the B. The branching ratio is estimated to be about 5×10^{-4}. This effect may be observable but about 10^7 B events will be necessary.

8.3 Other New Particles

In this last section a few problems will be discussed which got a lot of attention in the past, but which seem to move out of the LEP energy range, mainly because of new limits produced recently by CDF at the TEVATRON [9]. Some of them shall nevertheless be mentioned although there is no theoretical motivation for most of them, but one has to keep an open eye to unexpected effects.

The **number of fermion families** is not understood theoretically and hence there might be a fourth family of quarks and leptons. This question will hopefully be settled very soon by a measurement of the Z-width. Otherwise a fourth family would show up in the ρ-value which would differ from 1. Recent results from CDF gave a value of $\rho = 1.001 \pm 0.017$ as deduced from the masses of Z and W.

New **heavy gauge bosons** have been considered but recent limits from CDF indicate that their masses would be larger than 400 GeV for a Z' and 380 GeV for a W' and thus outside the range of LEP.

The **compositeness** of fermions can be studied in the process [15] $e^+e^- \rightarrow e^+e^-\gamma$ and one would be sensitive for a cut-off parameter Λ of about 250 GeV. The recent preliminary result from CDF indicates $\Lambda > 950$ GeV.

Various **non-SM heavy leptons**, charged and neutral, have been discussed [69, 15], but we shall not follow up this topic here any further.

At all new accelerators a hunt for free quarks has been carried out. This will certainly be done also at LEP although no specific experiment has been proposed for this purpose. However, some of the detectors have an excellent possibility to measure the ionisation of the produced particles which should make it possible to find free quarks if they exist in the accessible energy range.

The existence of magnetic monopoles is a topic which has attracted considerable attention due to their association with some cosmological models. Recently a special LEP experiment has been approved [6] which will look for highly ionising particles, in particular monopoles (see Sect. 3).

References

1 Previous theoretical work has been summarized in the report *Physics at LEP*, CERN 86-02, February 1986, ed. by J. Ellis and R. Peccei, Vol. I and II. In this article we shall not reproduce all the original references, but preferentially those which appeared after the publication of the review reports

2 Physics at higher energies has been discussed at the *ECFA Workshop on LEP 200*, CERN 87-08, ECFA 87/108, June 1987, ed. by A. Boehm and W. Hoogland, Vol. I and II

3 The physics with polarized beams has extensively been studied at a workshop *Polarization at LEP*, CERN 88-06, September 1988, ed. by G. Alexander et al., Vol. I and II

4 J.M. Jowett: CERN LEP TH 85-04 (1985)

5 F. Dydak: Academic Training Lecture, CERN, February 1987

6 CERN-Bologna-Harvard-Montreal Collaboration: CERN/LEP 89-2 and 89-9

7 R.D. Peccei: "Physics at LEP", DESY 86-163 (1986) and "Physics of the Standard Model", DESY 89-60 (1989)

8 W.F.L. Hollik: DESY 88-106 (or in *Polarization at LEP*, Vol. I, p. 83) and DESY 88-188 (1988)

9 *Lepton-Photon Symposium*, SLAC, August 1989

10 U. Amaldi et al.: Phys. Rev. **D36** (1987) 1385

11 G. Burgers: In *Polarization at LEP*, CERN 88-06, Vol. I, p.121

12 T. Riemann and M. Sachwitz: Phys. Lett. **B212** (1988) 488

13 A. Blondel et al.: In *Physics at LEP*, CERN 86-02, Vol. II, p. 35

14 F. Dydak: "Study Group on Electroweak Radiative Corrections", CERN 87-08 (1987)

15 "Five LEP working groups 1989", CERN, to be published. Rapporteur G. Altarelli and C. Verzegnassi, June 1989. Convenors for "Rare Z-Decays": E. Glover, J. van der Bij; for "Higgs Search at LEP 100": P.J. Franzini, P. Taxil; for "Supersymmetry Searches": R. Barbieri; for "Compositeness": F. Boudjema, F.M. Renard; for "New Bosons": R. Gatto

16 G. Bonvicini and L. Trentadue: Univ. Michigan UM-HE-88-36 (1988) (to be published in Nucl. Phys. B)

17 D. Bardin, M. Bilenky and G. Micelmacher: CERN-DELPHI 88-43 PHYS 27 (1988)

18 Z. Was and S. Jadach: MPI-Pae/PTh 33/89 and 34/89, München (1989)

19 S.K. Moenig: DELPHI 88-1 PHYS 23

20 A. Sirlin: Phys. Rev. **D22** (1980) 971

21 T. Riemann and M.Sachwitz: Phys. Lett. **B212** (1988) 488;
T. Riemann, M. Sachwitz and D.Y. Bardin: Inst. f. Hochenergiephysik, Zeuthen, PHE 88-11

22 W. Marciano and A.Sirlin: Phys. Rev. **D12** (1980) 2695; **D29** (1984) 945;
M. Consoli, S. Lo Presti and L. Maiani: Nucl. Phys. **B223** (1983) 474;
Z. Hioki: Prog. Theor. Phys. **68** (1982) 2134; Nucl. Phys. **B229** (1983) 284

23 G. Burgers and W. Hollik: In *Polarization at LEP*, CERN 88-06, Vol. I

24 R. Leiste et al.: Preprint; Inst. f. Hochenergiephysik, Zeuthen, PHE 89-02 and *Workshop on Electroweak Radiative Corrections*, Ringberg, April 1989

25 D.Y.Bardin et al.: Inst. f. Hochenergiephysik, Zeuthen, PHE 89-05

26 S.G. Gorishny, A.L.Kataev and S.A. Larin: Preprint JINR Dubna and INR Moscow (1988), quoted in Ref. [8]

27 E. Longo: In *Polarization at LEP*, CERN 88-06, Vol. I, p. 304

28 R. Kleiss: CERN preprint, June 1989

29 S.N. Ganguli and A. Gurtu: Tata Inst.Fund.Res. TIFR-EHEP/89/2

30 G. Barbiellini et al.: In *Physics at LEP*, CERN 86-02, Vol. II, p. 1 and several articles in *Polarization at LEP*, CERN 88-06, Vol. I

31 F.M. Renard: In *Polarization at LEP*, CERN 88-06, p. 197

32 D. Treille: In *Polarization at LEP*, CERN 88-06, p. 265

33 J.K. Drees et al.: DELPHI 88-53 PHYS 29 (1988)

34 P. Baillon, G. Barbiellini and D. Treille: In *Physics at LEP*, CERN 86-02, p. 172

35 A. Blondel et al.: Montpellier PM/87-14 (1987)

36 J. Drees et al.: In *Polarization at LEP*, CERN 88-06, Vol. I, p. 317

37 P.J. Dornan: In *Polarization at LEP*, CERN 88-06, Vol. I, p. 344;
A. Djouadi, J.H. Kuehn and P.M. Zerwas: Max-Planck-Inst. for Physics and Astrophysics, München MPI-PAE/PTh 48/1989

38 J. Chaveau: In *Physics at LEP*, CERN 86-02, Vol. I, p. 177

39 Z. Was: In *Polarization at LEP*, CERN 88-06, Vol. I, p. 250

40 C. Matteuzzi: DELPHI 88-14 PHYS 24
41 E. Simopoulou: In *Physics at LEP*, CERN 86-02, Vol. I, p. 197
42 A. Lopez and F. Bianchi: DELPHI 88-74 PHYS 33
43 M. Chen: "L 3 – The Precision Lepton Photon Detector", MIT Report
44 A. Ali: In *Physics at LEP*, CERN 86-02, Vol. II, p. 81
45 W. de Boer: DESY 89-067 (1989)
46 P.N. Burrows: DESY 88-122 and Particles and Fields **41** (1988) 375
47 A. Ali: In *Physics at LEP*, CERN 86-02, Vol. II, p. 109
48 G. Rudolph: In *Physics at LEP*, CERN 86-02, Vol. II, p. 150
49 H. Daum, H. Meyer and J. Bueger: Particles and Fields **8** (1981) 167
50 K.J.F. Gaemers and J.A.M. Vermaseren: Particles and Fields **7** (1980) 81
51 J.G. Koerner, G. Schierholz and J. Willrodt: Nucl. Phys. **B185** (1981) 365
52 O. Nachtmann and A. Reiter: Particles and Fields **16** (1982) 45
53 N.K. Falck, D. Graudenz and G. Kramer: DESY 88-186 (1988) and DESY 89-027 (1989)
54 H. Baer et al.: In *Physics at LEP*, CERN 86-02, Vol. I, p. 349
55 M. Chen et al.: Phys. Rep. **C159** (1988) 203
56 C. Wetterich: DESY 87-154 (1987)
57 J.F. Gunion et al.: "The Higgs Hunter's Guide", University of California at Davis Report UCD-89-4
58 See e.g., J. Ellis, M.K. Gaillard and D.V. Nanopoulos: Nucl. Phys. **B106** (1976) 292
59 J. Lee-Franzini: XXIV Int.Conf. on High Energy Physics, Munich 1988
60 P.O. Hulth and R.C. Shellard: DELPHI 88-91 PHYS 34
61 See e.g., W. Buchmueller et al.: In *Physics at LEP*, CERN 86-02, Vol. I, p. 203
62 J.H. Kuehn and P.M. Zerwas: Max-Planck-Inst. for Physics and Astrophysics, München MPI-PAE/PTh 52/87
63 K. Wacker et al.: "Proposal for an Electron Positron Collider for Heavy Flavour Particle Physics and Synchrotron Radiation", Paul Scherrer Institut PSI-PR-88-09 (1988), see also German Abstract
64 P.J. Dornan: In *Polarization at LEP*, CERN 88-06, Vol. I, p. 344
65 See e.g., H. Schroeder: DESY 88-101 (1988)
66 D. London: DESY 88-137 (1988)
67 S. Ting: private communication 1989 and L 3 reports to LEP Experiments Committee
68 A. Ali: In *Physics at LEP*, CERN 86-02, Vol. I, p. 262;
B. Lusignoli: Particles and Fields **41** (1989) 645;
D. Krawzsyk: DESY 88-163 (1988)
69 see e.g., H. Baer et al: In *Physics at LEP*, CERN 86-02, Vol. I, p. 424

The HERA Project

R. Felst

DESY, D-2000 Hamburg 52, Fed. Rep. of Germany

1. Introduction

The DESY Laboratory in Hamburg is presently constructing the storage ring facility HERA (Hadron Electron Ring Accelerator) which will collide 30 GeV electrons with 820 GeV protons. This accelerator complex will extend by about two orders of magnitude the experimental reach of spacelike four-momentum transfer Q^2 into a region where γ, Z and W exchange contribute with similar strength. This will allow to probe the proton structure in greater depth penetrating to distances as small as 10^{-16} to 10^{-17} cm and to test the Standard Model (S.M.) in this new kinematic region. It will allow searches for internal structure of quarks and leptons, for new exchange currents, for supersymmetric partners of the known fermions, as well as for leptoquarks in the mass range up to 300 GeV, typically.

Before speculating about possible new phenomena, we will try to extrapolate our present knowledge to the HERA energies. These topics are discussed in Sect. 3 after a short description of the machine in Sect. 2. The demands on the detector systems at HERA, and its realisations by the two general purpose facilities H1 and ZEUS are presented in Sect. 4, finally the status of the project is summarized in Sect. 5.

2. The Machine

The HERA accelerator complex is sketched in Fig. 1. It consists of an electron and a proton storage ring located in the same tunnel. A tunnel cross section is shown in Fig. 2. The main parameters of the two machines are listed in Table 1 [1]. The former e^+e^- storage ring PETRA is modified to accelerate electrons as well as protons before injection into HERA.

The electron ring is – apart from the relatively large number of circulating bunches – a standard machine similar to PETRA and LEP. Its maximum energy is determined by the energy losses due to synchrotron radiation. Recently superconducting acceleration structures were developed by DESY and its series production started in industry. These cavities yield higher acceleration fields of about 5 MV/m and are less power consuming than the normal conducting structures originally proposed, and will allow a maximum electron beam energy of about 35 GeV.

Physics for a New Generation Editors: H. Latal · H. Mitter

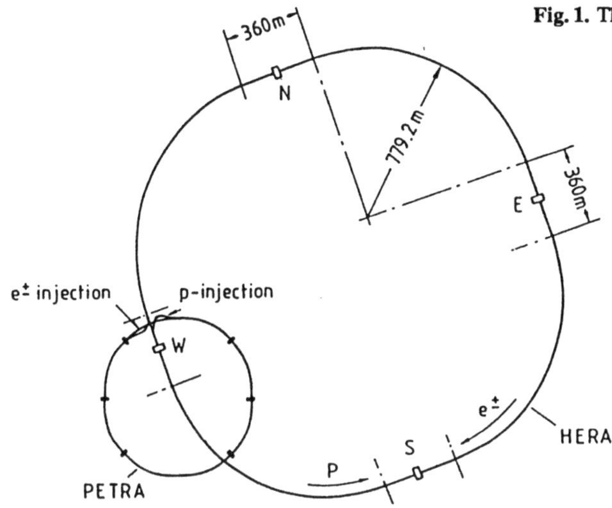

Fig. 1. The HERA accelerator

Table 1. Parameter of HERA

	p – ring		e – ring	units
Nominal energy	820		30	GeV
c.m. energy		314		GeV
Q^2_{max}		98.400		GeV^2
Luminosity		$1.5 \cdot 10^{31}$		cm^{2}s^{-1}
Polarization time			28	min
Number of interaction points		4		
Crossing angle		0		mrad
Free space for experiments		±5.5		m
Circumference		6336		m
Length of straight sections		360		m
Bending radius	588		608	m
Magnetic field	4.65		0.165	T
Energy range	300–820		10–33	GeV
Injection energy	40		14	GeV
Circulating current	160		58	mA
Total number of particles	$2.1 \cdot 10^{13}$		$0.8 \cdot 10^{13}$	
Number of bunches		200		
Number of bunch buckets		220		
Time between crossings		96		ns
Emittance $(\varepsilon_x/\varepsilon_y)$	0.71/0.71		3.4/0.7	10^{-8} m
Beta function (β^*_x/β^*_y)	10/1.0		2/0.7	m
Beam tune shift (Q_x, Q_y)	0.0026/0.0014		0.023/0.026	
Beam size at crossing σ^*_x	0.27		0.26	mm
Beam size at crossing σ^*_y	0.08		0.07	mm
Beam size at crossing σ^*_z	11		0.8	cm
Energy loss / turn	$1.4 \cdot 10^{-10}$		127	MeV
Critical energy	10^{-6}		111	keV
Max. circumf. voltage	0.2/2.4		260	MV
Total RF power	1		13.2	MW
RF frequency	52.033/208.13		499.667	MHz
Filling time	20		15	min

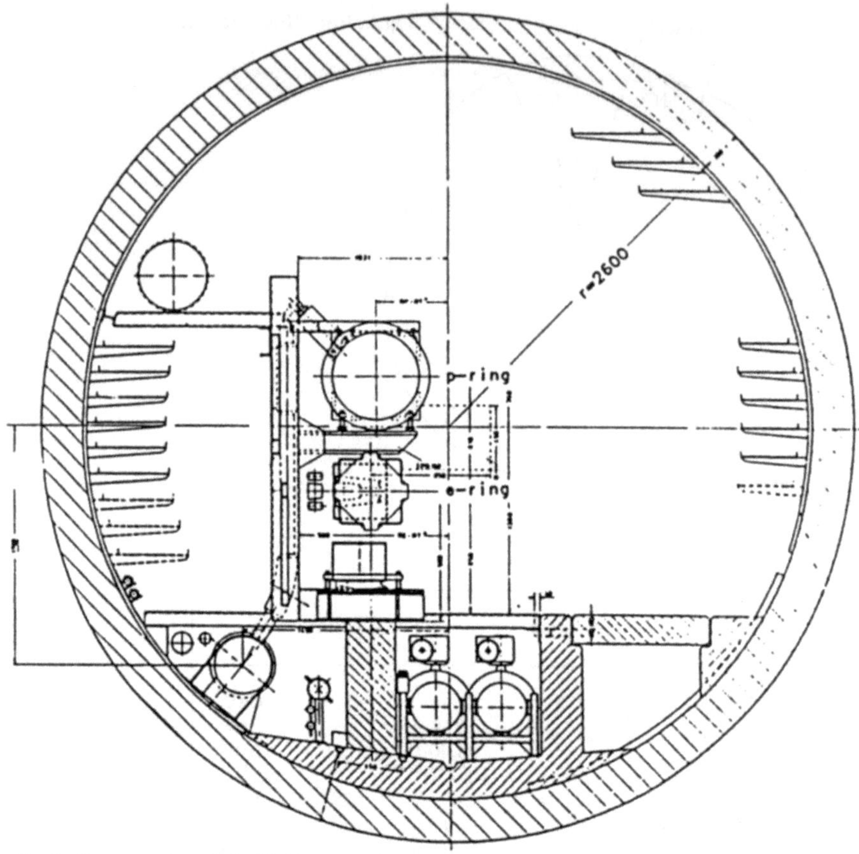

Fig. 2. Cross section of the HERA tunnel

The energy of the proton ring is limited by the strength of the magnetic bending field. Superconducting magnets were developed, following the pioneering work at FNAL, to achieve the highest possible energies for the diameter of the ring. We will shortly sketch these new accelerator magnets.

2.1 Superconducting Magnets

For high field superconducting magnets the field quality is determined by the current density distribution, whereas for a normal magnet it is essentially defined by the shape of the iron poles. Fig. 3 shows the cross section of the central part of the superconducting dipole and quadrupole magnets. The wedge shaped cable used for the dipole magnet is shown in an enlarged view. It consists of 24 strands each of which contains 1230 Niobium-Titanium filaments of 14 μm diameter embedded in a copper matrix. The arrangement of these conductors has been carefully tuned to achieve the necessary field quality. The conductor position has to be accurate to 10 – 20 μm [2]. This is especially difficult, as strong magnetic forces act on the conductor – the overall radial force in the horizontal plane amounts to about 10^6 N/m at 5 T – which are taken by the nonmagnetic collars clamped onto the coil.

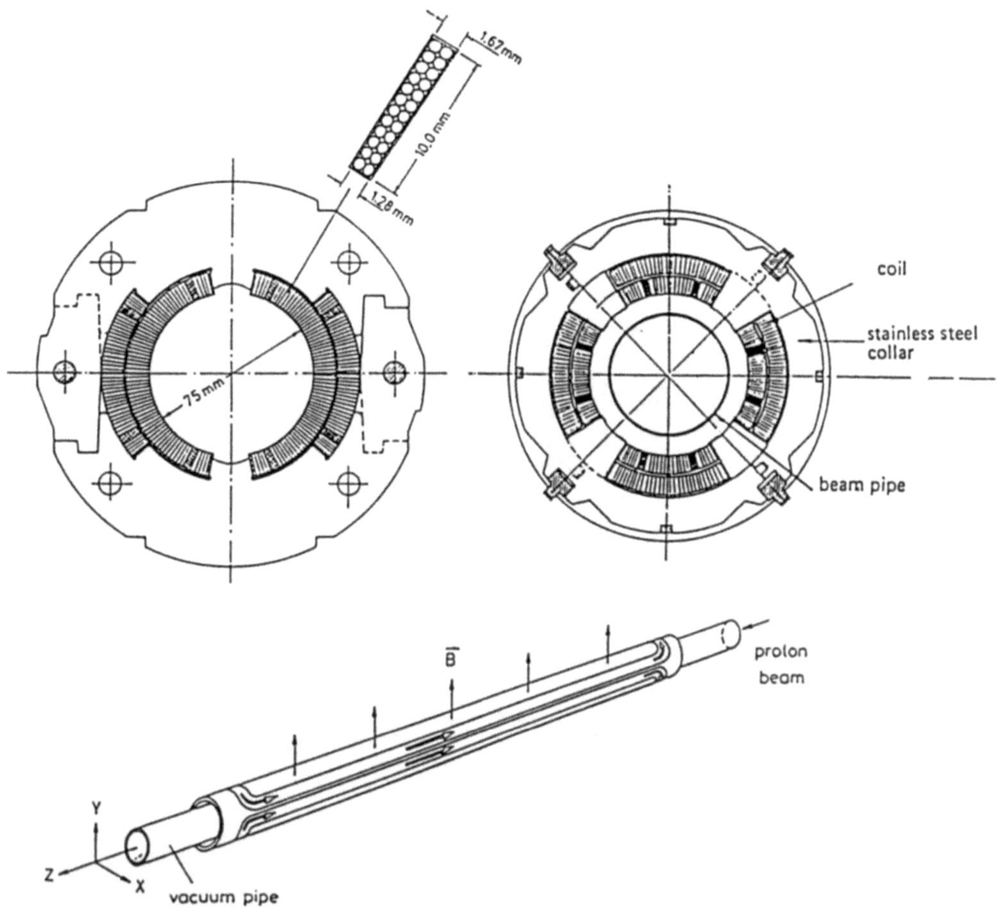

Fig. 3. The conductor arrangement of the superconducting HERA dipole and quadrupole magnets. The collars compressing the conductors to its precise position are also shown. The dipole magnet was developed by DESY together with industry and the quadrupole magnet by Saclay

The cross section of the complete dipole magnet is shown in Fig. 4; the cryogenic and thermally isolating parts of the quadrupole being similar. In the HERA design the beam pipe and the iron yoke surrounding the collars, which greatly reduce the fringe field outside the coil and enhances the field on the beam axis by about 20%, are at liquid helium temperatures.

The originally proposed HERA dipole magnets followed the Tevatron design, which had the iron yoke outside the cryostat at room temperature. In this case the distance between the coil and the iron yoke is larger and the enhancement of the central field by the iron yoke is only minor. It is therefore expected that the actual HERA magnets with its cold iron yoke will allow maximum proton energies in excess of the design value.

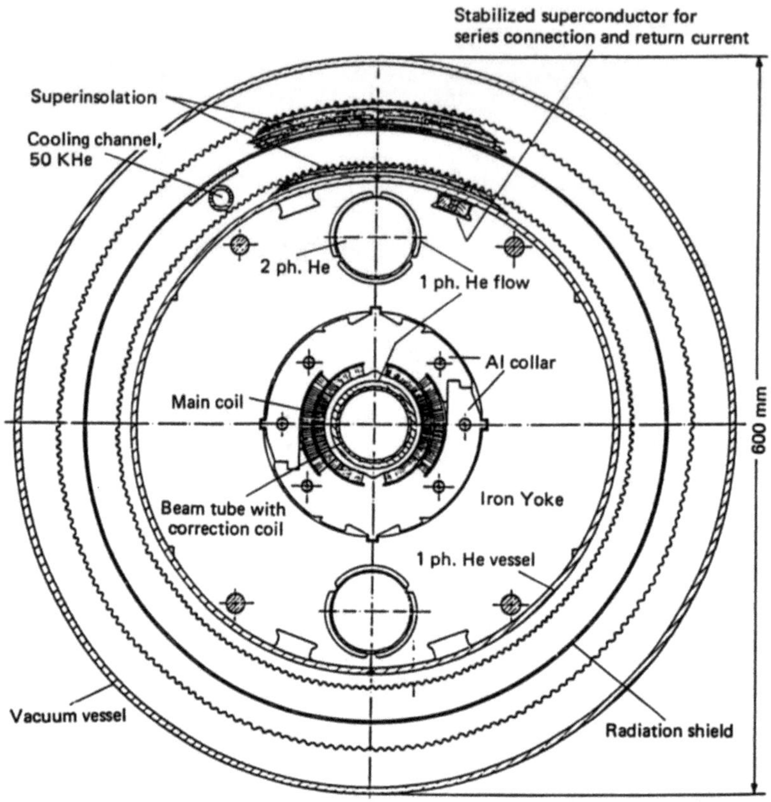

Fig. 4. Cross section of the HERA dipole magnet

2.2 Interaction Region and Spin Rotator

The HERA ring has 4 straight sections in which the beams can collide, and space for experimental facilities is provided around the 4 possible interaction regions. Initially, however, only in two of the straight sections will the beams be bent as sketched in Fig. 5, such that electrons and protons collide head on at the two interaction points. The two big detector facilities, H1 and ZEUS, are being assembled around these interaction regions.

A special magnet arrangement is inserted into the electron beam line in order to rotate the electron spin. In an ideal planar storage ring electrons become transversely polarized due to the Sokolov-Ternov mechanism [3] according to

$$P(t) = 0.98\left(1 - e^{-t/\tau}\right).$$

The polarization time τ is given by

$$\tau[s] = 98\frac{R^3[\mathrm{m}]}{E^5[\mathrm{GeV}]}\frac{< R >}{R}$$

where E is the electron energy, R the magnetic bending radius and $< R >$ the average radius. For HERA $\tau = 28$ min at 30 GeV. If in a transverse magnetic field

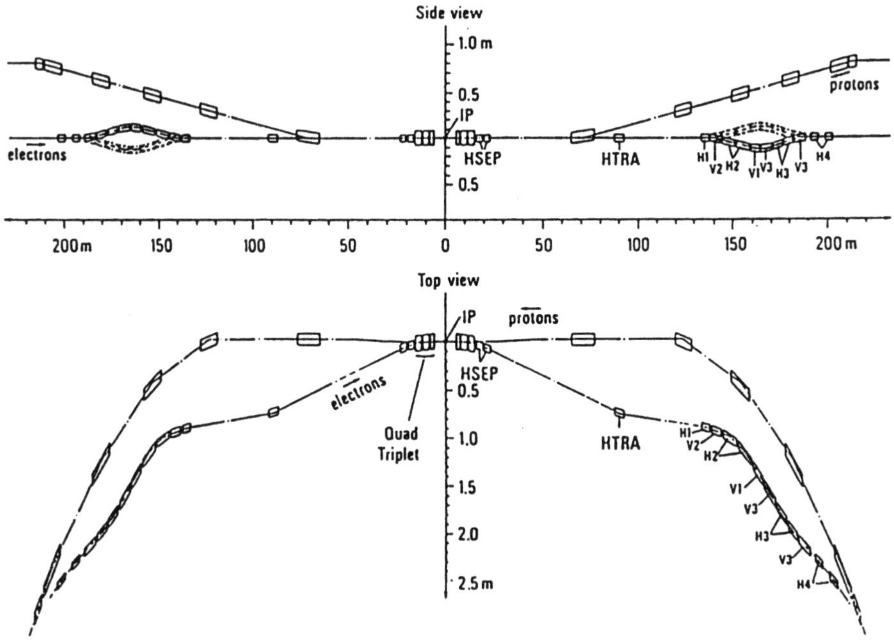

Fig. 5. Machine magnet arrangement in the straight sections to provide head on e–p collisions. Note the scale difference in the transverse and longitudinal direction

the electron momentum is rotated by an angle ϕ, the electron spin is rotated around the same axis by an angle

$$\theta = \frac{E[\mathrm{GeV}]}{440.65[\mathrm{MeV}]} \cdot \phi = \nu \cdot \phi,$$

due to the difference between the Lamor and the Cyclotron frequency; i.e. ν is the number of spin precessions per revolution. The special magnets shown in Fig. 5 rotate the spin by carefully tuned vertical and horizontal bending such that states of definite helicity are obtained at the interaction point [4]. The schematic principle of this spin rotator is shown in Fig. 6. Polarisation has been observed at electron positron storage rings but it is more difficult to obtain with increasing energies. This is due to the focussing fields present in a real machine and to the depolarising resonances at

$$\nu = n \pm n_z Q_z \pm n_x Q_x \pm n_s Q_s,$$

where n is an integer, $n_{z,x,s} = 0, 1$ and Q_z, Q_x are the number of vertical and horizontal betatron oscillations and Q_s the number of synchrotron oscillations per revolution. The energy distance of these resonances is essentially independent of E whereas the energy spread of the beam increases with E. A special effort will be made at HERA to tune the machine for maximum polarisation and the spin rotator sketched in Fig. 6 is prepared to become active, when transverse polarisation has been observed.

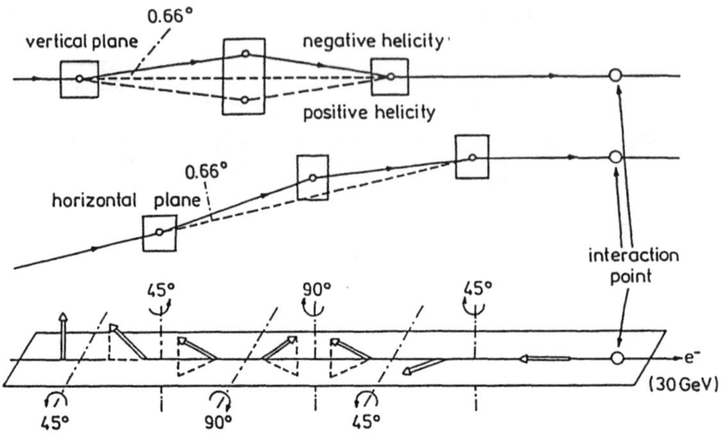

Fig. 6. Schematic principle of HERA spin rotator. The position of the magnets is shown in Fig. 5

3. HERA Physics

A main topic of the physics at HERA is electron scattering. It is a well known means of structure analysis ranging from atomic and nuclear physics to the famous lepton scattering experiments at SLAC and elsewhere revealing the parton structure of the nucleon. As HERA will enable us to enlarge the range of spacelike four momentum transfers Q^2 by about two orders of magnitude and will allow to penetrate to distances as small as 10^{-16}to10^{-17} cm, one is hoping that new structures will become apparent. Before discussing some possible new structures, we will extrapolate our present knowledge into this new kinematic region and mention the further insights to be gained on proton structure and the S.M.

But the physics at HERA is not limited to deep inelastic scattering. As electrons are surrounded by almost real photons and protons are made of quarks and gluons, HERA can also be looked at as a machine in which the constituents collide. The luminosity for these constituent collisions relative to the e - p luminosity is given in Fig. 7 as a function of the constituent c.m. energy. For $\gamma-$ q collisions as example:

$$dL^{\gamma \mathrm{q}} = G_{\mathrm{e}}^{\gamma}(x_1) G_{\mathrm{p}}^{\mathrm{q}}(x_2) dx_1 dx_2$$

where $G_{\mathrm{e}}^{\gamma}(x)$ is the number density of photons with momentum xp_{e} given by

$$G_{\mathrm{e}}^{\gamma}(x) = \frac{\alpha}{2\pi} \ln\left(\frac{s}{4m_{\mathrm{e}}^2}\right) \frac{1+(1-x)^2}{x}$$

in Weizsäcker-Williams approximation, and $G_{\mathrm{p}}^{\mathrm{q}}(x)$ the analogously defined proton quark density at low Q^2, known experimentally. With $\hat{s} = M^2 = (x_1 p_{\mathrm{e}} + x_2 p_{\mathrm{p}})^2 \simeq x_1 x_2 s$, i.e. M the $\gamma-$ q center of mass energy, and $\tau = M^2/s$, one obtains

$$\frac{dL^{\gamma \mathrm{q}}}{dM} = \frac{2\tau}{M} \int_{\tau}^{1} \frac{1}{x} G_{\mathrm{e}}^{\gamma}(x) G_{\mathrm{p}}^{\mathrm{q}}\left(\frac{\tau}{x}\right) dx$$

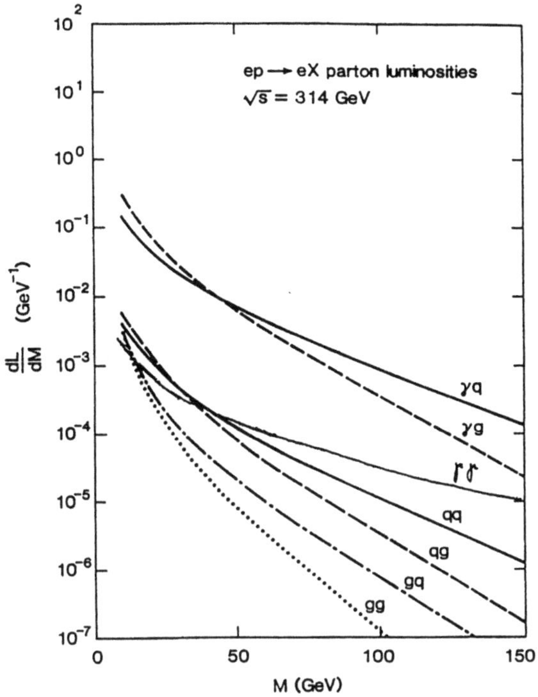

Fig. 7. Luminosity distribution for parton-parton scattering in e–p collisions with $Q^2 \cong 0$ (from [5])

which is plotted in Fig. 7. The contribution of the hard scattering subprocess $\gamma\,q \to q\,g$ for example to the low Q^2 cross section is then given by

$$\sigma^{\mathrm{ep}\to\mathrm{eX}} = \sum_{q} \int dM \frac{dL^{\gamma g}}{dM} \sigma^{\gamma q \to qg}(M).$$

In this sense HERA enables one to perform photoproduction experiments equivalent to fixed target photon energies well in excess of 10 TeV. These experiments will allow important tests of QCD – for details see [5, 6] – and will allow production studies of known and new heavy particles, as will be discussed in Sect. 3b.

Although the partonic densities G_e^q and G_e^g of the electron are considerably lower than the photon density as of order α^2, hadronic subprocesses often are of similar strength as photonic ones [5], because of their larger scattering cross sections.

3.1 Deep Inelastic Scattering

3.1.1 The Standard Prejudice

The e–p scattering kinematics is not quite as simple as the kinematics of e^+e^-–annihilation. Fig. 8 reminds the reader of the standard notation.

The following Lorentz invariants are used to define the kinematics:

$$(p_e + P)^2 = s \simeq 4E_e E_p$$
$$(p_e - p_\ell)^2 = q^2 \equiv -Q^2 \simeq -4E_e E_\ell \sin^2\theta/2,$$

Fig. 8

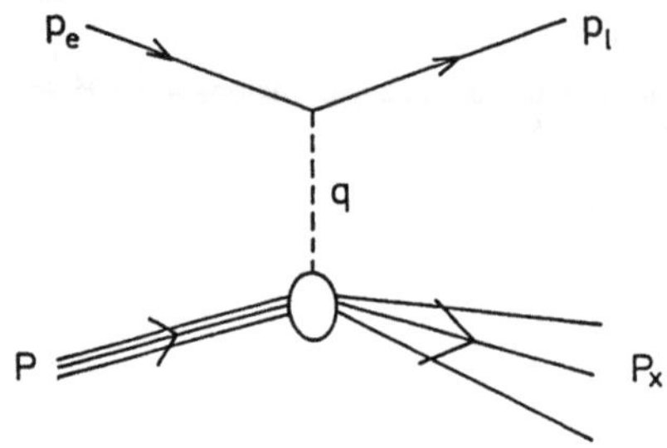

Fig. 9

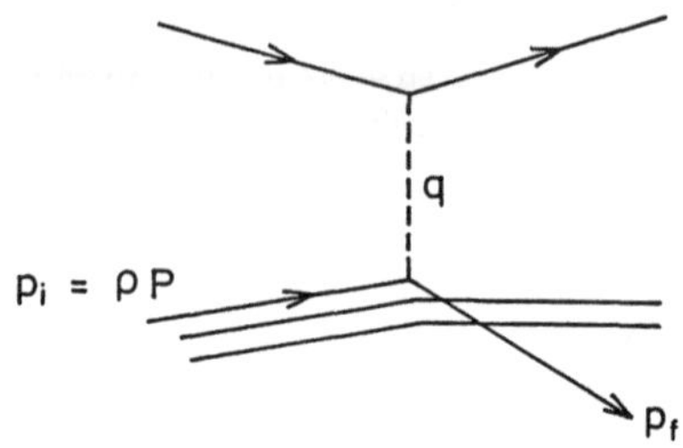

Fig. 8. Definition of four momenta in e–p scattering. P and p_e are the four momenta of the incoming proton and electron, respectively, and p_ℓ and P_X the ones of the outgoing lepton and the outgoing hadrons

Fig. 9. e–p scattering in the parton model. The incoming and outgoing momenta of the interacting parton are p_i and p_f

$$P \cdot q = m_p \nu \simeq 2E_p(E_e - E_\ell \cos^2 \theta/2)$$

where the approximative relations hold in systems with $\boldsymbol{p}_e$ and $\boldsymbol{P}$ being collinear if masses are neglected. One usually defines dimensionless ratios

$$x = \frac{Q^2}{2P \cdot q} = \frac{Q^2}{2m_p \nu} \qquad \text{and}$$

$$y = \frac{2P \cdot q}{s} = \frac{\nu}{\nu_{max}}$$

which are limited to values between 0 and 1. The mass W of the hadronic system is given by

$$W^2 = (q + P)^2 = Q^2 \frac{1 - x}{x} + m_p^2.$$

Describing the proton as a system of quasifree partons, with one of which the photon interacts (Fig. 9), one can interpret x, neglecting quark masses, as the fraction of the proton momentum carried by the interacting parton:

$$\begin{aligned} 0 = p_f^2 = (p_i + q)^2 &= p_i^2 + 2p_i \cdot q + q^2 \\ &= 2\rho P \cdot q - Q^2 \end{aligned}$$

$$\longrightarrow \rho = \frac{Q^2}{2P \cdot q} \equiv x$$

In the S.M. the neutral current contribution to e–p scattering is given by the amplitudes sketched in Fig. 10 leading to the cross section:

$$\frac{d^2\sigma^{NC}(e^\mp p)}{dx dQ^2} = \frac{4\pi\alpha^2}{xQ^4} \left\{ y^2 x F_1 + (1 - y)F_2 \pm (y - \frac{y^2}{2})F_3 \right\} \tag{1}$$

where $F_i = F_i(x, Q^2)$. The structure functions F_i could in principle be calculated by QCD, in practice, however, only their Q^2 dependence is predicted by the Altarelli-Parisi evolution equations [7]: For F_2 as an example

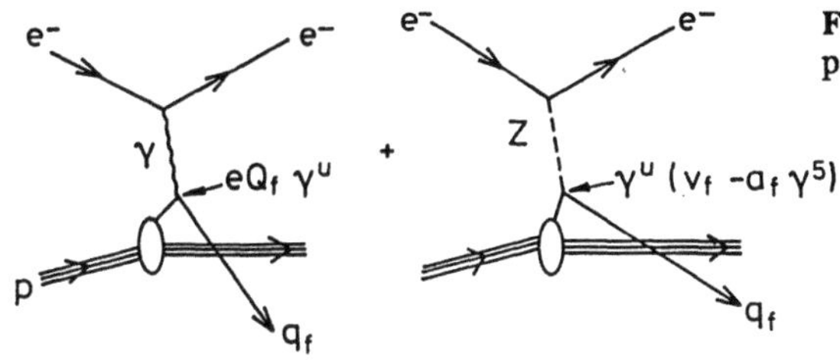

Fig. 10. Diagrams of γ and Z-exchange in e–p scattering

$$\frac{dF_2\,(x,Q^2)}{d\ln Q^2} = \frac{\alpha_s(Q^2)}{2\pi}\int_x^1 \{P_{qq}(x/z)F_2(z,Q^2)+ +2zN_f P_{qg}(x/z)G(z,Q^2)\}\,\frac{x}{z^2}dz. \tag{2}$$

The scaling violation, i.e. the Q^2 dependence of the F_i, is due to QCD corrections to the quark parton model. The first term in the integral of Eq.(2) takes account of the process sketched in Fig. 11a and the second term, which involves the gluon structure function $G(x,Q^2)$ of the proton, of Fig. 11b. The integral over the first term is negative, leading to a decrease of $F_2(x,Q^2)$ with increasing Q^2, while the second term, which dominates at small x, makes a positive contribution i.e. an increase of F_2 with increasing Q^2. The sum of the two terms leads to the behaviour shown in Fig. 12.

For high Q^2 and $x \gtrsim 0.2$ the Callan-Gross relation is a good approximation

$$F_L \equiv F_2 - 2xF_1 \simeq \frac{<k_T>^2}{Q^2} \simeq 0,$$

where $<k_T>$ is the average quark momentum component transverse to the beam direction. Eq.(1) then simplifies to

$$\frac{d^2\sigma^{NC}(e^\mp)}{dx\,dQ^2} = \frac{2\pi\alpha^2}{xQ^4}\left\{[1+(1-y)^2]F_2 \pm [1-(1+y)^2]xF_3\right\}. \tag{3}$$

The structure functions F_2 and F_3 can be expressed by the quark density distributions $q_f(x,Q^2)$ of flavour f. Referring separately to the different helicity states L and R of the incident lepton

$$\begin{aligned} F_2^{L,R}(x,Q^2) &= \sum_f \{xq_f(x,Q^2) + x\bar{q}_f(x,Q^2)\}\,A_f^{L,R}(Q^2) \\ xF_3^{L,R}(x,Q^2) &= \sum_f \{xq_f(x,Q^2) - x\bar{q}_f(x,Q^2)\}\,B_f^{L,R}(Q^2) \end{aligned} \tag{4}$$

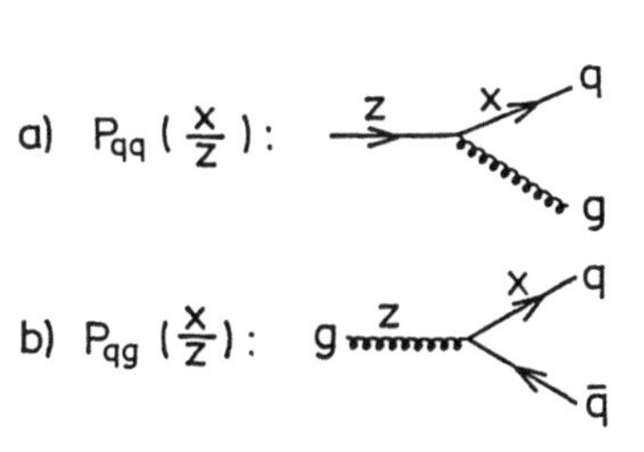

Fig. 11a,b. Processes described by the Altarelli-Parisi splitting functions. **(a)** A quark of momentum fraction z emits a gluon leading to fraction x. **(b)** A gluon of fraction z splits into a quark of fraction x and an antiquark

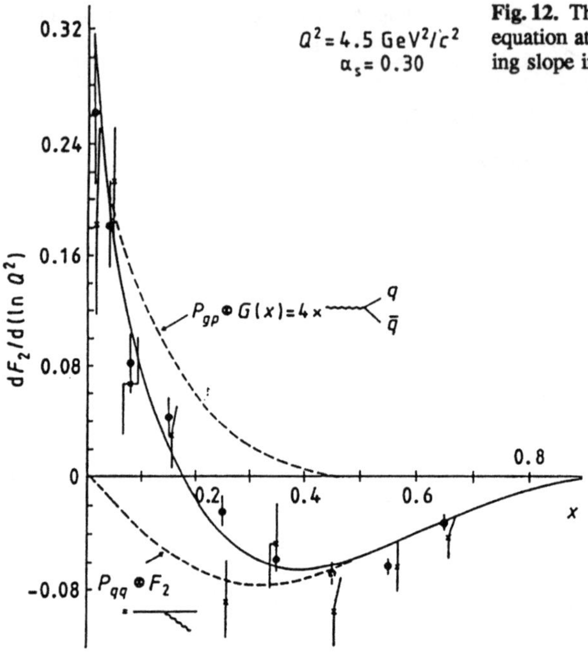

Fig. 12. The two contributions of the evolution equation at low Q^2 and comparison of the resulting slope in Q^2 with data (from [8])

with

$$A_f^{\mathrm{L,R}} = Q_f^2 - 2Q_f(v_\mathrm{e} \pm a_\mathrm{e})v_f\chi + (v_\mathrm{e} \pm a_\mathrm{e})^2(v_f^2 + a_f^2)\chi^2$$
$$B_f^{\mathrm{L,R}} = \mp 2Q_f(v_\mathrm{e} \pm a_\mathrm{e})a_f\chi \pm 2(v_\mathrm{e} \pm a_\mathrm{e})^2 v_f a_f \chi^2$$

and

Q_f = electric charge of quark flavour f in units of $|e|$,
$a_f = 2T_{3\mathrm{L}} - Q_f/|Q_f|$
$v_f = Q_f/|Q_f| - 4Q_f \sin^2\theta_\mathrm{W}$
θ_W = weak mixing angle

$$\chi = \frac{1}{16\sin^2\theta_\mathrm{W}\cos^2\theta_\mathrm{W}} \cdot \frac{Q^2}{Q^2 + M_\mathrm{Z}^2}$$
$$= \frac{G_\mathrm{F}M_\mathrm{Z}^2}{8\sqrt{2}\pi\alpha} \cdot \frac{Q^2}{Q^2 + M_\mathrm{Z}^2}$$

For incident positrons one has to make the following substitutions:

$$\mathrm{e}^- \to \mathrm{e}^+ : F_2^{\mathrm{L,R}} \to F_2^{\mathrm{R,L}}, xF_3^{\mathrm{L,R}} \to -xF_3^{\mathrm{L,R}}.$$

The Q_f^2 term in $A_f^{\mathrm{L,R}}$ describes the photon exchange, the term proportional χ^2 the Z-exchange, and the term proportional χ is due to the interference of the two amplitudes. Only the interference and Z-exchange term contribute to $B_f^{\mathrm{L,R}}$. The expected relative sizes of these contributions as a function of Q^2 are given in Fig. 13.

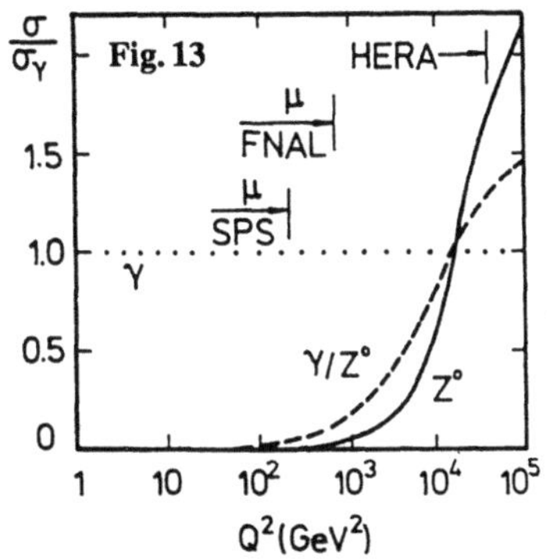

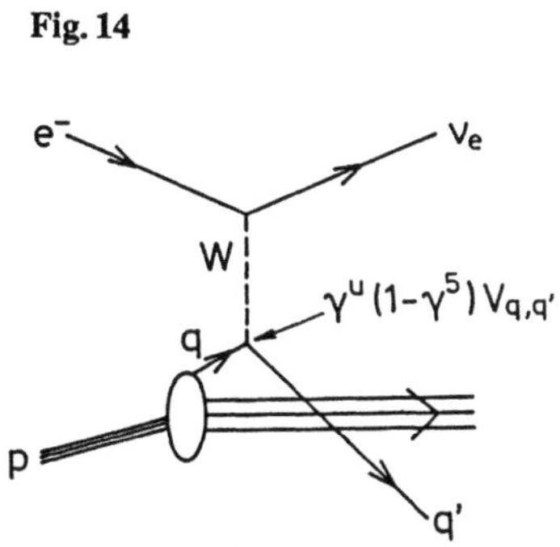

Fig. 13. Relative size of the cross sections from γ and Z^0 exchange and their interference (from [9]). The upper reaches of the different accelerators are also indicated

Fig. 14. Diagram for charged current process in ep-collision

For low Q^2 the Q_f^2 term dominates and F_3 can be neglected

$$\frac{d^2\sigma^{\mathrm{NC}}(\mathrm{e}^{\pm}\mathrm{p})}{dx\, dQ^2} \simeq \frac{2\pi\alpha^2}{xQ^4}\left\{1+(1-y)^2\right\} F_2^{\mathrm{em}}$$
$$= \frac{2\pi\alpha^2}{xQ^4}\left\{1+(1-y)^2\right\} \sum_f Q_f^2(xq_f + x\bar{q}_f)\,. \tag{5}$$

In an analogous way we can write for the charged current process Fig. 14:

$$\frac{d^2\sigma^{\mathrm{CC}}(\mathrm{e}_{\mathrm{L}}^-\mathrm{p})}{dx\, dQ^2} = \frac{G_{\mathrm{F}}^2}{\pi x(1+Q^2/M_{\mathrm{W}}^2)^2}\left\{y^2 xF_1^{\mathrm{CC}} + (1-y)F_2^{\mathrm{CC}} + (y - y^2/2)xF_3^{\mathrm{CC}}\right\}$$
$$= \frac{\pi\alpha^2}{2x\sin^4\theta_{\mathrm{W}}(Q^2+M_{\mathrm{W}}^2)^2}\left\{y^2 xF_1^{\mathrm{CC}} + (1-y)F_2^{\mathrm{CC}} + (y - y^2/2)xF_3^{\mathrm{CC}}\right\}$$
$$\frac{d^2\sigma^{\mathrm{CC}}(\mathrm{e}_{\mathrm{R}}^-\mathrm{p})}{dx\, dQ^2} = 0\,. \tag{6}$$

At larger x and Q^2 the Callan-Gross relation is again expected to hold $F_{\mathrm{L}} = F_1 - 2xF_2 \simeq 0$ and one obtains

$$\frac{d^2\sigma^{\mathrm{CC}}(\mathrm{e}_{\mathrm{L}}^-\mathrm{p})}{dx\, dQ^2} = \frac{\pi\alpha^2}{4x\sin^4\theta_{\mathrm{W}}(Q^2+M_{\mathrm{W}}^2)^2}\left\{[1+(1-y)^2]F_2^{\mathrm{CC}} + [1-(1-y)^2]xF_3^{\mathrm{CC}}\right\} \tag{7}$$

Expressing F_2^{CC} and F_3^{CC} in terms of the quark density function $q_f(x,Q^2)$ one gets

$$\begin{aligned} F_2^{\mathrm{CC}}(\mathrm{e}^-\mathrm{p}) &= \sum_{i,j}\left\{x\left|V_{u_i d_j}\right|^2 u_i + x\left|V_{u_j d_i}\right|^2 \bar{d}_i\right\} \\ xF_3^{\mathrm{CC}}(\mathrm{e}^-\mathrm{p}) &= \sum_{i,j}\left\{x\left|V_{u_i d_j}\right|^2 u_i - x\left|V_{u_j d_i}\right|^2 \bar{d}_i\right\} \end{aligned} \tag{8}$$

where u_i and d_i stand for a $Q_f = 2/3$ and $Q_f = -1/3$ quark, respectively, and $V_{u_i d_j}$ are elements of the Kobayashi-Maskawa matrix. For incident positrons: $e^- \to e^+$: $L \to R$, $R \to L$, $u \to \bar{u}, \bar{d} \to d$.

Considering only u, d, s and c quarks and using the unitary relations

$$\sum_i \left|V_{u_i d_j}\right|^2 = \sum_j \left|V_{u_i d_j}\right|^2 = 1$$

one gets:

$$\frac{d^2\sigma^{\mathrm{CC}}(\mathrm{ep})}{dx\, dQ^2} \simeq \frac{\pi\alpha^2}{2\sin^4\theta_{\mathrm{W}}(Q^2 + M_{\mathrm{W}}^2)^2} \times \begin{cases} (u+c) + (1-y)^2(\bar{d}+\bar{s}) & \text{for } \mathrm{e}^{-\mathrm{L}} \\ (\bar{u}+\bar{c}) + (1-y)^2(d+s) & \text{for } \mathrm{e}^{+\mathrm{R}} \\ 0 & \text{for } \mathrm{e}^{-\mathrm{R}},\ \mathrm{e}^{+\mathrm{L}} \end{cases}$$

Using parton distribution functions determined by various groups from low Q^2 lepton scattering data – to better constrain the gluon distribution, which enters only indirectly via the Q^2 evolution into lepton scattering, data on pp$\to \gamma$X (gq $\to \gamma$ q) were taken into account as well – and extrapolating the data to HERA energies with the help of Eq.(2), one obtains the predicted rates shown in Fig. 15. For $x, y \gtrsim 0.2$ the charged and neutral current rates are about similar. Fig. 16 shows a compilation of F_2 measurements on proton targets and its extrapolation to HERA energies. An estimate of the uncertainties of the extrapolation can be obtained from Fig. 17 where the relative differences of several extrapolations to $Q^2 = 10^4\mathrm{GeV}^2$ are shown. Unfortunately the most accurate μ-scattering experiments EMC [14] and BCDMS [15] measure cross sections differing by about 15% at low x, as shown in Fig. 16. This leads to differences in the deduced parton density distribution [16] at $Q^2 = 20\,\mathrm{GeV}^2$ of up to 20%.

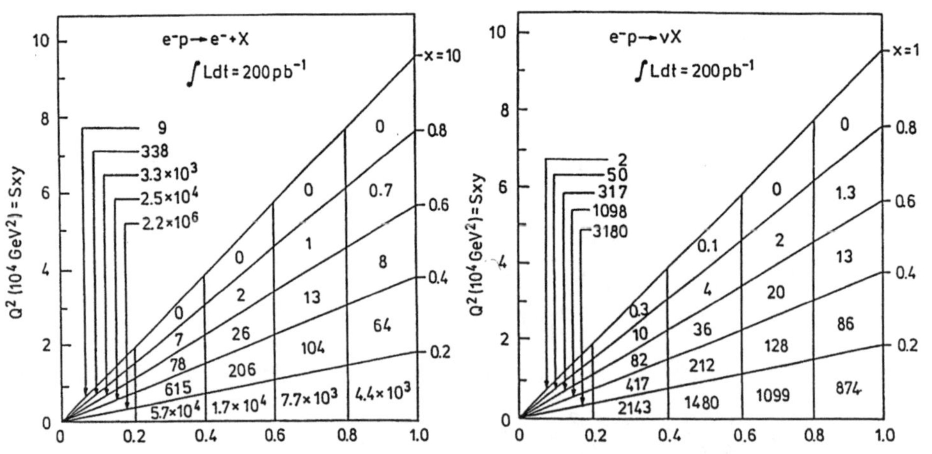

Fig. 15. Event rates predicted for $\int L\, dt = 200\,\mathrm{pb}^{-1}$ at $\sqrt{s} = 314\,\mathrm{GeV}$ for $x > 0.01$ and $y > 0.01$ using the Duke and Owens [10] parton distribution parametrization

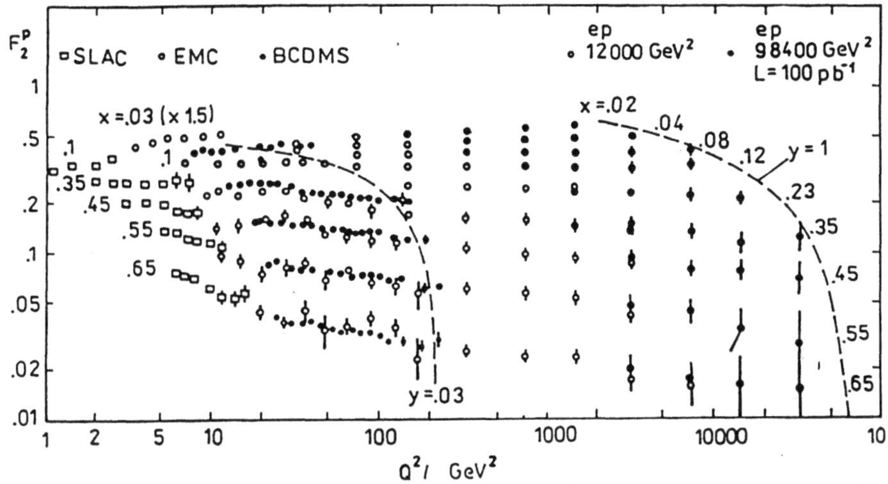

Fig. 16. Proton structure function F_2 from electron and muon scattering and its extrapolation to HERA energies of $\sqrt{s} = 110$ GeV and 314 GeV, using the parametrisation of [10] (from [11])

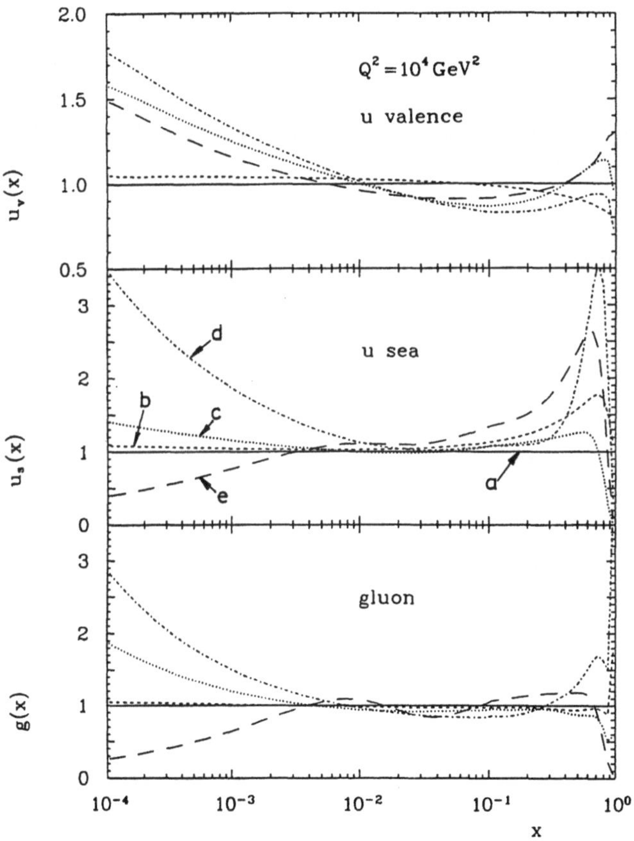

Fig. 17. Comparison of some parton density parametrizations extrapolated to $Q^2 = 10^4\,\mathrm{GeV}^2$, normalized to set 1 of [12]. Curve *b* is from set 2 of [12], curve *c* from set 1 of [10], curve *d* from set 2 of [10] and curve *e* from [13] (from [9])

3.1.2 Tests of the Standard Model at HERA

Deep inelastic scattering will be one of the main experiments at HERA and important confirmations of the S.M. can be obtained. The weak mixing angle $\sin^2\theta_W$ will be determined from the ratio of charged and neutral current rates, i.e. Eq.(6) and Eq.(1). In this ratio measurement several systematic uncertainties cancel and a precision of about 0.002 has been estimated to be obtainable for $\sin^2\theta_W$ [17]. From the neutral current data alone, i.e. Eq.(3) and Eq.(4), the estimated uncertainties amount to about 0.009 [18]. These measurements of $\sin^2\theta_W$ will be complementary to those at SLC and LEP, since they are obtained at large spacelike momentum transfers.

The experimental determination of the Q^2 evolution of the parton distributions is an important further test of QCD. Target mass or higher twist effects, which could mimic the dependence observed at lower Q^2, are proportional to $1/Q^2$ and expected to be negligible at HERA energies. Figs.18–20 which result from a computer experiment [19] and take fragmentation and detector acceptance into account, may provide an estimate of what can be achieved in the first years of data taking.

Figs. 18, 19 show that already rather global cross section distributions give a clear indication of QCD effects, whereas a separation of the various parton distributions is considerably more difficult as Fig. 20 indicates.

HERA also covers the hitherto unexplored region of very low x down to 10^{-4}. It is the region where the many soft gluons within the proton become important and where we reach the limit of perturbative QCD. Extrapolations of the various fits into this region differ considerably and HERA can, with a modest amount of luminosity, determine the structure functions there (Fig. 21).

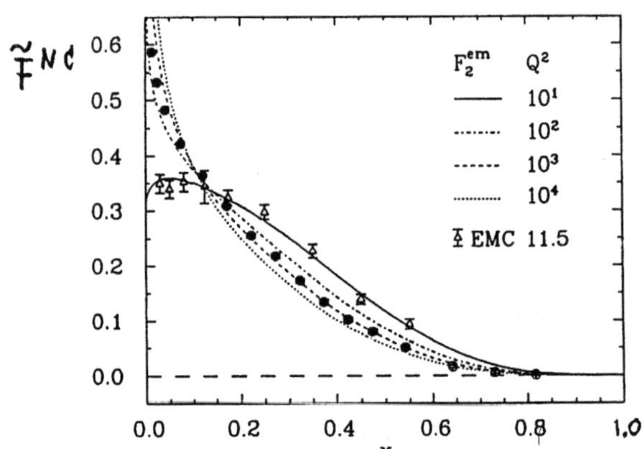

Fig. 18. Monte Carlo simulation of a neutral current cross section measurement for incident electrons. Plotted is the quantity $\tilde{F}^{NC} = d\sigma^{NC}/(dx\,dQ^2)\cdot(x\,Q^4)/[2\pi\alpha^2\left\{1+(1-y)^2\right\}] \simeq F_2^{em}$. The data are averaged over $0.03 \leq y \leq 1$ and correspond to $\int L\ dt = 200\ \mathrm{nb}^{-1}$ at $\sqrt{s} = 314$ GeV. Low Q^2 measurements of F_2 and the Q^2 dependence of F_2 as used for the data generation are shown (from [19])

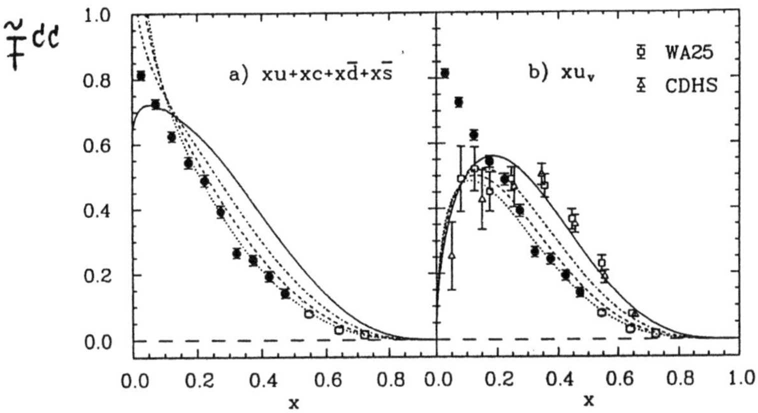

Fig. 19a,b. Monte Carlo simulation of charged current data. Same conditions as for Fig. 18 (from [19]). Plotted is the quantity $\tilde{F}^{CC} = d^2\sigma^{CC} \cdot 4x \sin^4 \theta_W \cdot \left(Q^2 + M_W^2\right)^2 / \left(dx\, dQ^2 \cdot \pi\alpha^2\right)$. The curves show **(a)** the $x\left(u + c + \bar{d} + \bar{s}\right)$ and **(b)** the valence quark distribution xu_v for $Q^2 = 10$ (*full*), 10^2 (*dash-dotted*), 10^3 (*dashed*) and 10^4 (*dotted*) GeV^2, as evaluated from the input distributions of the simulation

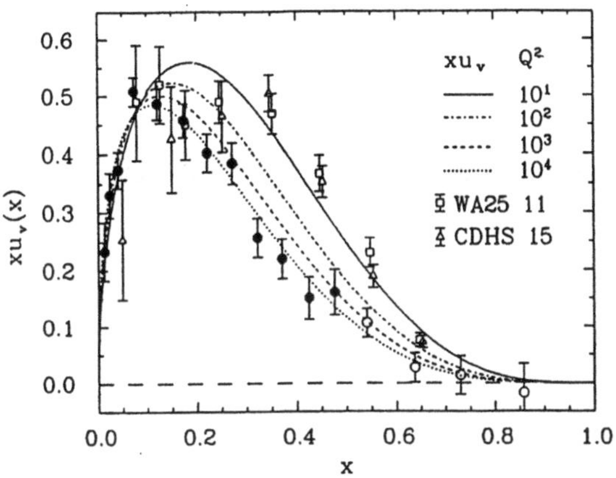

Fig. 20. Valence quark momentum distribution extracted from charged and neutral current data of a Monte Carlo experiment at $\sqrt{s} = 314$ GeV. The input data correspond to $\int L\, dt = 400\,\mathrm{pb}^{-1}$ shared equally between e^+ and e^-. The data have been averaged over $0.003 \leq y \leq 1$ (from [19])

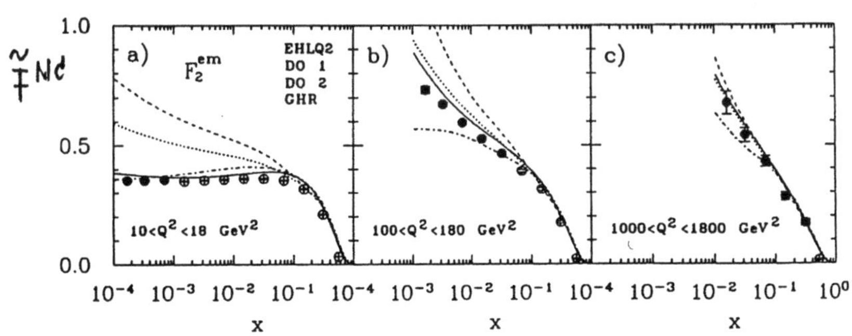

Fig. 21. Monte Carlo simulation of neutral current data at very low x (from [19]). The data correspond $\int L\, dt = 10$ pb. Otherwise same conditions as for Fig. 18. The curves represent different structure function extrapolations in this hitherto unexplored low x region, for references see Fig. 17

3.1.3 Speculations

The physics at HERA will be exciting if new structures will show up at these energies and in the following I sketch a few speculations about it.

a) Further Compositeness. Lacking imagination, we assume that nature repeats itself and quarks are made of further constituents, preons. To estimate possible observable effects at HERA we follow [20] and assume the preon distribution in the quarks to be the same as the quark distribution within the proton, roughly

$$Q(x) = \frac{2(1-x)^3}{x} \ .$$

Convoluting quark and preon distribution we obtain the preon distribution within the proton

$$P(x) = \frac{4}{x}\int_x^1 \frac{dy}{y}(1-y)^3\left(1-\frac{x}{y}\right)^3 .$$

The corresponding quark and preon momentum distributions, which are shown in Fig. 22, are used to construct F_2. For $Q^2 \ll \Lambda^2$, where Λ is the preon scale we have (Eq.(5) in scaling limit)

$$F_2(x, Q^2) \simeq \frac{5}{18}\, xQ(x),$$

where the factor 5/18 represents the $\langle Q_f^2 \rangle$ of u and d quarks. Assuming that the transition between the quark and preon region is described by a form factor

$$f\,(Q^2/\Lambda^2) = \left(1 + \frac{Q^2}{\Lambda^2}\right)^{-1}$$

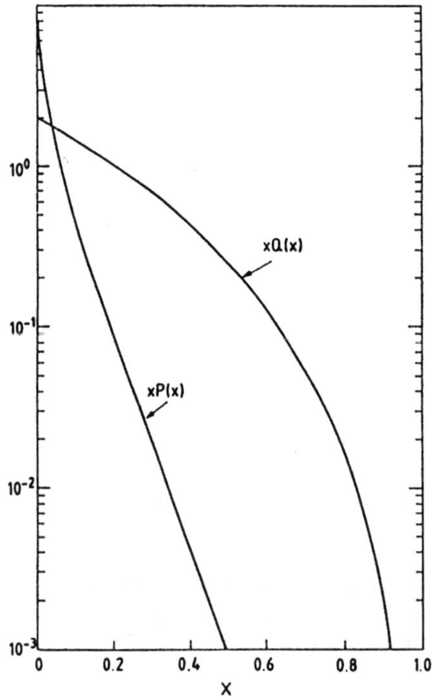

Fig. 22. The quark and preon momentum distribution $Q(x)$ and $P(x)$ discussed in the text (from [20])

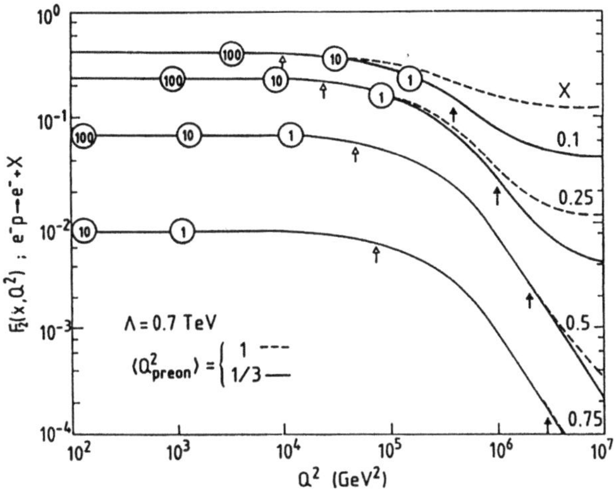

Fig. 23. The toy model structure function $F_2(x, Q^2)$ discussed in the text for Λ = 700 GeV. The kinematic limits of HERA and a LEP-LHC-collider are indicated by arrows. The encircled numbers give the events per $\int L\, dt \simeq 8\,\mathrm{pb}^{-1}$ (from [20])

one makes the following ansatz:

$$F_2(x, Q^2) = \frac{5}{18} \left\{ f^2 \left(\frac{Q^2}{\Lambda^2} \right) xQ(x) + \langle Q^2_{\mathrm{Pr}} \rangle \left[1 - f^2 \left(\frac{Q^2}{\Lambda^2} \right) \right] xP(x) \right\}$$

where Q_{Pr} is the preon charge. Numerical values of F_2 given in Fig. 23 for Λ = 700 GeV show a steep drop at $Q^2 \simeq \Lambda^2$. The kinematic boundaries of HERA are indicated. Values of Λ up to 700 GeV seem to be observable at HERA. At present, similarly direct limits on quark formfactors obtained from $e^+e^- \to$ hadrons amount to $\Lambda \gtrsim 250$ GeV.

If electron and quarks possess substructure one expects, however, that possible evidence may first show up by a new contact interaction. This is analogous to the first evidence for strong interactions obtained by Rutherford and coworkers in 1918 from the observation of anomalous scattering of α particles from light nuclei. For estimates one defines an effective Lagrangian by adding a chirality conserving contact term to the Standard Model

$$L_{\mathrm{eff}} = L_{\mathrm{SM}} \pm \frac{g^2}{\Lambda^2_\pm} \sum_{i,j} C_{ij} \left(\bar{\ell}_i \gamma_\mu \ell_i \right) \left(\bar{q}_j \gamma^\mu q_j \right)$$

where the sum extends over the various chirality structures i, j = R,L and $C_{ij} = 0, 1$. One usually sets $g^2/(4\pi) = 1$ leaving $\Lambda_\pm$ as the only free parameter.

Fig. 24 shows the neutral current cross sections expected from L_{eff} with $\Lambda_\pm$ = 3 TeV for incident electrons and positrons. The data have been limited to $Q^2 >$ 1000 GeV^2, $0.1 \geq x \geq 0.82$ and $y > 0.01$ and are normalized to the S.M. prediction. HERA experiments will be sensitive up to $\Lambda_\pm$ values of about 7 TeV, depending somewhat on the chiral structure.

There exist of course more stringent but model dependent limits on $\Lambda_\pm$. From the absence of lepton or baryon number violating decays such as $\mu \to e\gamma$, 3e or

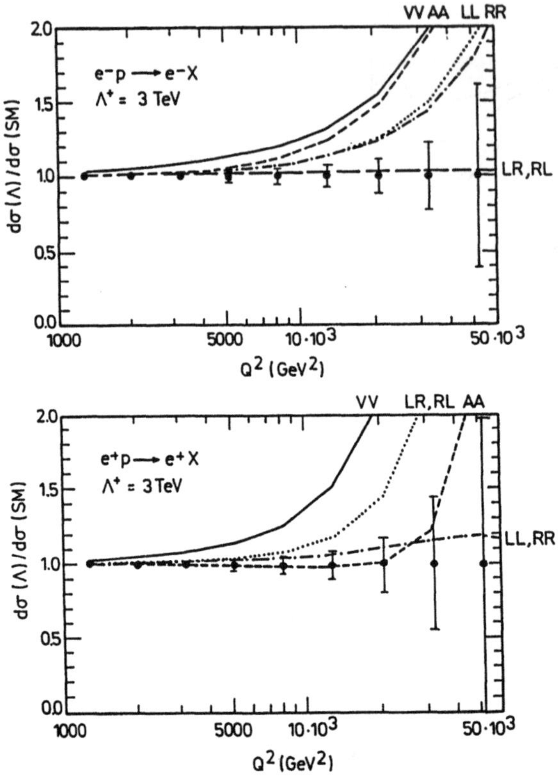

Fig. 24. Neutral current (e^-p) and (e^+p) cross sections from a contact interaction with $\Lambda_+ = 3$ TeV as a function of Q^2. The cross sections are normalized to S.M. expectation. The indicated errors correspond to $\int L\, dt = 100\,\text{pb}^{-1}$ and $\sqrt{s} = 314$ GeV (from [21])

$K^0 \to \mu\,e$ and proton decay, one obtains limits of $\Lambda_\pm > 10 - 10^3$ TeV and even much larger $\left(\Lambda_\pm > 10^{11}\,\text{TeV}\right)$ for proton decay. These decays, however, might be suppressed by flavour or family selection rules.

It seems hard to imagine a selection rule, which suppresses flavour diagonal terms $\left(g^2/\Lambda_\pm^2\right)\left(\bar{f}\gamma_\mu f\right)\left(\bar{f}\gamma^\mu f\right)$. Presently the best limits on these are obtained from Bhabha scattering $e^+e^- \to e^+e^-$ amounting to $\Lambda \gtrsim 2$–$3\,\text{TeV}$.

b) New Currents. New structures may also show up in the gauge boson sector and many models have been proposed with additional W's and Z's. To exemplify the HERA possibilities we consider an extension of the S.M., in which parity invariance is restored above a certain unification scale [22].

In this L,R symmetric model the S.M. gauge group is enlarged

$$SU(2)_\text{L} \times U(1)_Y \;\to SU(2)_\text{L} \times SU(2)_\text{R} \times U(1)_{\text{B}-\text{L}}$$

and yields for the electric charges

$$Q_f = T_{3\text{L}} + \frac{Y}{2} \to Q = T_{3\text{L}} + T_{3\text{R}} + \frac{B-L}{2}.$$

Fig. 25

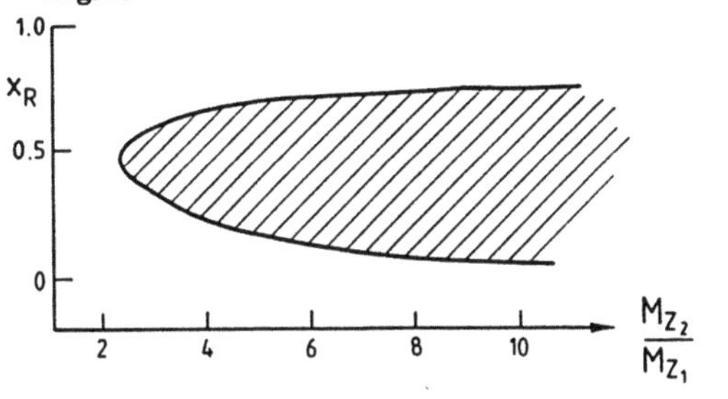

Fig. 26

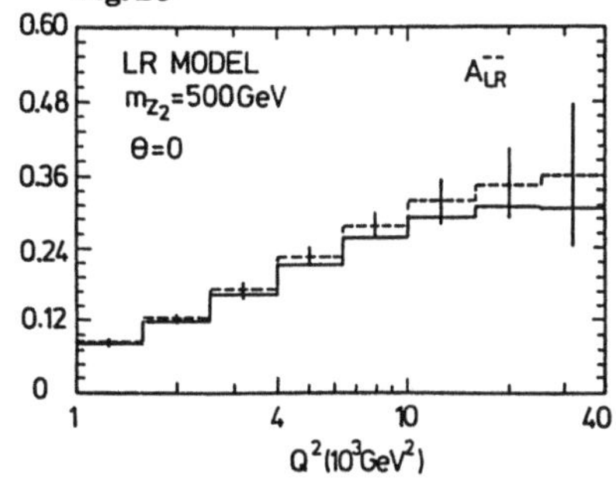

Fig. 25. Constraints on model parameters x_R and M_{Z_2} discussed in the text. The allowed region is shaded (from [20])

Fig. 26. The polarization asymmetry A_{LR}^{--} for the left-right symmetric model with M_{Z_2} = 500 GeV (*full line*) and for the S.M. prediction (*dashed line*). The error bars correspond to $\int L\ dt$ = 125 pb^{-1} per beam. The polarization is assumed to be 100% (from [23])

This enlargement leads to three new gauge bosons Z', W'$^\pm$, which couple to right handed fermions. The mass eigenstates are in general mixtures

$$W_1 = W \cos\rho + W' \sin\rho \qquad \text{and}$$
$$W_2 = -W \sin\rho + W' \cos\rho$$

$$Z_1 = Z \cos\theta + Z' \sin\theta$$
$$Z_2 = -Z \sin\theta + Z' \cos\theta\ ;$$

further parameters of the model are $x_L = \sin^2\theta_W$ and the corresponding x_R.

Identifying W_1 and Z_1 with the known gauge bosons one obtains [20] the lowest limits

$$\rho < 0.004\ ,\ M_{W_2} \geq 300\ \text{GeV},$$

and Fig. 25 gives the limits for x_R and M_{Z_2}. For a really symmetric model, i.e. $x_L = x_R$, one obtains $M_{Z_2} > 4M_{Z_1}$ and $\theta < 0.03$.

In order to estimate the sensitivity of the HERA experiments for the new gauge boson, the polarisation asymmetry

$$A_{L,R}^{--} = \frac{d\sigma^{NC}(e_L^- p)/dQ^2 - d\sigma^{NC}(e_R^- p)/dQ^2}{d\sigma^{NC}(e_L^- p)/dQ^2 + d\sigma^{NC}(e_R^- p)/dQ^2}$$

for M_{Z_2} = 500 GeV, $x_L = x_R$ and 100% polarized beams is shown in Fig. 26.

The function

$$\chi^2(m_2) = \sum_{Q^2}\left(\frac{F(m_2) - F_{SM}}{\delta F}\right)^2$$

deduced from a computer experiment assuming 80% polarization and $\int L\ dt$ = 250 pb^{-1} in total is shown in Fig. 27. The polarization asymmetries A_{LR}^{--} and A_{RR}^{-+} are considerably more sensitive to right handed currents than the cross sections for unpolarised beams.

Fig. 27

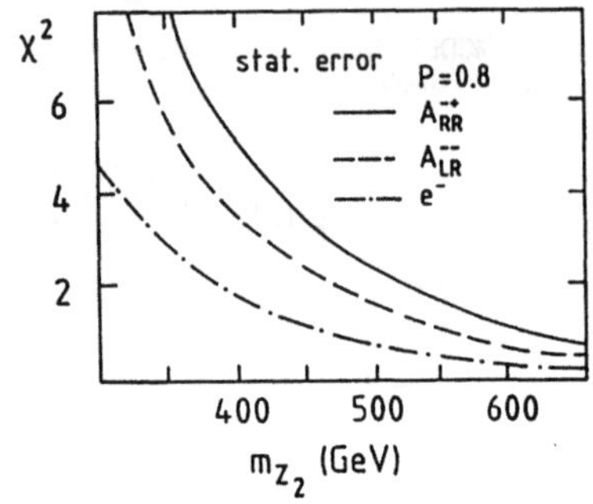

Fig. 28

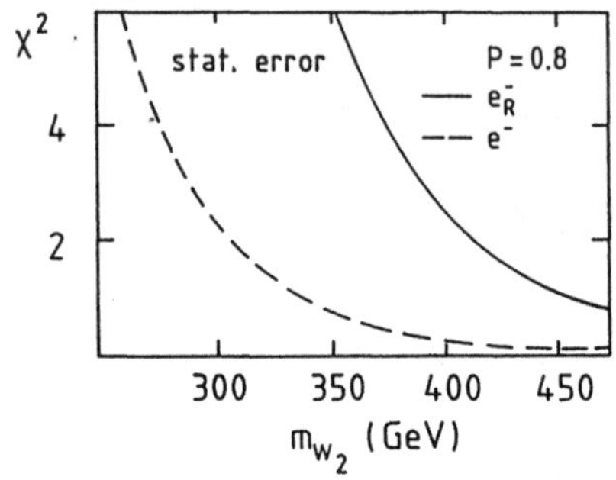

Fig. 27. The quantity $\chi^2(m)$ defined in the text for the polarisation asymmetries A_{RR}^{-+} and A_{LR}^{--} assuming a polarisation of 80% and for an unpolarised electron beam. The statistical errors correspond to $\int L\,dt = 250\,\mathrm{pb}^{-1}$ in total (from [23])

Fig. 28. The $\chi^2(m)$ function defined in the text for the (e_R^- p) charged current process assuming 80% and 0% polarisation. The statistical errors are based on $\int L\,dt = 250\,\mathrm{pb}^{-1}$ in total (from [23])

The W' could be detected via the process

$$e_R^- + p \rightarrow \nu_R + X$$

if the mass of the right handed neutrino is sufficiently low. If ν_R were a Majorana particle sufficiently heavy it would decay with equal probability to

$$\begin{aligned} \nu_R &\rightarrow e^- + X \\ &\rightarrow e^+ + X \end{aligned}$$

leading to clear experimental signatures. If it were a Dirac particle it would escape detection. Fig. 28 shows the $\chi^2(m)$ curves for the charged current cross section e_R^- p with $P = 80\%$ and the less sensitive one for unpolarized beams.

Though difficult, HERA will be able to look with reasonable sensitivity for new W' and Z' in the mass range m(W'), m(Z') $\lesssim 500$ GeV. A positive result would be a spectacular surprise.

3.2 Particle Production

The possibilities at HERA to study heavy quarks and search for new particles are shortly sketched. We start with the latter but discuss the top search under heavy particle production in Sect. 3.2.2.

3.2.1 New Particles

a) Fundamental Scalars – the mystery of the S.M. The dominant diagrams for the standard Higgs particle H^0 production at HERA are sketched in Fig. 29 leading to the rates given in Table 2.

The cross sections are rather small and considering the low detection efficiencies for an H^0 decay, HERA is not likely to contribute to its eventual discovery. Also for the discovery of non standard $H^\pm$ particles e^+e^--annihilation seems to be a considerably better process.

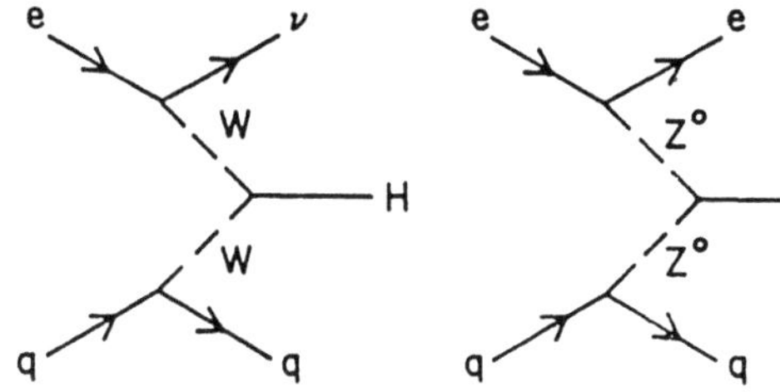

Fig. 29. Diagrams for Higgs particle production at HERA

Table 2. The number of standard Higgs particles produced with $\int L\, dt = 1000\,\mathrm{pb}^{-1}$ (from [24])

m_{H}[GeV]	10	20	30	40	50	60
# of H^0	25	19	15	10	8	6

b) Supersymmetric Particles. Supersymmetry is one of the main line of thoughts beyond the S.M.. As discussed in the lecture of Bartl [25] it predicts in its standard form the existence of spin 0 partners to leptons (sleptons $\tilde{\ell}_{\mathrm{L,R}}$) and quarks (squarks $\tilde{\mathrm{q}}_{\mathrm{L,R}}$), of spin 1/2 partners of the gluons (gluinos $\tilde{\mathrm{g}}$) the electroweak gauge bosons (gauginos $\tilde{\gamma}, \tilde{\mathrm{Z}}, \tilde{\mathrm{W}}^{\pm}$) and to the H^0 (higgsinos $\tilde{\mathrm{H}}_1^0, \tilde{\mathrm{H}}_1^-, \tilde{\mathrm{H}}_2^+, \tilde{\mathrm{H}}_2^0$).

In most models SUSY particles must be produced in association. The most promising production process at HERA is $\mathrm{e\,p} \rightarrow \tilde{\mathrm{e}}\,\tilde{\mathrm{q}}\,\mathrm{X}$ according to the diagram in Fig. 30a. The mass eigenstates of the exchanged neutralinos might be a mixture of the gauginos and the higgsinos. This is also the case for the chargino exchange in Fig. 30b, which, however, is more difficult to detect experimentally if the $\tilde{\gamma}$ is the lightest supersymmetric particle (LSP).

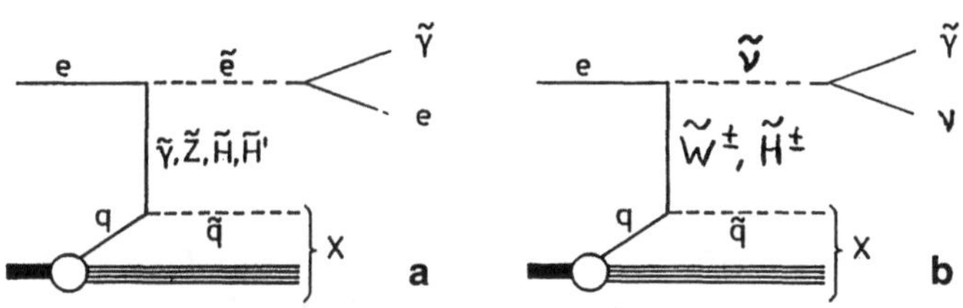

Fig. 30. Squark and slepton production by (**a**) neutralino and (**b**) chargino exchange

The actual cross section varies with the neutralino mass mixing; apart from that it depends essentially only on the sum of masses $m_{\tilde{\mathrm{e}}} + m_{\tilde{\mathrm{q}}}$. The production cross section is given in Fig. 31 for some representative mixing scenarios.

Event simulations show that one can detect $\tilde{\mathrm{e}}\,\tilde{\mathrm{q}}$ production, assuming the photino in the LSP and escapes detection, on an event to event basis with an efficiency of about 50%, provided the photino is not too heavy. In this way squark and slepton masses up to 150 GeV might be detectable at HERA. Table 3 gives some present lower mass limits of sparticles for comparison.

c) Leptoquarks. In models of quark-lepton substructure like technicolor, but also in models of grand unification and in some superstring theories leptoquarks, usually

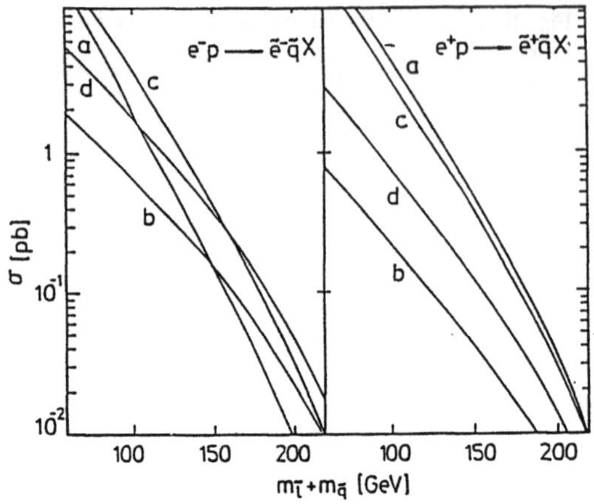

Fig. 31. Selectron-squark production cross sections for some representative neutralino mixing scenarios. The mass of the lightest neutralino varies between 0 for curve a and 100 GeV for curve d (from [42], see also [25])

Table 3. Lower mass limits for sparticles

sparticle	lower mass limit
squark	60 GeV
gluino	60 GeV
selectron	65 GeV if $m_{\tilde{\gamma}} \simeq 0$
Wino	23 GeV
Zino	23 GeV

color triplett bosons of fractional charge carrying both baryon and lepton quantum numbers, quite naturally occur. Table 4, taken from [26] gives a list of scalar and vector leptoquarks. We will consider only the S_1 particle as an example and will omit in the following the index 1.

The s-channel production amplitudes of the S are sketched in Fig. 32. The S production cross section for incident $e^{\pm}$ on protons is given in Fig. 33 for quark lepton couplings $\lambda_R = 0$ and $\lambda_L = 0.3, 0.1$ as a function of m_s. The cross sections are remarkably large. The S is relatively narrow

$$\Gamma_S = \frac{1}{16\pi}\left(2\lambda_L^2 + \lambda_R^2\right) m_s \simeq 0.7\text{GeV}\frac{2\lambda_L^2 + \lambda_R^2}{0.18} \cdot \frac{m_s}{200\text{GeV}}$$

and its Beit-Wigner amplitude adds coherently to the amplitudes of the S.M.

$$|As|^2_{L,R} = \frac{1}{2}\frac{\lambda_{L,R}^2\left(\lambda_L^2 + \lambda_R^2\right)\hat{s}^2}{\left(\hat{s} - m_s^2\right)^2 + m_s^2\Gamma_s^2}$$

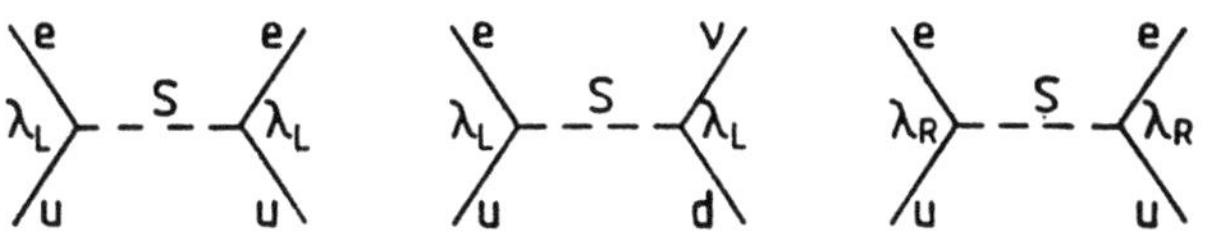

Fig. 32. Lepton quark fusion diagrams for S production

Table 4. Quantum numbers of scalar and vector leptoquarks with $SU(3)_C \times SU(2)_L \times U(1)_Y$ invariant couplings to quark-lepton pairs ($Q_{em} = T_3 + Y$) (from [26])

	Spin	F=3B+L	$SU(3)_C$	$SU(2)_L$	$U(1)_Y$
S_1	0	2	3	1	-1/3
$\tilde{S}_1$	0	2	3	1	-4/3
S_3	0	2	3	3	-1/3
V_2	1	2	3	2	-5/6
$\tilde{V}_2$	1	2	3	2	1/6
R_2	0	0	3*	2	-7/6
$\tilde{R}_2$	0	0	3*	2	-1/6
U_1	1	0	3*	1	-2/3
$\tilde{U}_1$	1	0	3*	1	-5/3
U_3	1	0	3*	3	-2/3

where $\hat{s} = x\,s$. This leads to a spectacular signal in the x-distribution as shown in Fig. 34. The regions of λ and m_s in which a signal would be seen at HERA are shown in Fig. 35.

The existing limits on leptoquarks are:

1. $m_s \gtrsim 10^{11}$ TeV for baryon and lepton number violating couplings from the proton lifetime.
2. for baryon and leptonnumber conserving but flavour violating couplings from

$$\mu\mathrm{N} \to \mathrm{eN} : m_s \gtrsim |\lambda_\mathrm{L}| \cdot 150\mathrm{TeV}$$
$$\mathrm{K}^+ \to \pi^+\nu\nu : m_s \gtrsim |\lambda_\mathrm{L}| \cdot 7\mathrm{TeV}$$

3. for flavour conserving couplings from

$$Br\left(\pi^+ \to \mathrm{e}^+\nu_\mathrm{e}\right) = 1.210^{-4} : m_s \gtrsim \sqrt{|\lambda_\mathrm{R} \cdot \lambda_\mathrm{L}|} \cdot 8\mathrm{TeV}$$

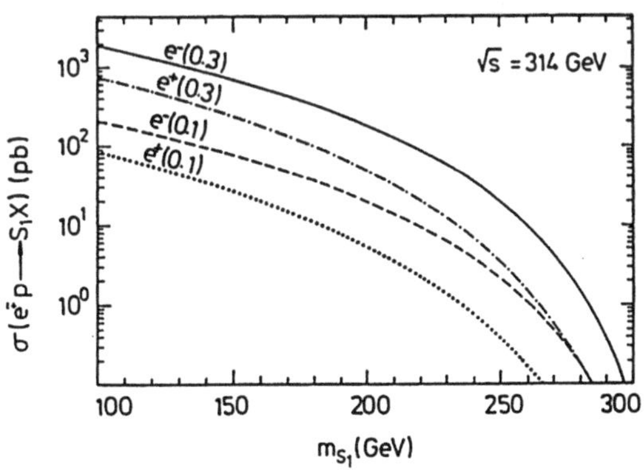

Fig. 33. Total cross section for S_1 leptoquark production in e^-p and e^+p collisions for $\lambda_R = 0$, $\lambda_L = 0.3$ and 0.1 (from [26])

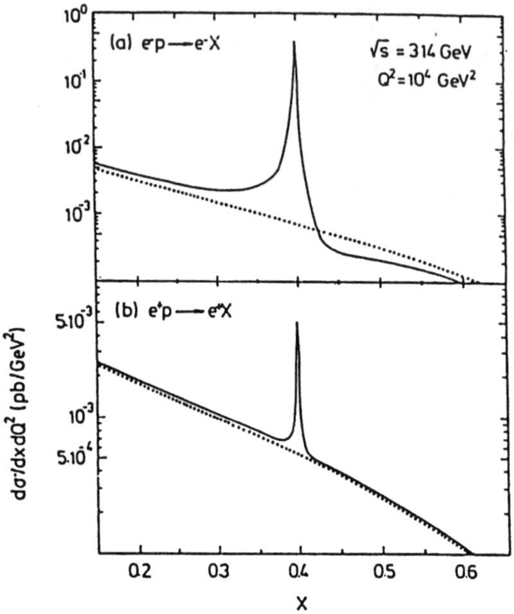

Fig. 34. Differential cross section for S_1 leptoquark production at $\sqrt{s}$ = 314 GeV and Q^2 = 10^4 GeV with m_s = 200 GeV and $\lambda_L = 0.3$, $\lambda_R = 0$ (from [26])

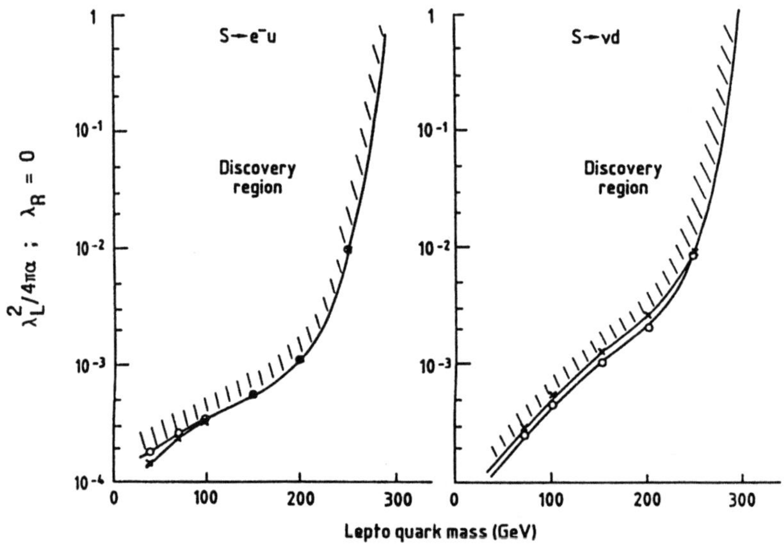

Fig. 35. Regions of λ and m_s in which HERA should be sensitive to possible leptoquarks if $\int L\, dt$ = 200pb^{-1} (from [27])

4. from the universality of β and μ decay $m_s \gtrsim \lambda_L \cdot 1.7$TeV
5. direct limits are:

$$m_s \geq 20.5\text{GeV from } e^+e^-$$
$$m_s \geq 33\,\text{GeV from } p\bar{p}$$

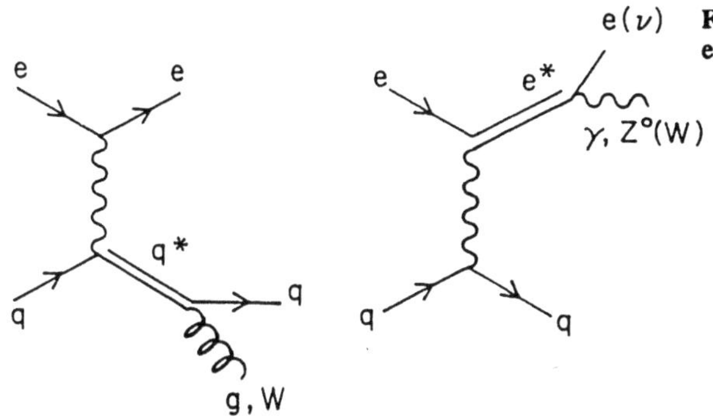

Fig. 36. Diagrams for the production of excited quarks and electrons

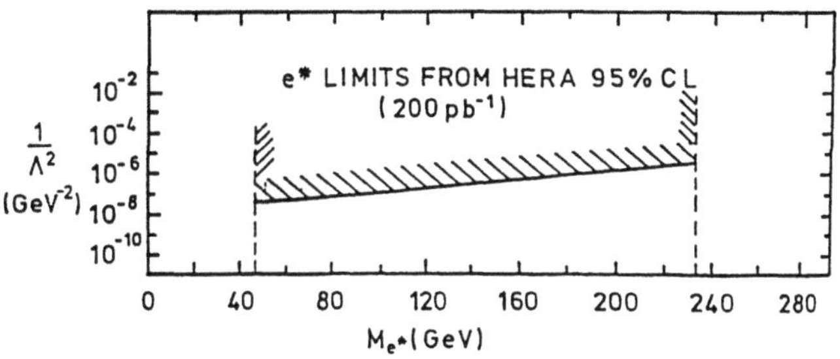

Fig. 37. Limits on m_{e*} and Λ obtainable with HERA at $\sqrt{s} = 314$ GeV (from [28])

Comparison with Fig. 35 shows that some room is left at HERA for spectacular discoveries.

d) Excited Quarks and Leptons. Excited quarks and leptons could be produced at HERA according to the diagrams in Fig. 36. Especially, sensitive searches for excited electrons e* can be performed. Assuming a magnetic coupling at the e e* γ - vertex of roughly speaking e/Λ HERA would be sensitive values of m_{e*} and Λ as sketched in Fig. 37. It is important to note that the elastic cross section e + p $\rightarrow$ e* + p is of similar size as the inelastic one e + p $\rightarrow$ e* + X and that background rejection is much easier for elastic production.

3.2.2 Heavy Quark Production

The dominant process for heavy quark production at HERA is photon-gluon fusion, Fig. 38b. The lowest order diagram Fig. 38a is not expected to contribute significantly since it involves a $\bar{d} \rightarrow \bar{c}, \bar{t}$ or u $\rightarrow$ b transition, which is strongly hindered. Fig. 39 shows the resulting cross section. The cross section for c- and b-production is large. Of order 10^7 c-quarks and 10^6 b-quarks ($\sim 10^{10}$ light quarks) are expected to be produced with $\int L\, dt = 100$ pb^{-1}, mostly at $x \lesssim 10^{-3}$, and it seems feasable to identify a fair fraction of these by their leptonic decays or eventually a measurement of their decay length. To separate b- from c-quarks dilepton decays will have to be used [29]. A measurement of heavy quark production, apart from studies of their decays, will test the model calculations and thus yield important information about the gluon distribution in the proton at low x. However, the most important issue of heavy quark physics is the search for the t-quark.

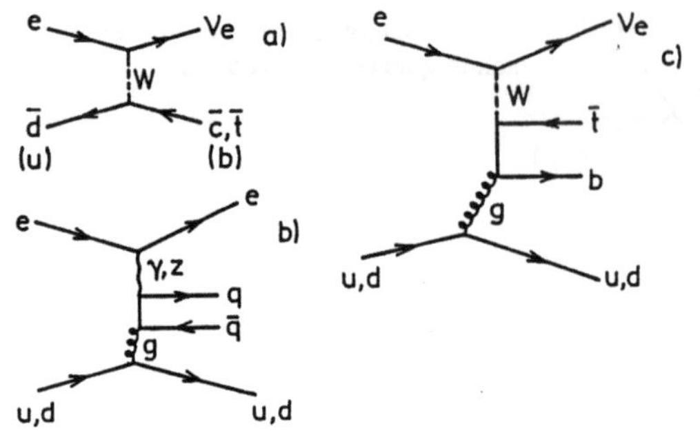

Fig. 38. Diagrams for heavy quark production

Top Search. For quark masses above 50 GeV the W-gluon fusion process, Fig. 38c, leading to a $\bar{t}b$ final state becomes important. The cross section is shown in Fig. 40 together with the other contributions. Next to leading order calculations indicate that the cross section might be higher by about 50% [32]. For m_t = 70 GeV and an integrated luminosity of 100 pb^{-1} about 10 t-quarks are produced. Will one be able to detect these events within a "background" 10^7 times more intense?

This question has recently been studied in [29] and especially in [33]. Both investigations are quite detailed in the event and background generation. In [29] emission of an additional gluon in electron-scattering of the proton constituents and in the diagrams of Fig. 38 is taken into account, but fragmentation effects are treated only globally. In [33] higher order QCD effects are approximated by using parton shower fragmentation schemes. Both investigations conclude that the background becomes negligible by requiring an isolated lepton, of large p_t, a large hadronic transverse energy ΣE_t, and a large transverse missing momentum p_t^{mis}. In Fig. 41

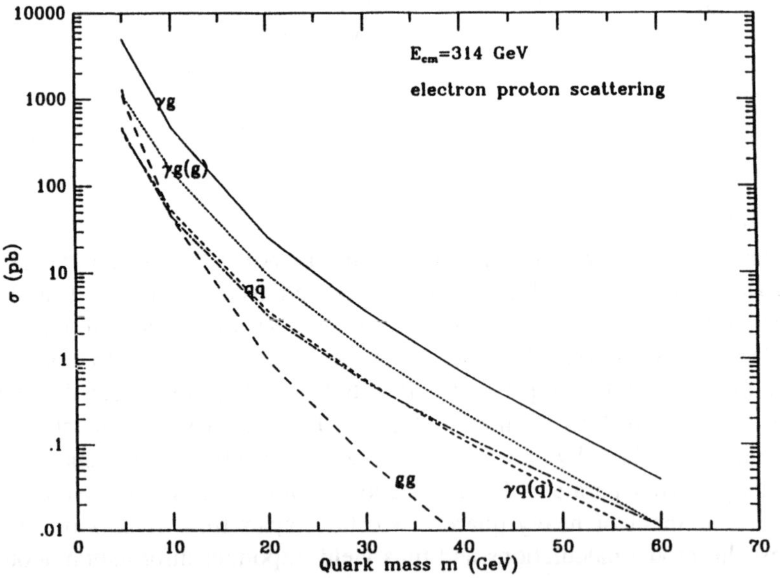

Fig. 39. Heavy quark production cross sections by the various electron and proton constituent collisions (from [30])

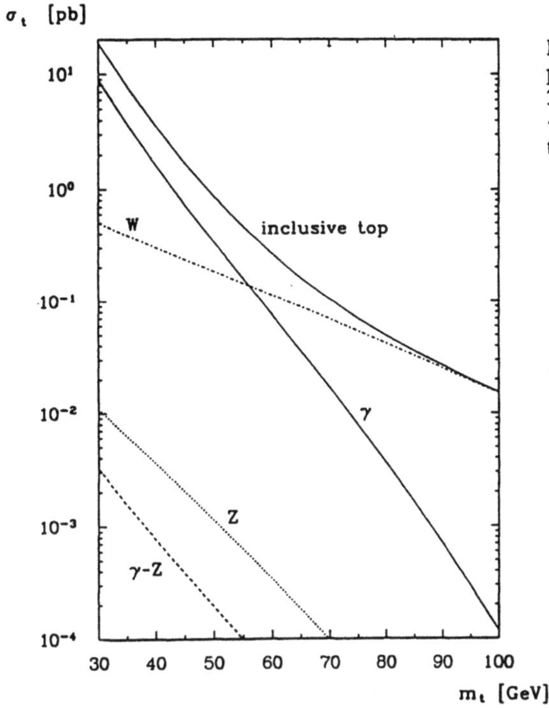

Fig. 40. Leading order cross section for t-production via boson gluon fusion at $\sqrt{s} = 314$ GeV. The contributions from W, Z and γ as well as the γ, Z interference contribution are separately shown (from [31])

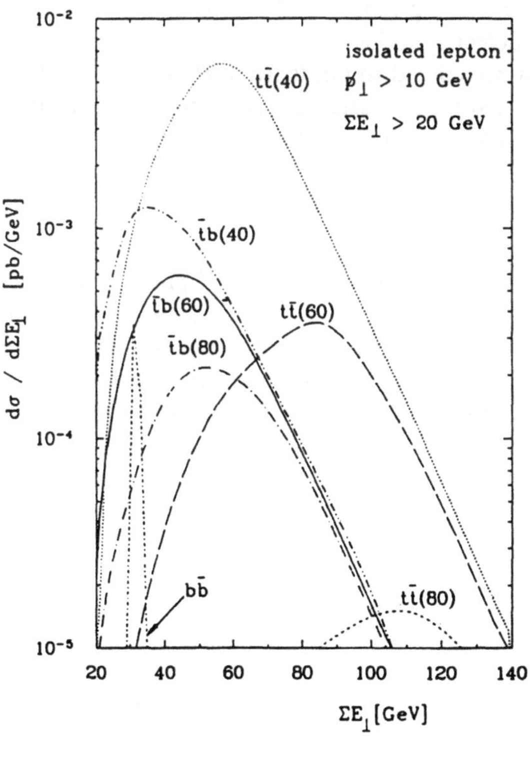

Fig. 41. Differential cross section $d\sigma/d\Sigma E_t$ in the total transverse energy for background and the top signal for three different top masses at $\sqrt{s} = 314$ GeV. The cuts applied are given in the figure (from [33])

Fig. 42

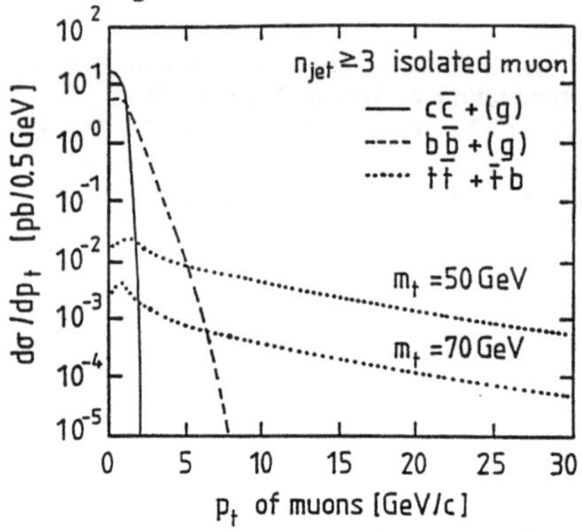

Fig. 43

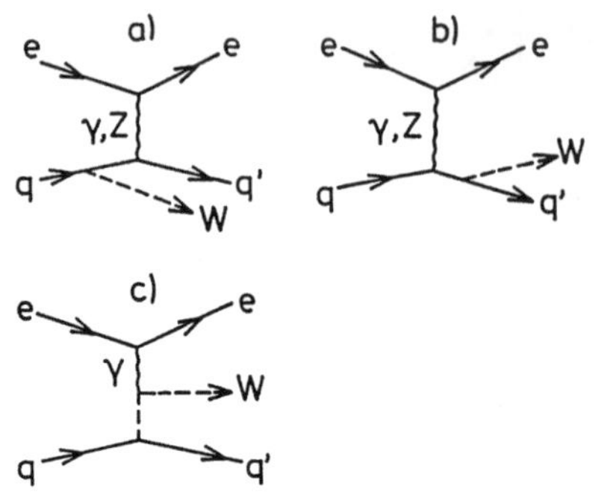

Fig. 42. Transverse energy distribution of single muon events for the background and t-production for two different t-quark masses at $\sqrt{s}$ = 314 GeV [29]

Fig. 43. The main diagrams for $W^{\pm}$ production in ep-collisions

the t-quark cross section and the remaining background is shown as function of ΣE_t, if apart from an isolated lepton a $p_t^{mis} > 10$ GeV is required. The only remaining background is from $b\bar{b}$ production and is negligible for $\Sigma E_t > 40$ GeV. The cut in p_t^{mis} is certainly problematic because of the detector resolution. But also without such a cut, the background can be separated as is demonstrated in Fig. 42. Here the cross section for the production of an isolated muon is shown as a function of the p_t of the muon for events with 3 or more jets. With such selection criteria the detection efficiency for t-production, including branching ratios, is about 10%.

One should be able at HERA to detect t-production if $m_t \lesssim 80$ GeV. A background of isolated high p_t leptons, not considered so far, is from W-production. This process is probably easy to discriminate since no quarks are produced in association but is interesting in itself and will be discussed in the following paragraph.

3.3 W-Production

At HERA W-production is expected to mainly proceed via low Q^2 photon exchange according to the diagrams in Fig. 43. Of special interest is the diagram c since it allows to probe the Wγ-vertex which is very sensitive to the non-Abelian gauge structure of the S.M. [34]. This vertex is defined by the magnetic dipole moment μ_W and the electric quadrupole moment Q_W of the W, which can be described by two free parameters κ and λ [35]:

$$\mu_W = \frac{e}{2M_W}(1 + \kappa + \lambda), Q_W = \frac{e}{M_W^2}(\lambda - \kappa)$$

In the S.M. at tree level $\kappa = 1, \lambda = 0$.

A good discrimination variable for diagram Fig. 43c is the transverse momentum of the jet initiated by the quark q'. Fig. 44 shows the transverse momentum distribution of q' for several values of κ and λ. The cross section for $p_t > 5$ GeV, though rather low, is strongly sensitive to couplings $\kappa \neq 1$. The detection of the W will

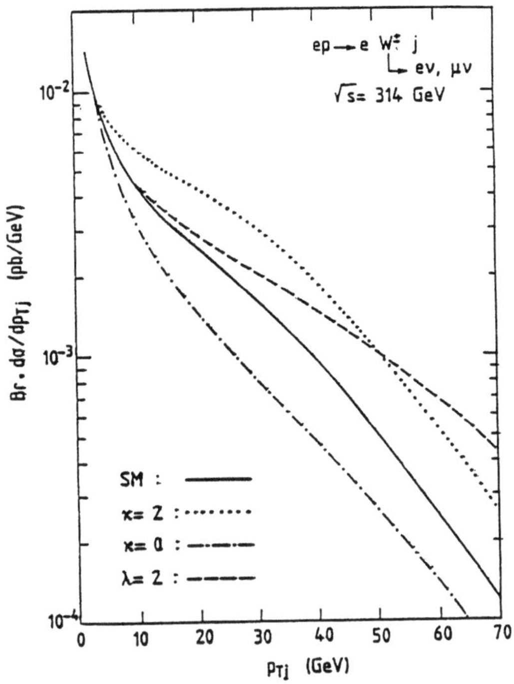

Fig. 44. The branching ratio for leptonic W decay multiplied with $d\sigma/dp_t$ of quark q' (from [36]

be possible by its leptonic decay. Estimates [36] show that at $\sqrt{s}$ = 314 GeV with $\int L\,dt = 200\,\mathrm{pb}^{-1}$ the quantity $\kappa - 1$ could be determined with a precision of about ± 0.5.

Closing this review of some selected topics of possible HERA physics it may be allowed to remind the reader that the measurement (1933) of the anomalous magnetic moment of the proton was the first evidence for the substructure of hadrons, more than 15 years before any more direct observations thereof were made. Evidence for an anomalous Wγ-coupling may perhaps be a first indication of the physics beyond the S.M..

4. Detection Facilities

4.1 General Requirements

In e-p collisions the hadronic c.m. energy and the momentum transfer of the virtual photon have to be determined from the final state particles. At an e^+e^-collider these quantities are determined by the initial state, i.e. the machine settings, and the requirements for a useful detector are less severe than at an e-p collider.

The range of the final state lepton momenta and of the struck quark, assuming parton model kinematics, are shown in Fig. 45, where also the lines of constant x and Q^2 are indicated. For NC processes Q^2 can be well determined by measuring the

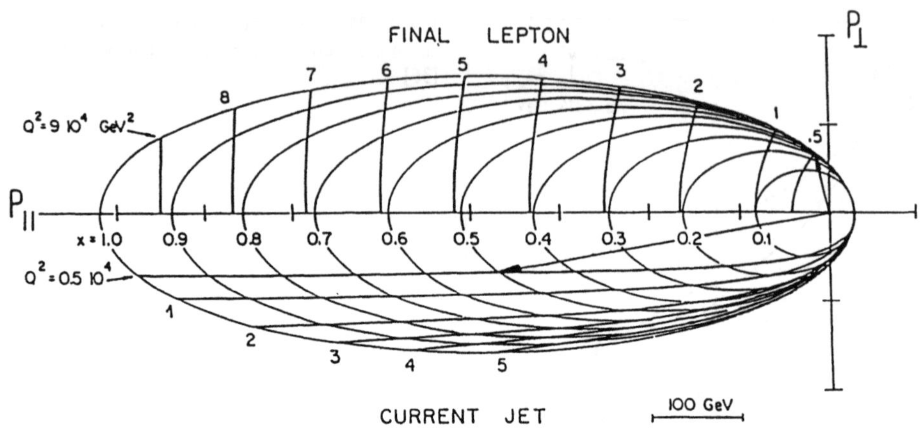

Fig. 45. Momentum distribution of the final state lepton (*upper half*) and, assuming parton model kinematics, of the struck quark. Lines of constant x and Q^2 are indicated. Connecting a (x, Q^2) point with the origin yields the laboratory momenta as shown by the example $x = 0.5, Q^2 = 5000$ GeV^2

energy of the scattered electron. Only at lower Q^2 the angular resolution becomes important,

$$\frac{\Delta Q^2}{Q^2} = \frac{\Delta E_l}{E_l} \oplus \cot\frac{\theta_l}{2}\Delta\theta_l,$$

where $\oplus$ stands for the quadratic sum. Measuring the scattered electron, the resolution in x becomes worse with increasing x and decreasing Q^2

$$\frac{\Delta x}{x} = \frac{1}{y}\frac{\Delta E_l}{E_l} \oplus \cot\frac{\theta_l}{2}\left(x\frac{E_\mathrm{p}}{E_\mathrm{e}} - 1\right)\Delta\theta_l.$$

The determination of the event kinematics is considerably more difficult for CC reactions. Reconstructing the current jet from the hadrons emitted under reasonable large angles and using the lower part of Fig. 45 to determine x and Q^2 is not sufficient because of multijet final states due to hard gluon emission and because of our incomplete knowledge of quark and gluon fragmentation. The expected energy flow of the final state hadrons for events at $x = 0.2$ and $y = 0.5$ GeV^2 is shown in Fig. 46. Indicated are the directions of the scattered lepton, the struck quark and the remnant diquark. The average energy flow is plotted for two different fragmentation models. The parton shower fragmentation model, which is based on QCD parton evolution in leading order, shows no clean separation between the quark and diquark region, whereas two sharply collimated jets are produced by the simple quark parton model fragmentation, which is characterized by fragments of limited transverse momentum along the direction of the primary partons.

An actual measurement of the hadronic energy flow would yield interesting insights to parton fragmentation, the basic parameters x and Q^2, however, one would like to determine in a manner independent of the fragmentation mechanism.

A scheme has been proposed by Jacquet and Blondel [37] which allows the determination of x and Q^2 for CC-processes without any assumption about the

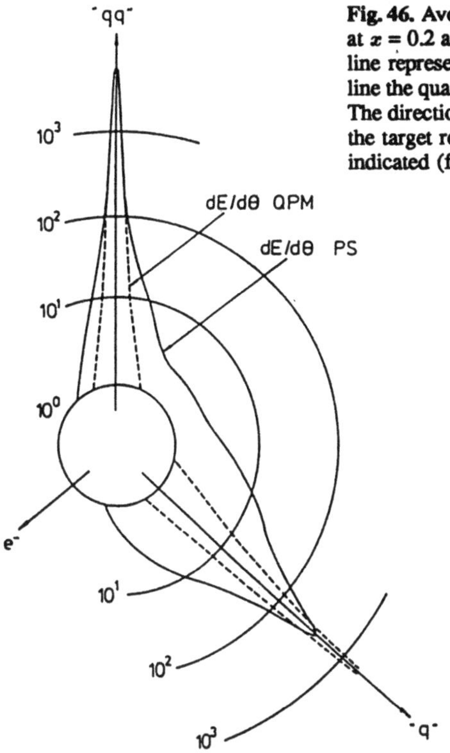

Fig. 46. Average energy flow in the event plane $dE/d\theta$ for events at $x = 0.2$ and $y = 0.5$ for the nominal HERA kinematics. The full line represents the parton shower model expectation , the dashed line the quark parton model result with independent fragmentation. The directions of the scattered lepton, the scattered parton 'q', and the target remnant 'qq' given by the parton model kinematics are indicated (from [38])

partonic structure of the final state. The scheme is quite insensitive to final state particles escaping detection by not leaving the beam pipe. Determining q from the hadronic vertex, one obtains

$$P \cdot q = P \cdot (P_X - P) = P \cdot P_X - m_p^2$$

where P_X is the total momentum of the final state hadrons $P_X = \sum_h p_h$, and if one neglects masses one gets, with $||$ being the momentum component in beam direction and $E_h^{||} = E_h p_h^{||} / |p_h|$,

$$P \cdot q = E_p E_X - |\boldsymbol{P}| \, P_X^{||} \simeq E_p \sum_h \left(E_h - E_h^{||} \right) .$$

It was noticed in [37] that the high momentum fragments of the diquark emitted in the proton beam direction and escaping detection through the beam pipe essentially do not contribute to the above sum since for these particles $E_h \simeq E_h^{||}$. A measurement of the hadronic energy flow thus determines, to good approximation in case of HERA,

$$y = \frac{2P \cdot q}{4 E_p E_e} \simeq \frac{\sum_h \left(E_h - E_h^{||} \right)}{2 E_e}$$

as well as

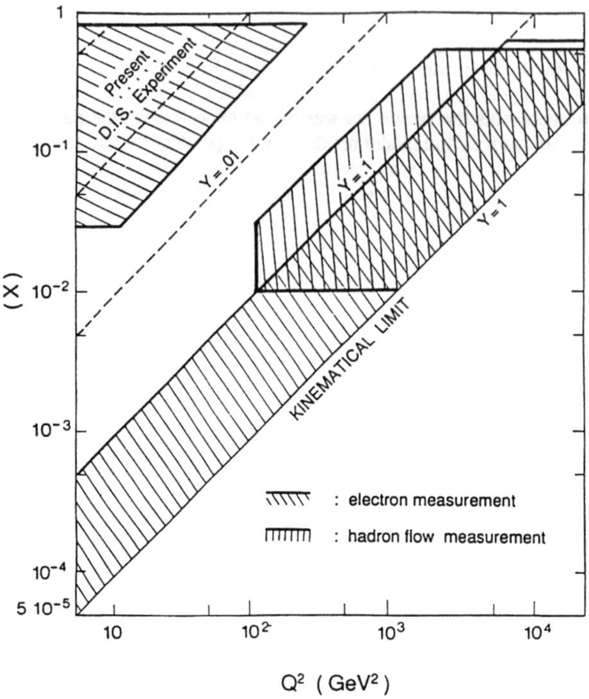

Fig. 47. Regions of the $x - Q^2$ plane where the systematic errors are below 10% for the detection of the scattered electron and for the measurement of hadronic energy flow with resolutions given in text. The region of the present fixed target experiments is also indicated (from [39])

$$Q^2 = \frac{(E_L \sin\theta_L)^2}{1-y} \simeq \frac{\left(\sum_{\mathrm{h}} E_{\mathrm{h}}^{\perp}\right)^2}{1-y}$$

and since $Q^2 = x \cdot y \cdot s$ also x.

To study deep inelastic NC and CC scattering at HERA a reasonable measurement of the scattered electron momentum is obviously needed as well as a good measurement of hadronic energy flow. It is also important that the detector resolution functions are well known and that they show no long tails, since otherwise the smearing corrections would be large and rather uncertain, as the cross section varies strongly over the available phase space.

Assuming a gaussian energy resolution of $\Delta E_{\mathrm{e}}/E_{\mathrm{e}} = 0.13/\sqrt{E_{\mathrm{e}}} \oplus 0.01$ for the electron and $\Delta E_{\mathrm{h}}/E_{\mathrm{h}} = 0.50/\sqrt{E_{\mathrm{h}}} \oplus 0.02$ for the hadronic energy flow, a detailed study [39] taking detector acceptance and resolution as well as cross section variations into account, shows that the cross section can be determined with systematic uncertainties of $\leq 10\%$ in the regions of x and Q^2 given in Fig. 47. These assumed resolutions will probably be achieved by the detector systems presently under construction. In order to provide an overlap with the range of the present deep inelastic scattering data, also indicated in Fig. 47, one has to run HERA at lower energies.

At HERA energies the hadronic energy flow is best measured by calorimetric means. For economical reasons one has to use sampling calorimeters in which absorber and detector material are interleaved. Normally these calorimeters have a different response to electrons or photons and to hadrons. In a nuclear cascade charged particles as well as π^0 are produced at random and the different response of a calorimeter to the electromagnetic and hadronic component usually limits the resolution achievable at high energies. To overcome this limitation one can either carefully choose the absorber and the detector material to yield the same response to e and π – as has been done for the ZEUS detector at HERA – or try to measure the electromagnetic component by the higher charge density of an electromagnetic shower and try to correct for it on an event to event basis by special weighting techniqes. The latter technique is persued by the H1-collaboration.

Apart from a good electromagnetic and hadronic energy flow measurement covering nearly the full solid angle – which is also important for the possible detection of new noninteracting particles like photinos via the missing momentum – the identification of charged leptons over the full momentum range is an essential requirement for a HERA detector. Charged leptons of high transverse momentum are often a good signature of new physics. Their charge determination requires a high magnetic field and a high precision track detector.

4.2 The H1-Detector

The H1-detector is shown in Figs. 48. Its main elements are: 1) A central tracking detector located in a solenoidal magnetic field of 1.2 T parallel to the beam direction, providing a momentum resolution of $\sigma(p)/p^2 \lesssim 0.003$ in the angular range $7° < \theta < 150°$ for isolated tracks and good pattern recognition. 2) An electromagnetic and hadronic calorimeter surrounding the tracking detector and covering the central and forward (in proton direction) angular range. The calorimetry is based on the liquid argon technique providing a stable, homogeneous and hermetic detector with good granularity for energy flow measurements and electron identification. The inner electromagnetic part extents of 20 to 30 radiation lengths and is made of lead absorber plates of 2.4 mm thickness leading to an energy resolution of $\sigma(E_e)/E_e \simeq 10\%/\sqrt{E_e} \oplus 1\%$. The hadron calorimeter is then continued with 19 mm thick stainless steel plates. The thickness of the total liquid argon calorimeter varies between 4.7 absorption length at 90° and 8 absorption length. The calorimeter is read out by about 45000 electronic channels and has a longitudinal segmentation varying between 5 and 10 sections. After weighting the hadronic energy resolution is $\sigma(E_h)/E_h \lesssim 55\%/\sqrt{E_h} \oplus 2\%$. 3) A large superconducting coil surrounding the liquid argon calorimeter at a radius of 3 m. 4) An iron yoke, which is built as the rear section of the hadronic calorimeter and acts as muon filter and as tail catcher for leaking hadronic showers. It has 10 iron plates, 7.5 cm thick, interspersed with layers of limited streamer tubes. 5) A compact copper silicon counter sandwich closes the calorimeter in the forward direction down to a polar angle of 0.7°. 6) A lead scintillator sandwich with photodiode readout records scattered electrons of low Q^2 events in the backward (electron beam) direction and will provide together with the tail catcher a rough measurement of the hadronic energy.

Muons are identified and measured in three layers of plastic streamer tube chambers inside, outside and in the middle of the yoke. In the forward direction the muon system is complemented by an iron toroid and layers of muon drift chambers. Electron identification is based on shower shape measurements in the electromagnetic

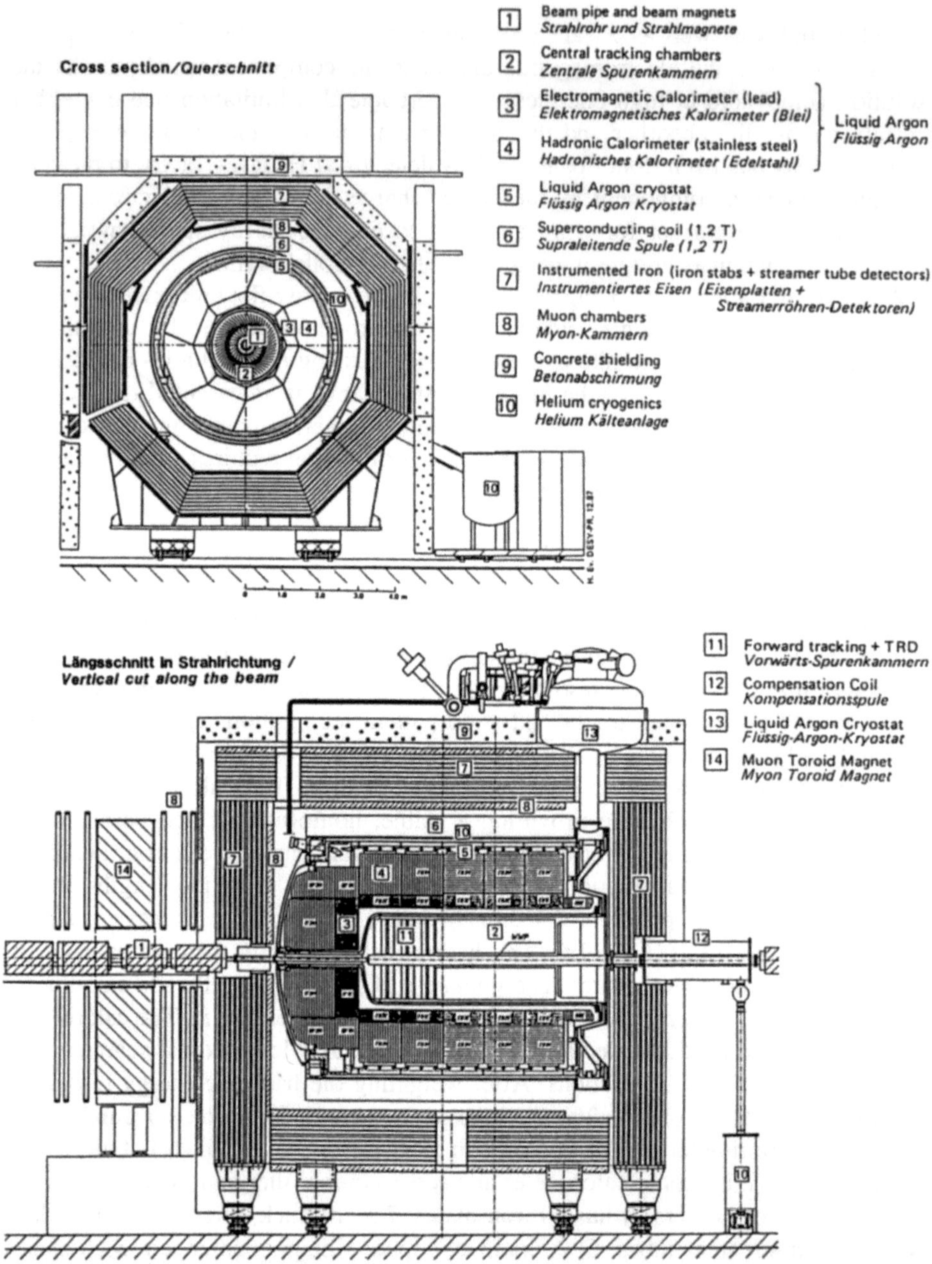

Fig. 48. Layout of the H1-detector. Vertical cuts transverse and along the beam direction are shown

calorimeter, on dE/dx measurements in the central drift chambers and on the use of transition radiation detectors in the forward direction. A pion rejection factor of $< 10^{-5}$ in the forward and 10^{-4} in the backward region is expected.

The liquid argon calorimeter has a different response to electrons and charged hadrons as shown in Fig. 49 for 50 GeV charged pions and electrons incident on a

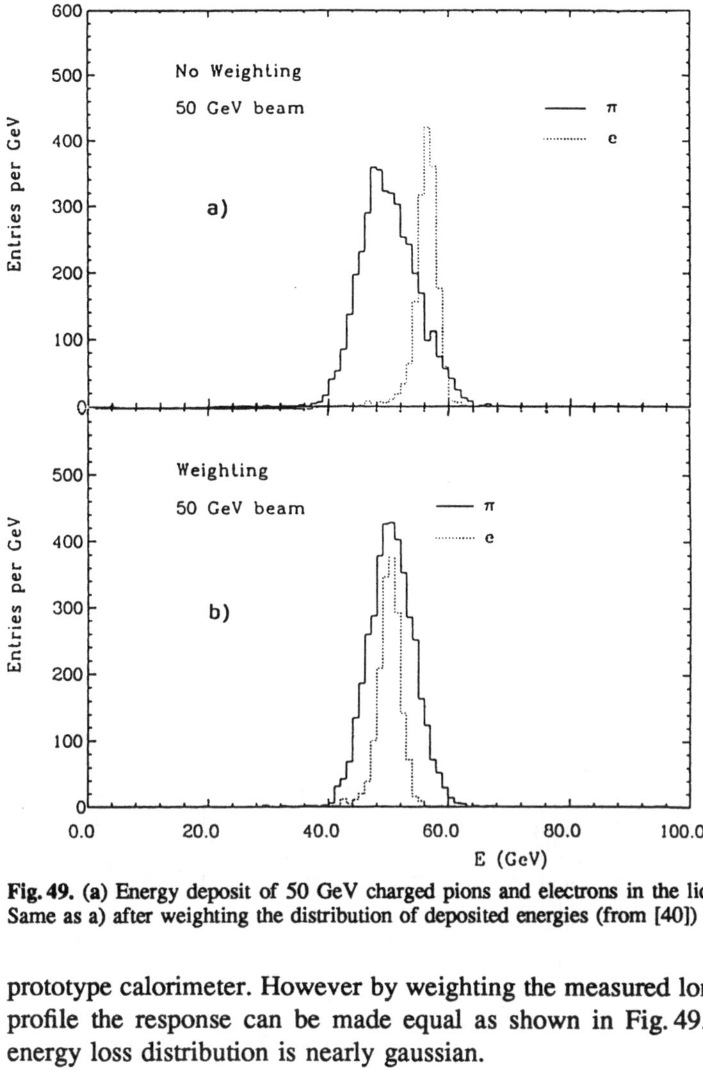

Fig. 49. (a) Energy deposit of 50 GeV charged pions and electrons in the liquid argon calorimeter. (b) Same as a) after weighting the distribution of deposited energies (from [40])

prototype calorimeter. However by weighting the measured longitudinal energy loss profile the response can be made equal as shown in Fig. 49. After weighting the energy loss distribution is nearly gaussian.

4.3 The ZEUS Detector

The ZEUS detector, shown in Fig. 50, is constructed to have the best possible hadronic energy resolution. A compensating calorimeter is used, i.e. calorimeter showing equal response to electrons and charged hadrons. Compensation is achieved by using 3.3 mm thick depleted uranium plates as an absorber and 2.6 mm thick scintillator plates as a detector. These parameters were chosen after an intensive test program [41]. The pulseheight spectra measured, show the equal response obtained for e and π with a prototype detector (Fig. 51).

The hadronic energy resolution is excellent $\sigma(E_h)/E_h = 35\%/\sqrt{E_h} \oplus 1\%$, whereas the resolution for electrons $\sigma(E_e)/E_e = 18\%/\sqrt{E_e} \oplus 1\%$ is only moderate, because of the relatively large sampling thickness. The scintillator plates are

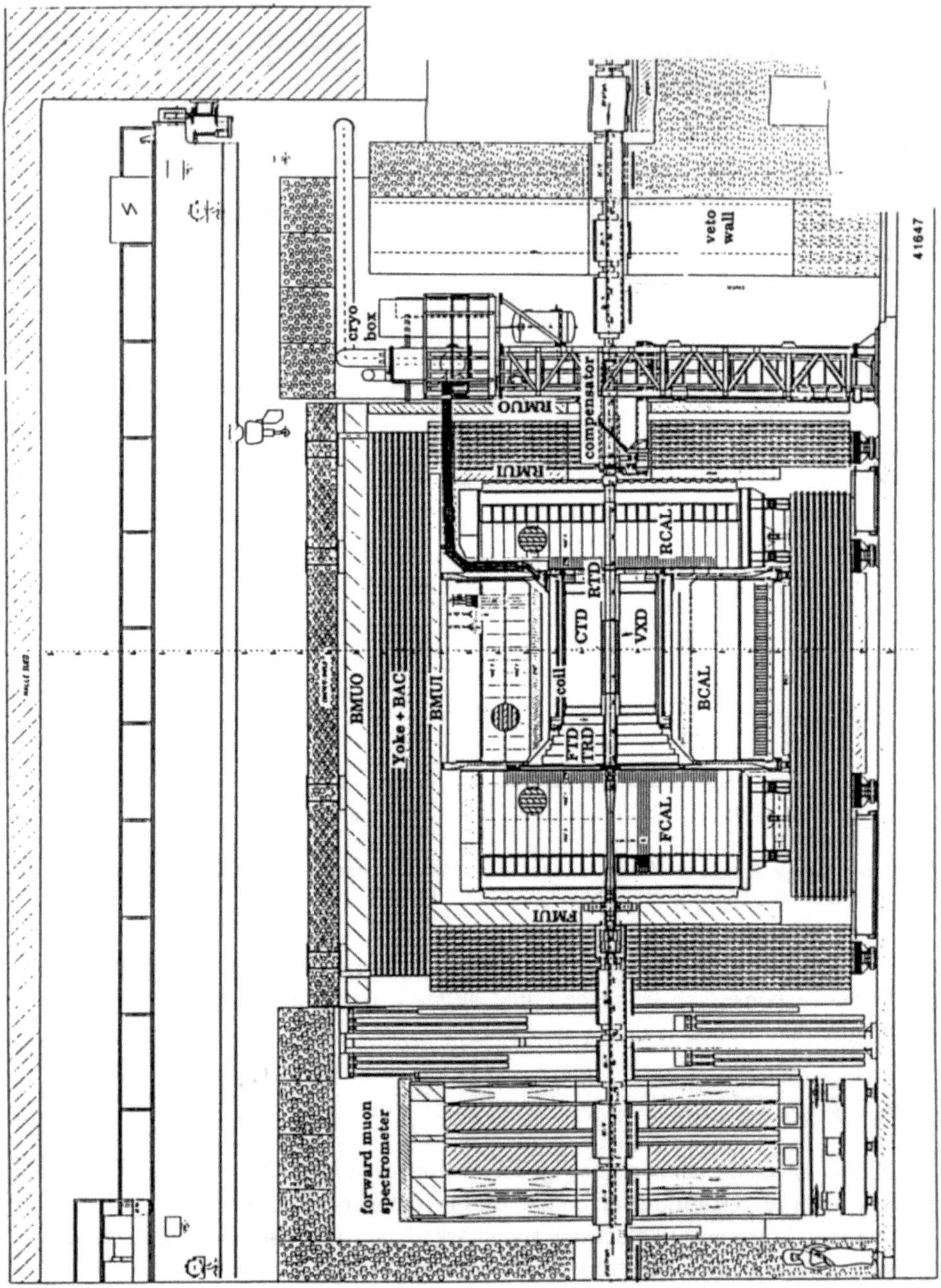

Fig. 50. Vertical cut of the ZEUS detector along the beam

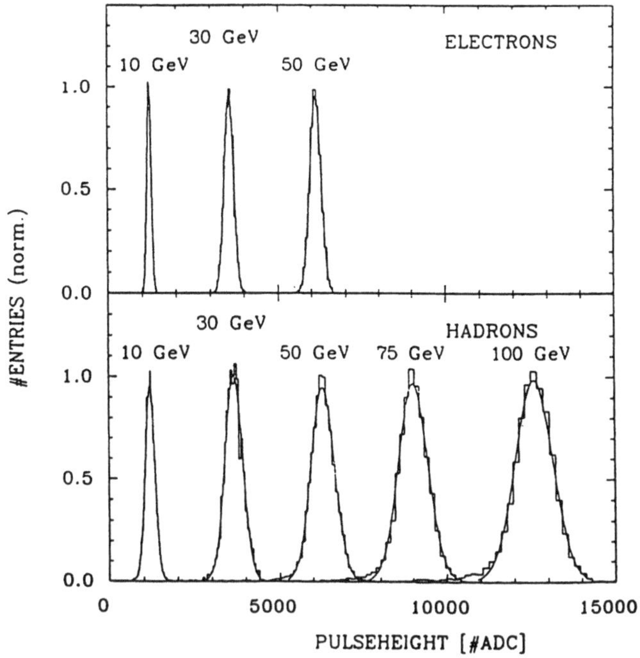

Fig. 51. Response of the uranium-scintillator test calorimeter for electrons and hadrons (from [41])

read out using wave lenght shifters, light guides and photo-multipliers. Longitudinally the calorimeter is segmented into an electromagnetic section containing the first 25 absorber plates followed by two hadronic sections. The system has a total of 13000 independent channels. The total absorption length varies from 7 interaction length in the forward direction to 4 in the backward. In order to improve the electron identification the ZEUS collaboration has proposed to insert two layers of silicon pad detectors at depths of 3 and 5 radiation lengths, improving the π rejection by about 2 orders of magnitude.

As photomultipliers are sensitive to magnetic fields, the calorimeter has to be installed outside of the superconducting coil, which is 0.9 radiation length thick and produces a field of 1.8 T.

The inner track detector is in principle similar to the one of H1. This is also true for the segmented iron structure serving as the flux return yoke, as a rough hadronic calorimeter to measure tails of the hadronic showers leaking out of the high resolution calorimeter and as a muon filter and spectrometer. In the forward direction the muon identification is performed by a toroidal spectrometer equipped with drift chambers.

ZEUS will be able to detect protons scattered in the very forward direction. The detectors reach down to 10σ of the proton beam size and consist of an elaborate set of Roman pots installed along the outgoing proton beam line between 20 m and 90 m downstream of the interaction point.

5. Status

The electron-proton collider HERA authorized 1984 is the first machine of this kind. At present, spring 1989, the civil engineering work has been completed and electrons have been stored for the first time in the HERA ring. The H^- linac and the 7.5 GeV proton synchrotron DESY III has been commissioned and the further acceleration of the protons in PETRA up to energies of 40 GeV will be tested in summer 89. Series production of the superconducting magnets is well under way and their installation in the HERA tunnel has commenced. Injection of protons into the first octant of the HERA proton ring is scheduled for the end of the year.

The optimization of the electron ring to store the multibunch design beam intensity of 60 mA will continue this summer. In the first test run only the single bunch design value of 0.3 mA was achieved.

The construction of the two general purpose detector facilities H1 and ZEUS is progressing well. The 2000 t iron yokes are assembled in the experimental halls, the superconducting coils are completed and tested, the construction of the tracking systems is well advanced, and mass production of the many calorimeter units is on its way after intensively testing prototype modules. In all cases the design resolution has been achieved by the prototypes.

Both detectors are expected to be ready for data taking by the end of 1990 when the first electrons and protons are scheduled to collide in HERA. These collisions will let us look deeper into the proton and, with a bit of luck, may allow us a glimpse into our microcosm beyond the Standard Model regime.

Acknowledgement. I would like to thank Professor Mitter and his colleagues for the invitation to Schladming to participate in this stimulating re- and preview of elementary particle physics in a beautiful surrounding and a friendly atmosphere. I am grateful to Mrs. S. Platz for editing the manuscript.

References

1 HERA proposal, DESY HERA 81/10
2 K-H. Meß, P. Schmüser: *Lectures on the CERN-Accelerator School, May 1988*, DESY HERA 89-11
3 A.A. Sokolov, I.M. Ternov: Sov. Phys. Doklady **8**, 1203 (1964)
4 J. Buon, K. Steffen: Nucl. Instr. Meth. **A245**, 248 (1986)
5 Z. Kunszt, W.J. Stirling: *Proceedings of the HERA Workshop October 87*, ed. R.D. Peccei, Vol.I, p. 331
6 A.C. Bawa, W.J. Stirling: *Proceedings of the HERA Workshop October 87*, ed. R.D. Peccei, Vol.I, p. 317
7 G. Altarelli, G. Parisi: Nucl. Phys. **B126**, 298 (1977)
8 F. Eisele: Rep. Prog. Phys. **49**, 233 (1986)
9 G. Ingelman et al.: *Proceedings of the HERA Workshop October 87*, ed. R.D. Peccei, Vol.I, p. 3
10 D.W. Duke, J.F. Owens: Phys. Rev. **D30**, 49 (1984)
11 J. Blümlein et al.: *Proceedings of the HERA Workshop October 87*, ed. R.D. Peccei, Vol.I, p. 67
12 E. Eichten et al.: Rev. Mod. Phys. **56**, 579 (1984); ibid. **58**, 1047 (1986)
13 M. Glück et al.: Z. Phys. **C13**, 119 (1982)
14 EMC-Collaboration – J.J. Aubert et al.: Nucl. Phys. **B272**, 158 (1986); ibid **B293**, 704 (1987)

15 BCDMS-Collaboration – A.C. Bonvenuti et al.: Phys. Lett. **195B**, 91; 97 (1987)
16 A.D. Martin et al.: *Proceedings of the HERA Workshop October 87*, ed. R.D. Peccei, Vol.I, p. 209
17 J. Blümlein et al.: *Proceedings of the HERA Workshop October 87*, Vol.II, p. 679
18 G. Cozzika et al.: *Proceedings of the HERA Workshop October 87*, Vol.II, p. 687
19 G. Ingelman, R. Rückl: *Proceedings of the HERA Workshop October 87*, Vol.I, p. 107
20 R.J. Cashmore et al.: Phys. Rep. **122**, 276 (1985)
21 H-M. Martyn: *Proceedings of the HERA Workshop October 87*, Vol.II, p. 801
22 For a review see R.N. Mohapatra: In *Quarks, Leptons and Beyond*, ed. by H. Fritzsch et al. (Plenum Press, New York 1985), p. 219
23 F. Cornet, R. Rückl: *Proceedings of the HERA Workshop October 87*, Vol.II, p. 771
24 K.J.F. Gaemers et al.: *Proceedings of the HERA Workshop October 87*, Vol.II, p. 739
25 For a review see the contribution of A. Bartl in these proceedings
26 W. Buchmüller et al.: Phys. Lett. **191B**, 442 (1987)
27 N. Harnew: *Proceedings of the HERA Workshop October 87*, Vol.II, p. 829
28 Ch. Berger et al.: *Proceedings of the HERA Workshop October 87*, Vol.II, p. 813
29 R.A. Eichler, Z. Kunszt: Nucl. Phys. **B308**, 799 (1988);
R.K. Ellis, Z. Kunszt: Nucl. Phys. **B303**, 653 (1988)
30 R.K. Ellis, Z. Kunszt: Nucl. Phys. **B303**, 653 (1988)
31 G.A. Schuler: Nucl. Phys. **B299**, 21 (1988)
32 For a review see the contribution of Z. Kunszt in these proceedings
33 G. Ingelman et al.: DESY 88/143; and *Proceedings of the HERA Workshop October 87*, Vol.I, p. 418
34 M. Böhm, A. Rosado: Z. Phys. **C39**, 275 (1988)
35 K.J.F. Gaemers, G.J. Gounaris: Z. Phys. **C1**, 259 (1979)
36 D. Atwood et al.: CERN-TH 5213/88; MAD/PH/444
37 A. Blondel, F. Jacquet: *Proceedings of the Study of an ep Facility for Europe*, ed. U. Amaldi (DESY 79/48 1979), p. 391
38 M. Bengtsson et al.: *Proceedings of the HERA Workshop October 87*, Vol.I, p. 149
39 J. Feltesse: *Proceedings of the HERA Workshop October 87*, Vol.I, p. 33
40 W. Braunschweig et al. – H1-Collaboration: Nucl. Instr. Meth. **A265**, 419 (1988)
41 R. Klanner: Nucl. Instr. Meth. **A265**, 200 (1988);
Zeus Calorimeter Group – G. d'Agostini et al.: NIKHEF - H/88-4
42 H. Komatsu, R. Rückl: Nucl. Phys. **B299**, 407 (1988)

Physics in p$\bar{\text{p}}$ Collisions

Felicitas Pauss

CERN, CH–1211 Genève, Switzerland

Abstract. A review of experimental results obtained by UA1 and UA2 before the upgrading of the Proton-Antiproton Collider at CERN is presented. Topics covered in these lectures are the physics of the Intermediate Vector Bosons, heavy-flavour production, B^0-$\bar{B}^0$ mixing, the search for the top quark, and searches for physics beyond the Standard Model. We give preliminary results from the 1988/89 Collider run at CERN [UA1, upgraded UA2 and improved CERN p$\bar{\text{p}}$ Collider (ACOL)]. The physics potential of the full upgrading is discussed as well. Also, where available, results from the Fermilab Tevatron Collider are quoted. And finally, from the results obtained we infer the possible features of physics at future hadron colliders (LHC and SSC).

1. Introduction

In a series of data-taking runs between 1982 and 1985, an integrated luminosity of close to 1 pb^{-1} was accumulated at the CERN p$\bar{\text{p}}$ Collider by each of the two major experiments, UA1 and UA2. The study of general event characteristics of hadron-hadron collisions at $\sqrt{s}$ = 630 GeV has provided revealing information about the physics of the Standard Model (SM). A long list of important results covering a wide range of physics topics has come from the analyses of these data, e.g. the detailed measurements of jet production and fragmentation properties, the production and decay characteristics of the charged and neutral Intermediate Vector Bosons (IVBs), heavy-flavour production, B^0-$\bar{B}^0$ mixing, etc. The main objective of all these studies is to know if the SM of electroweak and strong interactions is verified by the data, or if deviations are found which would hint at new physics.

Tests of the electroweak sector have been performed in measuring the SM parameters (Sect. 2). The strong sector (QCD) has been explored by studying jets, photons, the production of IVBs (Sect. 2), and the production of heavy flavour (Sect. 3). Within the minimal version of the SM there are three more elementary particles still waiting to be discovered, namely the ν_τ, the top (t) quark, and the neutral Higgs boson. Even though the ν_τ has not yet been discovered directly, there are indirect arguments for its existence, e.g. observation of the $W \to \tau\nu$ decay at the rate consistent with e-μ-τ universality of the weak charged-current coupling at $Q^2 = m_W^2$. The upper limit of 1.2×10^{-3} (90% CL) on $BR(B^0 \to \mu^+\mu^- X)$ is strong indirect evidence that the t-quark exists. At the CERN p$\bar{\text{p}}$ Collider, the main sources

Physics for a New Generation Editors: H. Latal · H. Mitter

of t-quarks are the decay $W \to t\bar{b}$, if kinematically allowed, and QCD production of $t\bar{t}$ pairs by gluon-gluon fusion or $q\bar{q}$ annihilation. At Fermilab, QCD $t\bar{t}$ production is the dominant production mechanism for all m_t. The search for the t-quark at CERN and Fermilab will be discussed in Sect. 4.

One of the most important discoveries still to be pursued is the experimental observation of the Higgs boson – which is believed to be the source of particle masses in the SM. The Higgs mass itself is, unfortunately, not predicted ($m_H \leq 1$ TeV/c^2), and the only 'safe' experimental limit on the Higgs mass is $m_H > 15$ MeV/c^2. There are two ways of searching for the Higgs at existing Colliders [1]: i) $p\bar{p} \to W^* \to WH + X$ (or $W \to W^* H$), i.e. the production of the Higgs with either a real or a virtual IVB; and ii) $p\bar{p} \to H + X$. In the first case the number of events expected is small, even for a very light Higgs. For example, for 1 pb^{-1}, one expects about four events (H + IVB) produced for $m_H < 1$ GeV/c^2 at $\sqrt{s} = 630$ GeV. A decay signature of $H \to e^+e^-$ ($\mu^+\mu^-, \pi^+\pi^-$, or K^+K^-) could lead to possible background-free detection of the Higgs, but, owing to the small branching ratios, a very high statistics W/Z data sample is required. At present $p\bar{p}$ Collider energies the dominant mechanism for Higgs production is through gluon-gluon fusion via a heavy quark loop. The Higgs decays into the heaviest fermion pair, i.e. into $b\bar{b}$, if $m_H > 10$ GeV/c^2. Even though in this case the rates are large enough to observe the Higgs up to masses of about 50 GeV/c^2 (about 100 GeV/c^2) at $\sqrt{s} = 630$ GeV (1.8 TeV), the background from QCD jets is about a factor of 10^5 above a possible $H \to b\bar{b}$ signal. It therefore seems very unlikely that the Higgs will be discovered at existing hadron colliders.

There is a general consensus, however, that the SM is not expected to be our ultimate theory. Theorists have proposed numerous possible solutions to the well-known problems of the SM. For example, compositeness [2] could solve the Flavour Problem, i.e. the understanding of the number of matter species. The Hierarchy Problem – the understanding of the origin of the different particle masses – could be solved by supersymmetry [3]. It is expected that new physics should show up for $m \leq O$(TeV/c^2). Unfortunately, the precise mass scale of this new physics is, as in the case of the Higgs, not known. Searches for physics beyond the SM can therefore provide limits on possible new mass scales of new physics (Sect. 5).

This review includes mostly data from the pre-ACOL era. These results have already been summarized and published in various conference proceedings. Therefore, instead of repeating details for the selected physics topics, the relevant results are stated and the interested reader is referred to the original publications.

The CERN $p\bar{p}$ Collider has been undergoing an important upgrading programme (addition of a new separate antiproton collector ring and six-bunch operation of the SPS). The performance reached during the first run of ACOL at the end of 1987 was far below design value ($L \approx 50$ nb^{-1} only). However, the following two runs (autumn 1988 and spring 1989) have been very successful, reaching an order of magnitude increase in luminosity. By end of June 1989, the total integrated luminosity delivered by the machine is 8.4 pb^{-1}, out of which 4.7 pb^{-1} (6.7 pb^{-1}) are recorded on tape by UA1 (UA2). Expectations from the present data samples will be discussed.

The electromagnetic (e.m.) calorimeters of the UA1 detector have been removed for the 1988 and 1989 runs in order to prepare for the installation of the new uranium-TMP calorimeter [4]. The muon detection capability has been improved by adding additional iron shielding in the forward region. The calorimetric measurements rely on the hadron calorimeters alone.

UA2 has been running since 1987 with the upgraded detector [5]. Full e.m. and hadronic calorimeter coverage has been achieved with the addition of new end-cap modules covering the angular regions $6° < \theta < 40°$ with respect to the beam directions. The electron identification is improved by a new central detector assembly consisting of a jet chamber, a silicon-pad detector, two layers of transition radiation detectors, and a scintillating-fibre detector.

The first physics run at $\sqrt{s}$ = 1.8 TeV $p\bar{p}$ collisions at the Fermilab Tevatron Collider took place in 1987, resulting in a delivered luminosity of $\sim$ 50 nb^{-1}, out of which $\sim$ 30 nb^{-1} have been collected by the Collider Detector at Fermilab (CDF) [6]. Published results from these data will be discussed in these lectures. The Tevatron started a long collider run in June 1988, which ended in June 1989, with excellent performance of the machine. The luminosity delivered is 9.5 pb^{-1}, with about 50% data-taking efficiency of the CDF detector (i.e. 4.7 pb^{-1}). Expectations from the present data sample will be discussed. The next Collider run at Fermilab is scheduled for the beginning of 1991.

Finally, some examples will be given of the possible features of physics at future hadron colliders (LHC and SSC), inferred from our present understanding of physics at hadron colliders.

2. Production and Decay of IVBs

2.1 IVB Masses and Standard Model Parameters

The data samples used in the UA1 analysis correspond to integrated luminosities of 136 nb^{-1} and 568 nb^{-1} at the Collider energies of $\sqrt{s}$ = 546 GeV and 630 GeV, respectively. The UA2 results are based on 142 nb^{-1} and 768 nb^{-1} at the two energies. The full statistics of UA1 [7] and UA2 [8] result in a total of about 650 $W \to (e, \mu, \tau)\nu$ decays and about 100 $Z \to e^+e^-$ or $\mu^+\mu^-$ decays, observed in both experiments. The major fraction of the statistics is dominated by the $W \to e\nu$ and $Z \to e^+e^-$ channels.

The W and Z masses are the two parameters that can be directly measured by the Collider experiments. The most precise values for m_W and m_Z obtained from the electron channel are [9]

$$m_W = 82.7 \pm 1.0 \pm 2.7 \text{ GeV}/c^2 \qquad \text{(UA1)},$$
$$m_W = 80.2 \pm 0.6 \pm 0.5 \pm 1.3 \text{ GeV}/c^2 \qquad \text{(UA2)},$$
$$m_Z = 93.1 \pm 1.0 \pm 3.1 \text{ GeV}/c^2 \qquad \text{(UA1)},$$
$$m_Z = 91.5 \pm 1.2 \pm 1.7 \text{ GeV}/c^2 \qquad \text{(UA2)}.$$

The first errors quoted are statistical, followed by the systematic uncertainties. In the case of m_W, UA2 quotes the systematic uncertainties separately for transverse-mass determination and energy scale. The errors are already dominated by the energy-scale uncertainties of the e.m. calorimeters (about ±3% for UA1 and about ±1.5% for UA2).

From a data sample corresponding to 25.3 nb^{-1} integrated luminosity at $\sqrt{s} = 1.8$ TeV, CDF obtained 22 $W \to e\nu$ events from an analysis with missing transverse energy $\not{E}_T$ larger than 25 GeV and with electron transverse energy larger than 15 GeV. A fit of the expected spectrum to the measured transverse mass distribution of the 22 events leads to [10]

$$m_W = 80.0 \pm 3.3 \pm 2.4 \text{ GeV/c}^2 \qquad \text{(CDF)},$$

where the first error is statistical, and the second error, which is systematic, is dominated by the uncertainty in the absolute energy scale.

One of the important tests of the electroweak model is the need for radiative corrections. The relation between the radiative correction (Δr) and m_W and m_Z is given by

$$(1 - \Delta r) = (A^2/m_W^2)/[1 - (m_W^2/m_Z^2)],$$

where $A = (\pi\alpha/\sqrt{2}G_F)^{1/2} = 37.2810 \pm 0.0003$ GeV. With this relation, we obtain [9]:

$$\Delta r = 0.036 \pm 0.100 \pm 0.067 \qquad \text{(UA1)},$$
$$\Delta r = 0.068 \pm 0.087 \pm 0.030 \qquad \text{(UA2)}.$$

A more precise value of Δr can be arrived at by using $\sin^2\theta_W = 0.233 \pm 0.003 \pm 0.005$ from a world average of neutral-current neutrino experiments as additional input [11]:

$$\Delta r = 0.127 \pm 0.023 \pm 0.060 \qquad \text{(UA1)},$$
$$\Delta r = 0.068 \pm 0.022 \pm 0.032 \qquad \text{(UA2)}.$$

The experimental results are summarized in Fig. 1, where correlations between the uncertainties of the m_Z and $m_Z - m_W$ measurements are shown. As can be seen from the error ellipse, the measurements of the IVB masses are not yet sensitive to the radiative corrections.

The one-loop radiative correction on the IVB masses has been calculated [12] to be

$$\Delta r = 0.0713 \pm 0.0013,$$

assuming that $m_t = 45$ GeV/c^2 and $m_H = 100$ GeV/c^2. The value of Δr decreases with increasing t-quark mass ($m_t \approx 250$ GeV/c^2 leads to $\Delta r = 0$). Hence a precise measurement of Δr provides information on m_t. A gobal fit to all neutral-current data and to the W and the Z mass would correspond to $m_t < 180$ GeV/c^2 (90% CL) for $m_H = 100$ GeV/c^2 and assuming that only three fermion families exist [13].

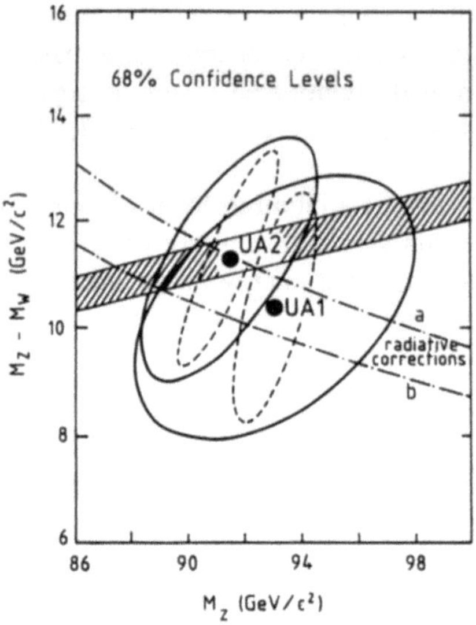

Fig. 1. Confidence contours (68%) in the $(m_Z - m_W)$ versus m_Z plane, taking into account the statistical error (*dashed line*) and with statistical and systematic errors combined in quadrature (*solid line*). The shaded region is allowed by the average of recent low-energy measurements of $\sin^2\theta_W$. The upper dashed-dotted curve (a) includes radiative corrections, the lower one (b) is without radiative corrections

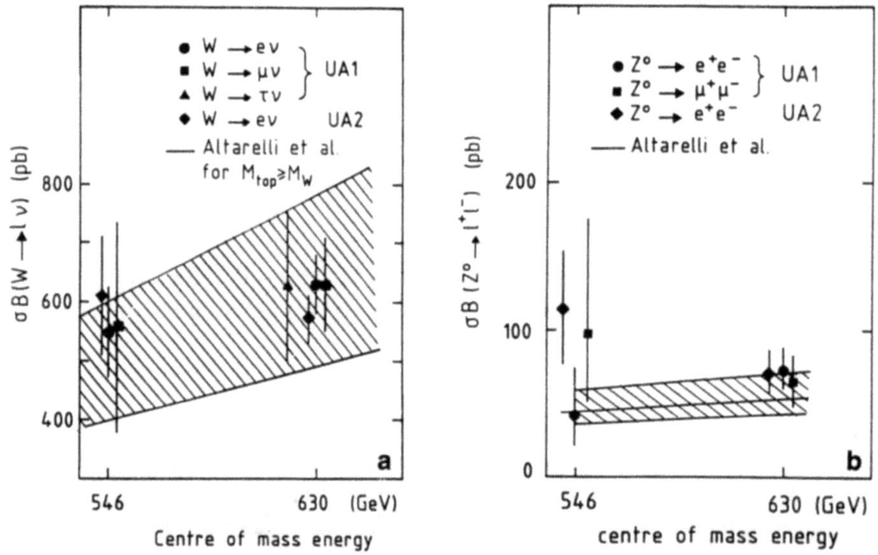

Fig. 2a,b. Partial cross-sections versus c.m. energy: (a) for $W \to \ell\nu$ production and (b) for $Z \to \ell\ell$ production. The theoretical predictions and their error bands on the total W and Z production cross-sections are taken from [14], and the leptonic branching ratios for $W \to \ell\nu$ and $Z \to \ell\ell$ have been calculated under the assumption that $m_t > (m_W - m_b)$

2.2 IVB Production Cross-Sections and e-μ-τ Universality

Figures 2a,b show the UA1 and UA2 values for $\sigma \cdot BR(W \to \ell\nu)$ and $\sigma \cdot BR(Z \to \ell^+\ell^-)$ at $\sqrt{s} = 546$ GeV and at $\sqrt{s} = 630$ GeV. The experimental errors are dominated by systematic uncertainties in the luminosity in the case of $W \to e\nu$ and by statistics in the case of $Z \to e^+e^-$ [9]. Also shown are the theoretical predictions and their error bands on the total W and Z production cross-sections taken from [14]. The

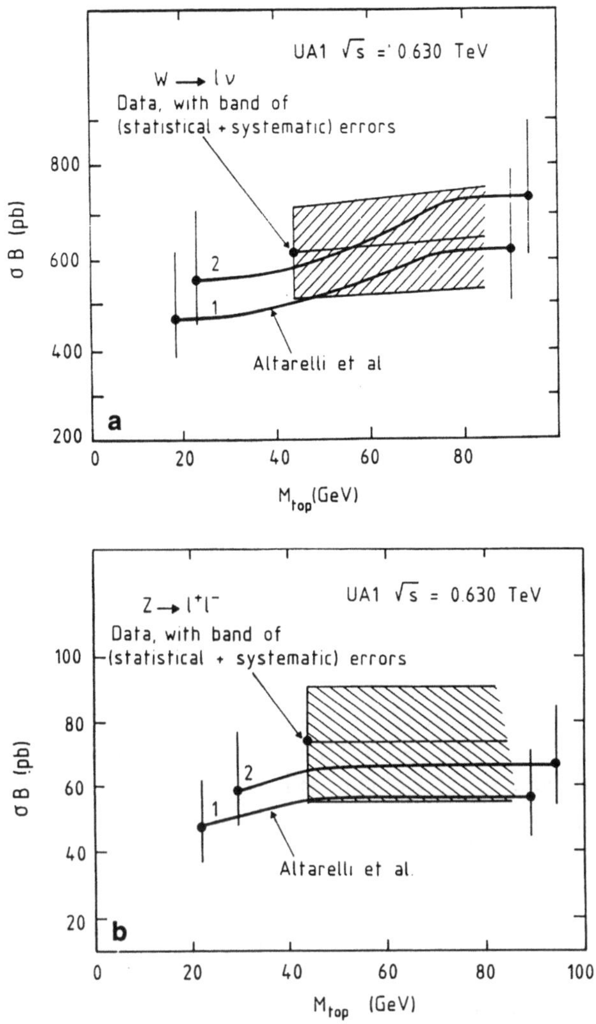

Fig. 3. Measurements and theoretical predictions for **(a)** $W \to \ell\nu$ and **(b)** $Z \to \ell\ell$ partial cross-sections as a function of m_t. The line labelled '1' represents the predictions [14] used in Figs. 2a,b. The line labelled '2' includes a 20% increase due to the BCDMS results on structure function parametrization [15]. The uncertainties are indicated by the error bars. The hatched band represents the measurements

leptonic branching ratios for $W \to \ell\nu$ and $Z \to \ell^+\ell^-$ have been calculated for $m_t > (m_W - m_b)$. The theoretical predictions are affected by several uncertainties, e.g. i) various choices of structure functions and Q^2-scale, ii) uncertainty in $BR(W \to \ell\nu)$ and $BR(Z \to \ell^+\ell^-)$ due to the as yet unknown t-quark mass. The sensitivity of the experimental results to the t-quark mass and to the choice of structure functions [15] is illustrated in Figs. 3a,b.

The W production cross-section at the Tevatron energy of 1.8 TeV is expected to be about a factor of 3 larger compared with the CERN Collider energy of 0.63 TeV. CDF has published the measurement of $\sigma \cdot BR(W \to e\nu)$ obtained from a

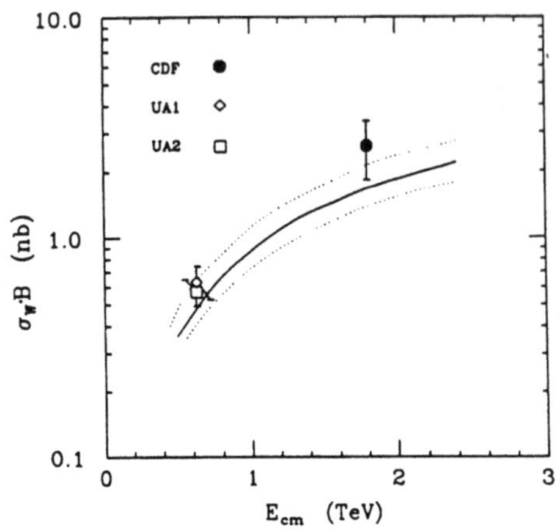

Fig. 4. Cross-section times branching ratio for W $\rightarrow$ eν versus E_{cm} measured by UA1, UA2 and CDF [10]. The prediction is from [14], adjusted for a W mass of 80 GeV/c^2. The dotted lines indicate the 1σ error limits for the theoretical curve

data sample of 25×3 nb^{-1} integrated luminosity. Using the 22 observed W $\rightarrow$ eν events and correcting for background and efficiency, CDF obtains [10]

$$\sigma \cdot BR = 2.6 \pm 0.6 \pm 0.5 \text{ nb}.$$

This result is shown together with the UA measurements in Fig. 4. The expected increase in W production is observed.

The UA1 data provide the first experimental verification of the universality of weak charged- and neutral-current coupling at $Q^2 = m_W^2 (m_Z^2)$. Defining the weak charged-coupling constants by $(g_i/g_j)^2 = \Gamma(\mathrm{W} \rightarrow \ell_i \nu_i)/\Gamma(\mathrm{W} \rightarrow \ell_j \nu_j)$, and similarly for the weak neutral-coupling constants k, UA1 obtains [9]

$$g_\mu / g_e = 1.00 \pm 0.07(\text{stat.}) \pm 0.04(\text{syst.}),$$
$$g_\tau / g_e = 1.01 \pm 0.10(\text{stat.}) \pm 0.06(\text{syst.}),$$
$$k_\mu / k_e = 1.02 \pm 0.15(\text{stat.}) \pm 0.04(\text{syst.}).$$

2.3 IVB Transverse-Momentum Distribution

In the framework of the QCD improved Drell-Yan model for quark-antiquark annihilation, the radiation of gluons from the incoming partons gives rise to a non-zero IVB transverse momentum and leads to the associated emission of high transverse-momentum jets. The normalized W transverse-momentum distributions obtained by UA1 and UA2 are shown in Fig. 5 [16]. The UA2 results are shown only for $p_T^W > 15$ GeV/c. The experimental data are compared with the theoretical predictions of [14]; also shown are the theoretical uncertainties indicated by the hatched band. The measurements agree well with the QCD prediction, except possibly at the very highest values of p_T^W. For $p_T^W > 60$ GeV/c, the theory predicts (0.8 ± 0.3) events for a total of three events observed (one event by UA2 and two events by UA1). The UA1 events have been discussed in detail in [17]. These events contain two high-p_T jets with an invariant mass compatible with the W mass, which makes their interpretation in terms of QCD correction to the W production somewhat unlikely. The analyses

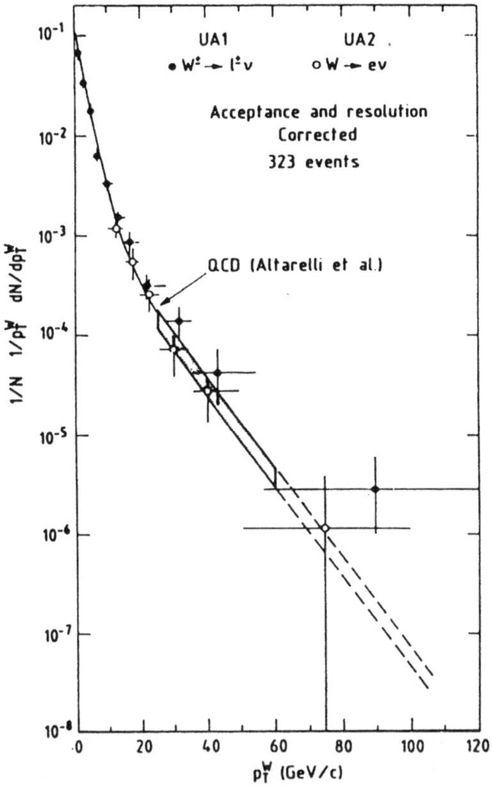

Fig. 5. W transverse-momentum distributions from UA1 (*full points*) and UA2 (*open points*). Only data with $p_T^W > 15$ GeV/c are shown for UA2. The curves are QCD predictions as reported in [17], based on calculations of [14] and extrapolating for high p_T^W

of the large new data samples at CERN and Fermilab are expected to clarify the question of a possible excess of high-p_T W's.

Figure 6 displays the inclusive p_T cross-sections for jets [18], direct photons [19], low-mass Drell-Yan $\mu^+\mu^-$ pairs [20], and W production [17], as measured by UA1. The QCD predictions are from Stirling [21], scaled by 1.5 for the jet cross-section, from Aurenche et al. [22] for direct photons, and the line describing the W cross-section is from Altarelli et al. [14]. For p_T^W, $p_T^\gamma \geq 50$ GeV/c, the effect of finite W and Z masses are reduced relative to photon production, and the absolute production cross-sections for W and γ become comparable. This is a direct manifestation of electroweak unification, the ratio of W and γ coupling to quarks being $\sin\theta_W$.

2.4 Measurement of α_s Using W + Jet Data

Jets produced in association with the W provide a means of measuring α_s. To first approximation, the yield of W + 1-jet events is proportional to α_s. UA2 used the measured one-jet to zero-jet ratio by comparing this experimental value with the QCD-predicted value obtained from Monte Carlo simulation of W production [23]. Requiring that the experimental ratio R_{exp} be reproduced by the Monte Carlo ratio $R_{MC}(\alpha_s)$, the following value for α_s was obtained:

$$\alpha_s(m_W^2) = 0.13 \pm 0.03(\text{stat.}) \pm 0.03(\text{syst.1}) \pm 0.02(\text{syst.2}),$$

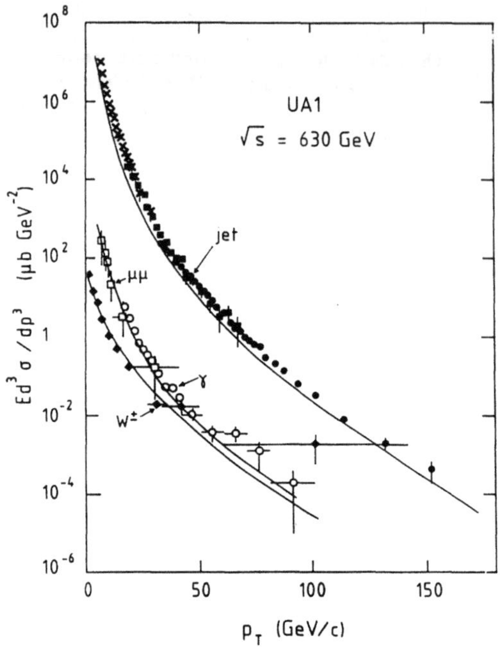

Fig. 6. Inclusive transverse-momentum distributions for W's, jets, direct photons, and nearly on-shell photons from low-mass dimuons as measured in the UA1 experiment [9]. The curves are QCD predictions as described in the text

where syst.1 represents the experimental systematic error and syst.2 is an estimate of the theoretical uncertainty. This measurement represents the first α_s determination that is (almost) free of K-factor effects in hadronic collisions. The result of this analysis is compared with other experimental determinations of α_s in Fig. 7 [23].

2.5 Number of Light Neutrino Species

The measured ratio $R = \sigma_W \cdot BR(W \to \ell\nu)/\sigma_Z \cdot BR(Z \to \ell^+\ell^-)$ depends on the t-quark mass and on the number of neutrino species via the branching ratios. The systematic measurement error largely cancels in the ratio R, and hence the available Z statistics limits the current precision on R. Combining all the available data of UA1 and UA2 (assuming lepton universality) results in [8, 24]

$$R = 8.4 \begin{array}{l} +1.2 \\ -0.9 \end{array}$$

$$R < 10.1 \qquad \text{at 90\% CL.}$$

The theoretical prediction for R depends on: i) the ratio of the total production cross-section $R_0 = \sigma_W/\sigma_Z$ (an important source of uncertainty comes from the choice of structure functions; recent calculations of R_0 give values between 3.0 and 3.4 with a typical error of 0.15); ii) the partial width of the IVBs, which are precisely determined by the SM; iii) the total decay widths of W and Z, which depend on the t-quark mass and on the number of light neutrino species (N_ν).

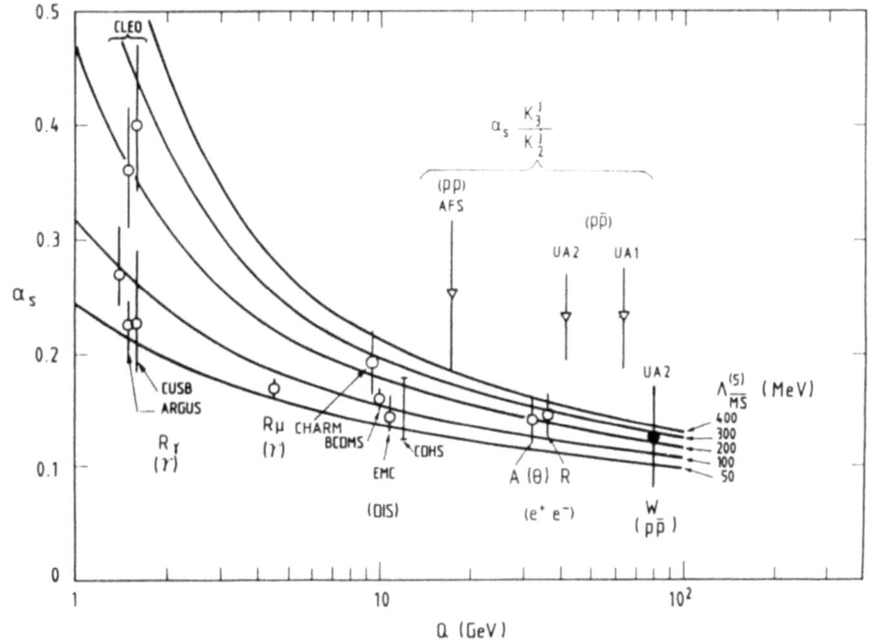

Fig. 7. Recent experimental determinations of the strong coupling constant compared with second-order QCD predictions [23]

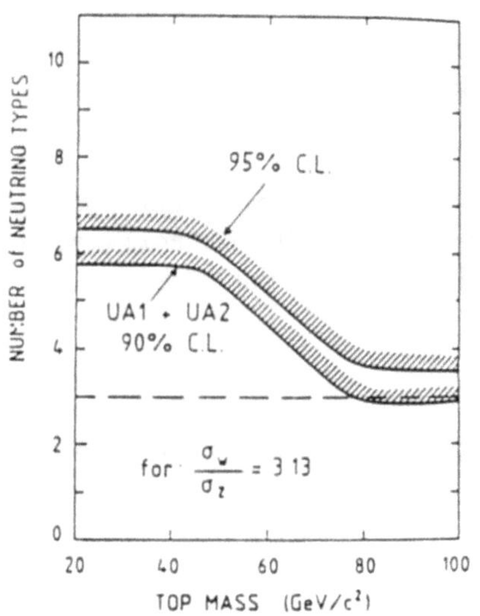

Fig. 8. The total number of light-neutrino species as a function of the t-quark mass, deduced from the combined UA1 and UA2 data

The combined (UA1 and UA2) limit on N_ν as a function of m_t is shown in Fig. 8 [24], using the conservative theoretical input for $\sigma_W/\sigma_Z = 3.13$. Assuming that there is no W decay into fourth-generation leptons and that $m_\nu < m_Z/2$, one obtains

$$N_\nu < 5.7 \qquad (90\% \text{ CL}).$$

If a new heavy charged lepton exists with a mass corresponding to the UA1 lower limit of 41 GeV/c^2, the limit obtained becomes less restrictive by one neutrino type. The data are not yet accurate enough to place an upper limit on the t-quark mass.

2.6 Outlook

Although the production cross-sections for W and Z increase with increasing $\sqrt{s}$, one expects that precision measurements in the W,Z sector will become more difficult at higher $\sqrt{s}$. For example, the factor of ~ 3 increase of σ_W and σ_Z at the Tevatron compared with that at CERN has to be compared with about a factor of 10 increase in QCD two-jet production ($p_T^{\mathrm{jet}} \approx 40$ GeV/c), thus resulting in a worse signal-to-background ratio at the Tevatron. In addition, the mean p_T^W at the Tevatron is expected to be almost twice that at CERN, thus significantly broadening the Jacobian peak observed in $W \to e\nu$ decay, and affecting the precision measurement of m_W.

From an integrated luminosity collected during the 1988/89 runs of 4.7 pb^{-1} for UA1, 6.7 pb^{-1} for UA2, and 4.7 pb^{-1} for CDF, a total of 1600 (5000) reconstructed $W \to e\nu$ decays and 200 (500) reconstructed $Z \to e^+e^-$ in UA2 (CDF) can be expected. UA1 (CDF) will have collected about 500 (2500) $W \to \mu\nu$ and about 100 (200) $Z \to \mu^+\mu^-$ decays.

With the existing data sample and the Collider runs in ≥ 1990, it should be possible to perform detailed tests of the SM, e.g. make a precise measurement of the W mass.

Using 10 pb^{-1} integrated luminosity UA2 estimated the expected error on the W and Z mass determination to be [25]

$$\delta m_W \approx \pm 0.22(\mathrm{stat.}) \pm 0.20(\mathrm{syst.1}) \pm 0.80(\mathrm{syst.2})\ \mathrm{GeV/c^2}$$
$$\delta m_Z \approx \pm 0.25(\mathrm{stat.}) \pm 0.92(\mathrm{syst.2})\ \mathrm{GeV/c^2}$$

where the errors are due to statistics (stat.), method (syst.1) and calibration (syst.2), assuming an uncertainty of $\pm 1\%$ on the absolute energy scale. This last error cancels in the measurement of $R = m_W/m_Z$, the expected precision on this ratio is

$$\delta(m_W/m_Z) \approx \pm 0.004.$$

The precise measurement of m_Z (to ± 50 MeV/c^2) expected from LEP allows to calibrate the mass scale of the hadron collider experiments. For UA2 this leads to an overall error on m_W (adding all contributions in quadrature) of

$$\delta m_W \approx \pm 0.37\ \mathrm{GeV/c^2}.$$

UA1 estimated the error on m_W, assuming 50 pb^{-1} and an absolute energy scale uncertainty of $\sim 0.5\%$ obtainable with the uranium-TMP calorimeter, to be

$$\delta m_W \approx \pm 0.10(\mathrm{stat.}) \pm 0.15(\mathrm{syst.})\ \mathrm{GeV/c^2},$$

again assuming a precise Z-mass value from LEP. This accuracy on m_W is unlikely to improve significantly until LEP operates above the threshold for W pair production.

At LHC and SSC energies, about 10^9 W's and Z^0's will be produced per year at a luminosity of 10^{33} cm^{-2} s^{-1}. This has to be compared with $\sim 10^7$ Z's produced in three years of running at LEP with $L = 1.7 \times 10^{31}$ cm^{-2}s^{-1}, and $\sim 10^4$ W^+W^- at LEP 200. Is it possible to use the LHC/SSC to search for rare W and Z decays, either to probe new physics or to perform precision measurements of the SM? For the Z, the very unfavourable signal-to-background ratio at hadron colliders is to be contrasted with the complete purity of the signal at e^+e^- Z factories; thus, looking for rare decays will very likely be the domain of LEP. However, observing the conventional decays $Z \rightarrow e^+e^-$ and $\mu^+\mu^-$ at hadron colliders is of great importance for testing perturbative QCD.

The same argument holds for the standard W-decays. Note that the calculations depend on parton distributions at small $x \approx 0.002$ [compared with $x \approx 0.15$ (0.04) at CERN (Fermilab) energies].

With $\sim 10^9$ W events produced per year, a branching ratio of $BR \geq 10^{-6} - 10^{-7}$ can in principle be reached. Looking for hadronic decay modes seems rather discouraging owing to the overwelming QCD-jet background. Leptonic decay modes are detectable, but it should be remembered that the W rapidity distribution is flat for $|y| \lesssim 4$ at SSC energies. For more details on possible detection of rare W-decays see [26].

It is clear that the sensitivity for observing interesting purely leptonic decay modes will be limited by large background – therefore rather challenging requirements for the detector performances are necessary.

3. Heavy-Flavour Production

3.1 Inclusive b-Quark Production

At the CERN Collider, the predicted cross-sections for charm and beauty production are about 100 mb and 10 mb, respectively, according to QCD estimates. These cross-sections need to be extracted from a large background of hadronic production due to $gg \rightarrow gg$. Furthermore, it is necessary to distinguish between charm and beauty production, which are expected to have comparable cross-sections at large p_T. Charm and beauty quarks can be tagged by the presence of one or more muons coming from the semileptonic decay of the heavy quarks. These muons are expected to be embedded in hadronic jets.

Inclusive muon production has been studied in UA1 for $p_T^\mu > 6$ GeV/c and within a pseudorapidity range $|\eta| < 1.5$ [27]. Figure 9 shows the p_T spectrum for muons after the decay background has been subtracted. The decays of charged π and K mesons amount to about 70% at $p_T^\mu = 6$ GeV/c and decrease to about 20% at $p_T^\mu = 20$ GeV/c. The solid curve is the prediction for the sum of all expected contributions listed in Fig. 9. The agreement with the data shown in this figure is very good. In the interval of $10 < p_T^\mu < 15$ GeV/c, heavy-flavour production is expected to be the dominant contribution. The harder fragmentation of the b-quark, compared with

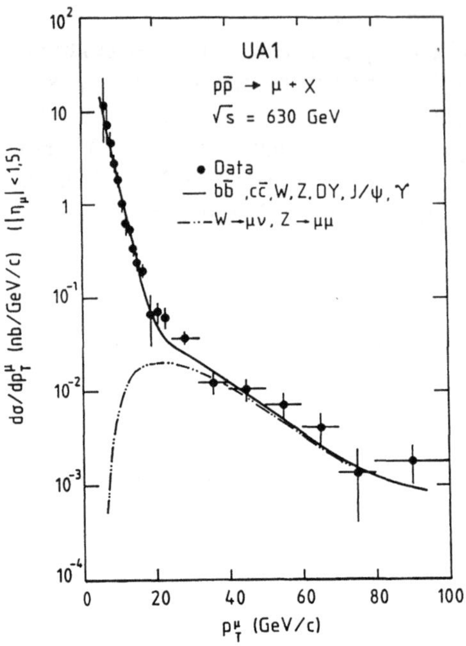

Fig. 9. Inclusive transverse-momentum distribution for muons, after decay background subtraction. Predictions made with ISAJET are shown for comparison [27]

charm, leads on the average to a harder p_T spectrum of muons; thus $b\bar{b}$ is favoured over $c\bar{c}$ once a p_T cut is applied to the muons.

An independent estimate of the beauty fraction is made by fitting the distribution of p_T^{rel}, the transverse momentum of the muons with respect to the nearest jet axis [27]. Figure 10 shows the p_T^{rel} distribution together with the fitted contributions from beauty, charm, and decay background, which for single-muon events results in

$$b\bar{b}/(b\bar{b} + c\bar{c}) = (76 \pm 12)\%.$$

The b-quark production cross-section has been evaluated using four different data sets [27]:

- J/ψ events: $p_T(J/\psi) > 5$ GeV/c;
- low-mass dimuons: $2m_\mu < m_{\mu\mu} < 6$ GeV/c^2; $p_T^{\mu 1} > 3$ GeV/c; $p_T^{\mu 2} > 3$ GeV/c;
- high-mass dimuons: $m_{\mu\mu} > 6$ GeV/c^2; $p_T^{\mu 1} > 3$ GeV/c; $p_T^{\mu 2} > 3$ GeV/c;
- single muons: $p_T^\mu > 6$ GeV/c.

The measured lepton cross-sections $\sigma(\mu)$ are related to the cross-sections for producing b-quarks, with quark transverse momentum above p_T^{min}, by the following expression:

$$\sigma(p_T^b > p_T^{min}) = \sigma(\mu)[\sigma_{MC}(p_T^b > p_T^{min})/\sigma_{MC}(\mu)],$$

where the cross-sections with subscript 'MC' are evaluated with the ISAJET Monte Carlo program [28]. The b-quark cross-section is then quoted for $p_T^b > p_T^{min}$, where p_T^{min} is defined in such a way that 90% of the muon events have $p_T^b > p_T^{min}$. Figure 11 shows the resulting inclusive b-quark cross-sections for six values of p_T^{min}, and

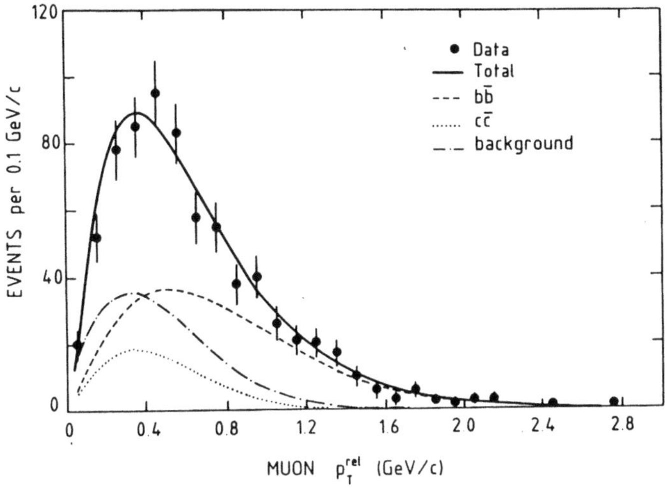

Fig. 10. Distribution of the transverse momentum relative to the nearby jet, for muons of $p_T > 10$ GeV/c in a sample of events with at least one jet of $E_T > 10$ GeV [27]

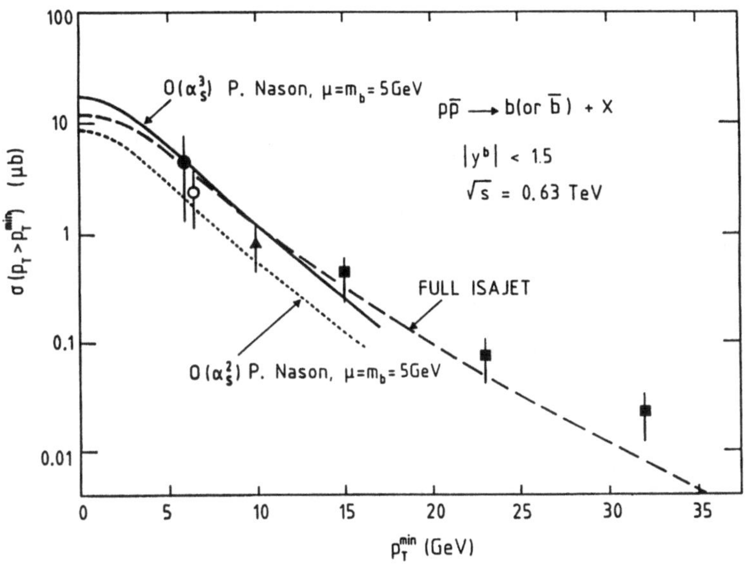

Fig. 11. Cross-section for b (or $\bar{b}$) quark production with transverse momentum above p_T^{min} and $|y| < 1.5$ [27]. The interpretation of the symbols is as follows: (■) single muons, (▲) low-mass dimuons, (o) high-mass dimuons, and (•) J/ψ. Also shown are predictions from ISAJET, as well as from second- and third-order QCD calculations [29]

for $|y^b| < 1.5$. The solid curve is the next-to-leading-order calculation of Nason et al. [29], and the dashed line corresponds to the full ISAJET prediction. The level of agreement between the cross-section measurements made with quite different techniques (using all available μ data samples) and the theoretical predictions are very satisfactory.

The total cross-section for beauty production is obtained by normalizing the $O(\alpha_s^3)$ prediction to the data and extrapolating to $p_T^{min} = 0$. Excluding the high-mass dimuon measurement as well as the measurement for $p_T^{min} > 15$ GeV/c results in [27]

$$\sigma(p\bar{p} \to b \text{ or } \bar{b}, |y| < 1.5) = 14.7 \pm 4.7\,\mu b.$$

Finally, extrapolating to all rapidities gives

$$\sigma(p\bar{p} \to b\bar{b}) = 10.2 \pm 3.3\,\mu b.$$

This can be compared with a recent theoretical calculation [30], which gives

$$\sigma(p\bar{p} \to b\bar{b}) = \left(12 \begin{array}{c} +7 \\ -4 \end{array}\right) \mu b \qquad \text{for} \quad m_b = 5\,\mathrm{GeV/c^2},$$

$$\sigma(p\bar{p} \to b\bar{b}) = \left(19 \begin{array}{c} +10 \\ -8 \end{array}\right) \mu b \qquad \text{for} \quad m_b = 4.5\,\mathrm{GeV/c^2}.$$

Within the large experimental and theoretical errors, there is good agreement between the QCD prediction and the UA1 data.

3.2 B^0-$\bar{B}^0$ Oscillations

Evidence for B^0-$\bar{B}^0$ oscillations has been reported by UA1 from the observation of an apparent excess of like-sign dimuon events [31]. Oscillation in the B_d^0-$\bar{B}_d^0$ system has been observed by ARGUS and CLEO [32].

An experimentally accessible quantity that measures the degree of oscillations at the Collider is the fraction of wrong-sign decays:

$$\chi = P(B^0 \to \bar{B}^0 \to \mu^-)/P(B^0 \to \mu^\pm).$$

The experimental result obtained by UA1 yields

$$\chi = 0.121 \pm 0.047,$$

which is an average over all beauty states. Oscillations can occur for the neutral-meson states B_d^0 and B_s^0. The corresponding mixing parameters χ_d and χ_s are related to χ via the expression

$$\chi = f_d\chi_d + f_s\chi_s,$$

where $f_{d(s)}$ is the fraction of beauty quarks hadronizing into $B_{d(s)}$ mesons. To obtain a limit on χ_d and χ_s from UA1 data, the values $f_d = 0.36$ and $f_s = 0.18$ have been used, based on the ratio of K^+/π^+ measured at the CERN Intersecting Storage Rings (ISR) [33]. Figure 12a shows the resulting ± 1 standard-deviation band obtained by UA1, compared with the weighted average of the ARGUS and CLEO results ($\chi_d = 0.162 \pm 0.039$). Figure 12b shows the corresponding combined 90% CL limit. The line in Fig. 12b indicates the upper limit on χ_d as a function of χ_s , which can be derived from existing limits on the Kobayashi-Maskawa (KM) matrix elements,

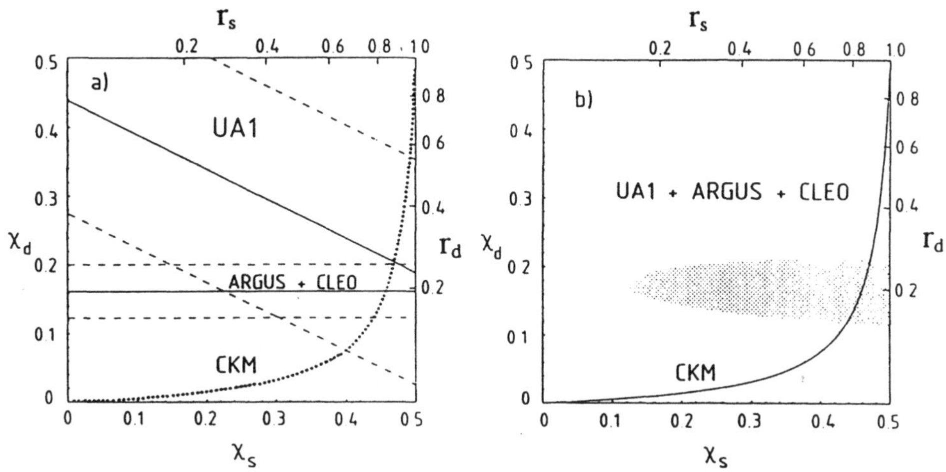

Fig. 12a,b. Beauty-oscillation results in the χ_s-χ_d plane: (a) one-standard-deviation bands, and (b) combined 90% CL limit. The region below the line is allowed by the present constraints from the KM matrix, assuming three generations [34]

assuming that we have a SM with three families. Combining this constraint with the experimental limits, only a small allowed region in the χ_s-χ_d plane remains, indicating almost maximal mixing in the B_s^0 channel [34].

4. Search for Heavy Quarks (t, b′)

In the framework of the SM, the t-quark should exist and its mass should be smaller than about 200 GeV/c^2 in order to be consistent with low-energy data. Thus, taking into account the lower limit on m_t from TRISTAN [35], the t-quark mass is expected to be in the range

$$28.0 < m_t \lesssim 180 \text{ GeV}/c^2.$$

An indirect lower mass limit of about 50 GeV/c^2 has also been obtained from analyses of B^0-$\bar{B}^0$ mixing [36].

The most promising channel for finding the t-quark at the CERN Collider is the decay $W \to t\bar{b}$, followed by the semileptonic decay of the t-quark. This channel dominates t-quark production at $\sqrt{s} = 630$ GeV if m_t is in the range 40–70 GeV/c^2. For m_t outside this range, direct $t\bar{t}$ production is the dominant process. The t-quark production cross-section for $t\bar{b}$ can be derived from the measurement $\sigma \cdot BR(W \to \ell\nu)$, taking into account the m_t dependence of the branching ratio for $W \to t\bar{b}$ and QCD corrections to the partial decay widths. The $t\bar{t}$ production cross-section can be estimated, using perturbative QCD, at the next-to-leading order and taking into account theoretical uncertainties. This leads, for example – assuming that m_t = 40 GeV/c^2 – to $\sigma(t\bar{b}) = 1.41$ nb and $\sigma(t\bar{t}) = 0.64$ nb at $\sqrt{s} = 0.630$ TeV. The production through the decay $Z \to t\bar{t}$ has a comparatively negligible cross-section and is not considered here.

UA1 has searched for heavy quarks using the distinctive signature of the semileptonic decay modes in the e and μ channels [37]. The experimental statistics available ($\sim 0.7\ \mathrm{pb}^{-1}$) allowed a systematic investigation of the region $m_t \lesssim 45\ \mathrm{GeV/c^2}$. A signal for the t-quark was searched for in the isolated $e/\mu + \geq 1$-jet channel. The characteristic topology of a t-quark decaying semileptonically allows other sources of charged leptons to be discriminated, such as W, Z, Drell-Yan, J/ψ, Υ, $b\bar{b}$, and $c\bar{c}$ – the main physics backgrounds to a new heavy-quark signal in the UA1 analysis.

Since no evidence for a t-quark signal was found in the data ($\sim 0.7\ \mathrm{pb}^{-1}$), limits have been placed on the t-quark production cross-section in $p\bar{p}$ collisions. A lower limit on m_t is obtained by comparing this experimental limit with the theoretical prediction of the cross-section, which is a function of m_t. The contribution from $W \to t\bar{b}$ decays is fixed by the measured $\sigma \cdot BR(W \to e\nu)$, therefore the result is a limit on $\sigma(t\bar{t})$. UA1 has published lower limits of 44 GeV/c² and 32 GeV/c² on the masses of the t and b′ (fourth-generation) quarks, respectively, at the 95% CL [37]. The results depend on the $O(\alpha_s)$ QCD prediction for the heavy-quark production cross-section. Using the $O(\alpha_s^3)$ calculation of Nason et al. [29] and the DFLM structure functions, Altarelli et al. [30] have revised these limits to

$$m_t > 41\ \mathrm{GeV/c^2} \qquad (95\%\ \mathrm{CL}),$$

$$m_{b'} > 34\ \mathrm{GeV/c^2} \qquad (95\%\ \mathrm{CL}),$$

as shown in Fig. 13. Since no contribution from $W \to t'\bar{b}'$ is included, the 95% CL upper limit on $\sigma(b'\bar{b}')$ is larger than the $\sigma(t\bar{t})$ limit.

To derive the b′-quark limit, it was assumed that the b′ is lighter than the t- and t′-quarks, so the b′ decays to a c-quark or a u-quark. Since there exists no lower bound on $V_{b'u}$ and $V_{b'c}$, these couplings could be very small, thus leading to long lifetimes of the b′. If the b′-quark has such a long lifetime, the quoted UA1 limit on $\sigma(b'\bar{b}')$ is no longer valid, because the selection of lepton candidates requires that the

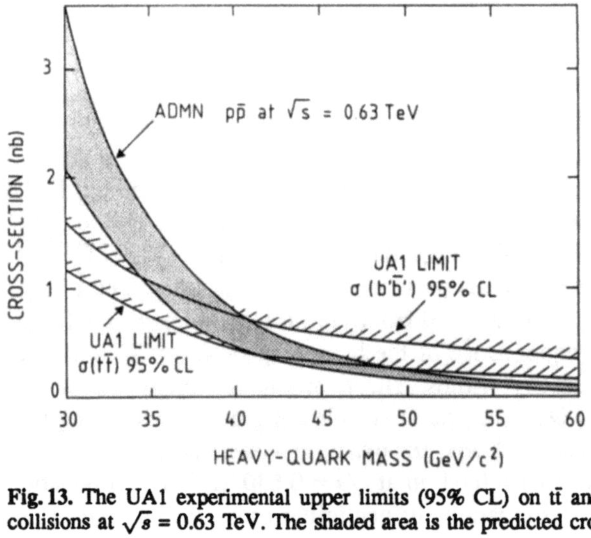

Fig. 13. The UA1 experimental upper limits (95% CL) on $t\bar{t}$ and $b'\bar{b}'$ production cross-sections in $p\bar{p}$ collisions at $\sqrt{s} = 0.63$ TeV. The shaded area is the predicted cross-section for heavy quarks, from [30]

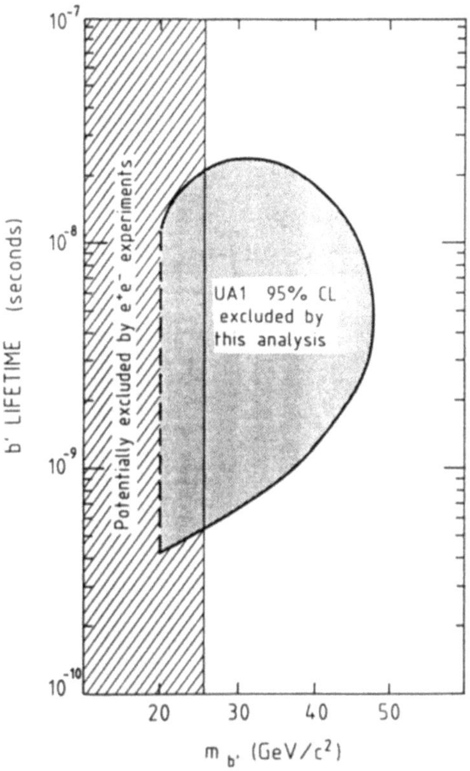

Fig. 14. The region excluded at 95% CL by UA1 in the $m_{b'}$-$\tau_{b'}$ plane, for a long-lived b′-quark [38]

lepton track points back to the primary vertex of the p$\bar{p}$ interaction. However, such a long-lived b′ provides an alternative signature: the b′, after travelling some distance, decays and produces a secondary vertex of charged particles in the UA1 central drift chamber and a high-momentum jet in the calorimeter. This event topology has been used to search for long-lived heavy quarks. The result of this search is plotted in Fig. 14 and details of the analysis can be found in [38].

4.1 Outlook

One of the most important analysis topics of the new data of UA1, UA2, and CDF ($\sim$ 5 pb^{-1}, $\sim$ 7 pb^{-1}, and $\sim$ 5 pb^{-1}, respectively), is the search for the t-quark.

The p$\bar{p}$ $\rightarrow$ t-quark production cross-section (from both t$\bar{b}$ and t$\bar{t}$) as a function of m_t at $\sqrt{s}$ = 0.63 and 1.8 TeV is shown in Fig. 15. At CERN energies the W $\rightarrow$ t$\bar{b}$ production dominates for $m_t \lesssim 70$ GeV/c^2, which has the advantage of a simpler topology than for t$\bar{t}$ and a more precisely known production rate, since normalization to the measured W $\rightarrow$ eν cross-section is possible. At Fermilab, QCD t$\bar{t}$ production is the dominant production mechanism for all m_t. From Fig. 15 it is evident that the Fermilab Collider has a substantial advantage over the CERN Collider in t-production rate. Especially at large m_t (e.g. m_t = 100 GeV/c^2), the t$\bar{t}$ cross-section increases by a factor of $\sim$ 40 between $\sqrt{s}$ = 0.63 and 1.8 TeV. If $m_t > m_W$, the interesting situation occurs that the t-quark decays into a real W + b and the process

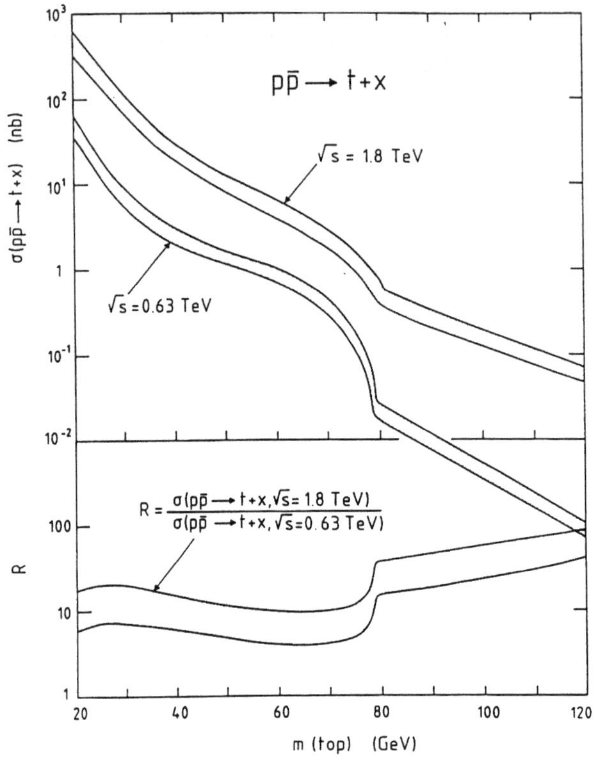

Fig. 15. The $p\bar{p} \to$ top production cross-section (from both $t\bar{t}$ and $t\bar{b}$) as a function of m_t at $\sqrt{s} = 0.63$ and 1.8 TeV

$p\bar{p} \to t\bar{t} + X$ leads to final states with two real W's and two b-quarks. The average p_T of each t-quark is $\sim m_t/2$. The W – because of its large mass – will carry most of the transverse momentum of the parent t-quark, and will therefore be produced at large p_T ($p_T^W \approx 45$ GeV/c). The $p\bar{p} \to t\bar{t} + X \to W^+W^-b\bar{b} + X$ final state has therefore a very distinctive experimental signature: large transverse momentum p_T^t, highly unbalanced transverse momenta of the two t-quark decay products ($p_T^W \gg p_T^b$), and unbalanced W decay products, characteristic of large p_T^W production.

Preliminary results on the search for the t-quark using the new data samples are available from all three Collider experiments [39].

The new data of UA1 allow the t-quark to be searched for in the muon channel only, since the e.m. calorimeters have been removed during the 1988/89 data-taking period. Applying analysis techniques similar to those discussed in Sect. 4 to the 1988 data (1.3 pb^{-1}), UA1 does not observe a signal for t-quark production in the single-muon and dimuon data samples. Combining the additional data from 1988 (μ+jet, $\mu\mu$) and data collected before 1988 (μ+jet, $\mu\mu$, e+jet, and e+μ), UA1 obtained a new, still preliminary, limit on m_t:

$$m_t > 56 \text{ GeV/c}^2 \quad (95\% \text{ CL}),$$
$$m_t > 60 \text{ GeV/c}^2 \quad (90\% \text{ CL}),$$

and for the b′-quark, the new UA1 lower limit on the b′ mass is

$$m_{b'} > 38 \text{ GeV}/c^2 \qquad (95\% \text{ CL}).$$

Including the new data from the 1989 run, UA1 expects a sensitivity to t-quark masses approaching the kinematic limit of the W channel.

UA2 also presented preliminary results from the 1988 data [39] on the t-quark search in the e-channel (2.5 pb^{-1}). The selection requires missing transverse momentum $\not{p}_T > 15$ GeV, $p_T^e > 11.5$ GeV and $15 < m_T(e\nu) < 50$ GeV/c^2 for the t signal region. An increase in acceptance is expected if the $\not{p}_T$ is reduced to 10 GeV; however, this requires more rejection power against background due to conversions and overlaps, i.e. tighter quality cuts on the electron candidate. So far the data are consistent with the expected background for both selections, where the dominant physics contribution comes from W + jet production. In 2.5 pb^{-1} of analysed data, no t-quark signal is observed in the mass range $30 < m_t < 60$ GeV/c^2. The corresponding confidence limits on m_t are not yet quoted. UA2 expects a sensitivity to t-quark masses approaching the kinematic limit in the W-channel from the full data sample including the 1989 run.

A status report on the t-quark search by CDF has been presented [39]. The search strategy at the Fermilab Tevatron is different from that at the CERN $p\bar{p}$ Collider; this is because the $W \to t\bar{b}$ contribution is small and $t\bar{t}$ can be observed in many channels. Owing to the large cross-section and large integrated luminosity, CDF can afford to look for $e^{\pm}\mu^{\mp}$ events from $t\bar{t}$ decay. This channel provides a very clear signature, no background from W + jets, and negligible background from $b\bar{b}$ due to p_T and isolation cuts. In a data sample corresponding to 2.0 pb^{-1}, CDF searched for events with $E_T^e > 15$ GeV and $p_T^\mu > 12$ GeV/c. The expectation from top in 2.0 pb^{-1} for this channel varies from 22 events ($m_t = 30$ GeV/c^2) to 1.8 events ($m_t = 70$ GeV/c^2). From zero events observed in the data, it was preliminarily concluded that $30 < m_t < 60$ GeV/c^2 is unlikely. The corresponding confidence limits or m_t are not yet quoted. From the analysis of the full data sample (~ 5 pb^{-1}), CDF should have a sensitivity to t-quark masses up to ~ 80 GeV/c^2, which could be extended to ~ 120 GeV/c^2 with 20 pb^{-1} (expected by the end of 1991).

Heavy-quark production at the LHC/SSC will be dominated by pair- production through gg fusion [40]. The total production cross-section for heavy quark pairs as a function of m_Q is plotted in Fig. 16a at $\sqrt{s} = 20$ TeV. The available centre-of-mass energy is an important parameter for heavy-quark production, e.g. for $m_Q = 0.5$ TeV/c^2, the cross-section decreases by a factor of ~ 10 if $\sqrt{s}$ decreases from 20 to 10 TeV. Assuming that $Q \to q\,W$, the most promising decay mode to be detected is $Q\bar{Q} \to (W \to \ell\nu) + 4$ jets. For one year of running at $L = 10^{33}$ $cm^{-2}\,s^{-1}$, a mass reach of [40]

$$m_Q \approx 0.8 \text{ TeV}/c^2 \qquad \text{at the LHC,}$$

$$m_Q \approx 1.0 \text{ TeV}/c^2 \qquad \text{at the SSC,}$$

can be expected provided that the main background from W + jet production is kept small. This can be achieved by cutting in a range of p_T^e values, for which the signal exceeds the inclusive $W \to e\nu$ background, as can be seen in Fig. 16b.

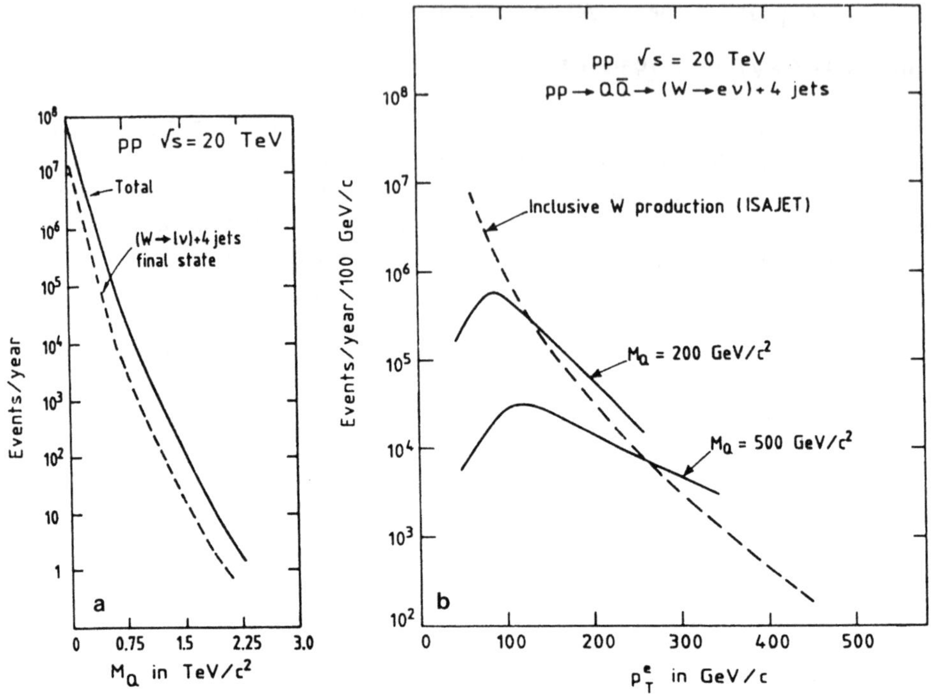

Fig. 16. (a) The total production cross-section for heavy-quark pairs as a function of the heavy-quark mass m_Q at $\sqrt{s}$ = 20 TeV. Also shown is the cross-section for pp → $Q\bar{Q}$ → (W → $\ell\nu$) + 4 jets, assuming that Q → W q [40]. (b) The inclusive electron transverse momentum p_T^e spectrum from W → eν decays (*dashed line*) and for the process pp → $Q\bar{Q}$ → (W → eν) + 4 jets, for two different mass values of the heavy-quark Q (*solid lines*) [40]

5. Searches for New Physics

5.1 Search for Heavy Leptons and the Number of Neutrino Species

An analysis of events with large missing transverse energy ($\not{E}_T$) due to the production of one or more energetic neutrinos or weakly interacting particles, can extend our understanding of the SM, and is a sensitive way to search for new physics. A fourth generation of quarks and leptons would contain a charged heavy lepton L, together with its associated ν_L. Assuming that there is a universal strength of coupling to W and Z particles, and that $m_L < m_W$, the heavy lepton can be produced in W decays. One then looks for a process p$\bar{p}$ → W + X (with W → Lν, and L → q$\bar{q}$ + ν). The experimental signature would be events with one or more jets plus large $\not{E}_T$. This decay pattern gives events that are qualitatively similar to W → $\tau\nu$ (τ → hadrons + ν), except for the phase-space factor and the typically larger invariant mass of the q$\bar{q}$ system. UA1 has used the $\not{E}_T$ + jet data sample – which has led to the identification of W → $\tau\nu$ decays – to place a limit on m_L [41]. Using the calculated heavy-lepton contribution and taking into account the existence of the additional neutrino (ν_L) coupling to the Z, it was found that

$$m_L > 41 \text{ GeV}/c^2 \qquad (90\% \text{ CL}).$$

Similarly, the rate of events from the process $p\bar{p} \to Z + X$ ($Z \to \nu\nu$) is proportional to the number of light-neutrino species coupling to the Z. If the Z is produced in association with an energetic jet (Z production at high p_T), such events are expected to appear as hadronic events with large values of $\not{E}_T$. The same data sample and method as were used for the heavy-lepton search have been employed by UA1 [41] to derive a limit on the number of light-neutrino species:

$$N_\nu \leq 10 \qquad (90\% \text{ CL}).$$

5.2 Searches for Supersymmetric Particles

As already mentioned, supersymmetric theories are attractive because they provide solutions for some of the fundamental problems of the SM. In supersymmetric theories, fermions and bosons are linked by a common symmetry, every particle having a supersymmetric partner, a 'sparticle'. The coupling of these new particles is well defined and therefore production cross-sections can be calculated. Supersymmetric particles are pair-produced (e.g. $p\bar{p} \to \tilde{g}\tilde{g}$), they decay into lighter SUSY particles (e.g. $\tilde{g} \to q\bar{q}\tilde{\gamma}$), and the lightest SUSY particle (LSP) is expected to be stable and non-interacting, and thus will escape detection. In many models, the LSP is assumed to be the photino ($\tilde{\gamma}$).

At hadron colliders the dominant final states containing SUSY particles are $\tilde{g}\tilde{g}$, $\tilde{g}\tilde{q}$, and $\tilde{q}\bar{\tilde{q}}$ [42]. In the case where $m_{\tilde{q}} > m_{\tilde{g}}$, the squark decays into $q\tilde{g}$ and the gluino decays into $q\bar{q}\tilde{\gamma}$; in the case where $m_{\tilde{q}} < m_{\tilde{g}}$, the decay modes are $\tilde{q} \to q\tilde{\gamma}$ and $\tilde{g} \to \bar{q}\tilde{q}$ or $(q\bar{\tilde{q}})$. In both cases, the event topology consists of hadronic jets (from the fragmentation of outgoing q and $\bar{q}$) and $\not{E}_T$ (from the outgoing $\tilde{\gamma}$'s). The number of jets depends upon the production subprocess and the $\tilde{q}$ and $\tilde{g}$ masses. When $m_{\tilde{q}} \ll m_{\tilde{g}}$, $\tilde{q}\bar{\tilde{q}}$ pair-production dominates the production of strongly interacting sparticles, whereas $\tilde{g}\tilde{g}$ dominates if $m_{\tilde{q}} \gg m_{\tilde{g}}$.

UA1 has performed a search for gluinos and squarks [43] using the jet(s) + $\not{E}_T$ signature (assuming a stable $\tilde{\gamma}$, which escapes detection). The domain of the ($m_{\tilde{g}}$, $m_{\tilde{q}}$) plane excluded by the UA1 analysis is shown in Fig. 17. In the case where $m_{\tilde{g}} \gg m_{\tilde{q}}$, the lower bound $m_{\tilde{q}} > 45$ GeV/c^2 is obtained. Similarly, if $m_{\tilde{q}} \gg m_{\tilde{g}}$, the lower bound $m_{\tilde{g}} > 53$ GeV/c^2 is found. For $m_{\tilde{g}} \approx m_{\tilde{q}}$, a more stringent limit of $\tilde{m} > 75$ GeV/c^2 is obtained owing to the increase in the number of events expected from $\tilde{g}\tilde{q}$ production. The results are valid for photino masses up to ~ 20 GeV/c^2. For $m_{\tilde{\gamma}} > 20$ GeV/c^2, however, the squark and gluino mass limits decrease rapidly.

In the search for $\tilde{g}$ and $\tilde{q}$ by UA2 [44] two cases were considered: i) the photino is the LSP and is therefore stable, and ii) the photino decays into $\tilde{H}\gamma$, where $\tilde{H}$ escapes detection. In the first case, a somewhat lower mass limit for $\tilde{q}$ and $\tilde{g}$ has been obtained, compared with that from UA1. In the case of an unstable photino,

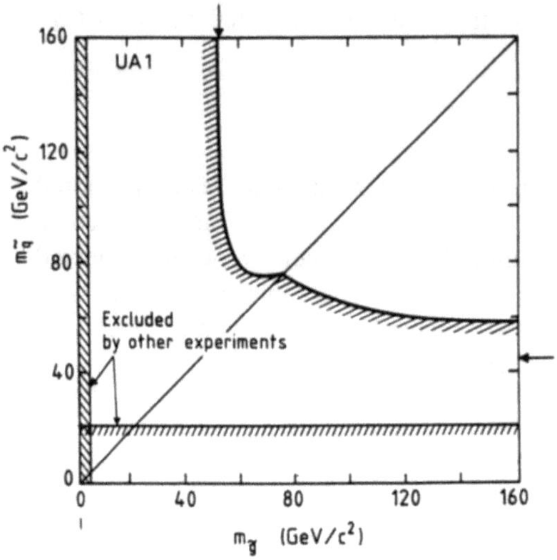

Fig. 17. Limits on squark and gluino masses (90% CL) obtained by UA1. The arrows indicate the asymptotic values of the 90% CL contours as the squark or gluino mass becomes infinitely large [43]

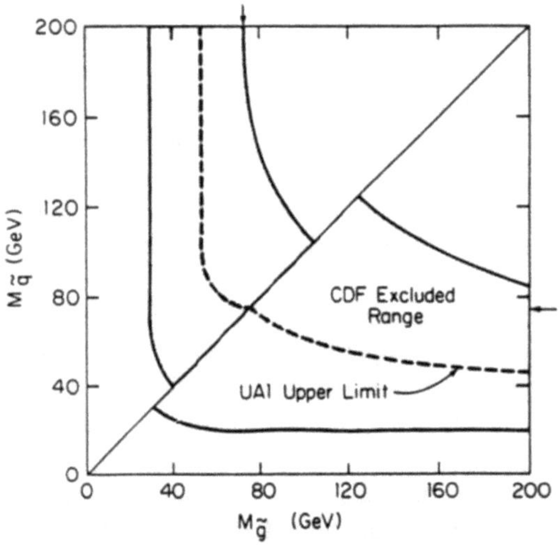

Fig. 18. Limits on squark and gluino masses (90% CL) obtained by CDF. The arrows indicate the asymptotic values of the 90% CL contours, where the $\tilde{g}$ or $\tilde{q}$ mass is very large. The lower-limit curve corresponds to an CDF acceptance of 0.1% [45]

the better granularity of the UA2 calorimeter allows an analysis of events containing two photons + jets in the final state. In this analysis, UA2 excludes gluino masses between 15 and 20 GeV/c² and squark masses between 9 and 46 GeV/c². Again, assuming $m_{\tilde g} \approx m_{\tilde q}$, a limit of $\tilde{m} > 60$ GeV/c² is obtained.

CDF has performed a search for gluinos and squarks using the jet(s) + $\not\!\!E_T$ signature (assuming a stable $\tilde{\gamma}$, which escapes detection). The data sample used corresponds to 25.3 nb^{-1} of integrated luminosity [45]. The domain in the $(m_{\tilde g}, m_{\tilde q})$ plane excluded by the CDF analysis is shown in Fig. 18. The asymptotic mass limits are $m_{\tilde g} > 73$ GeV/c² and $m_{\tilde q} > 74$ GeV/c² at 90% CL. The lower-limit curve in Fig. 18 has been chosen where the acceptance dropped to 0.1%. The quoted results are valid for photino masses up to 30 GeV/c² in the high-mass region.

Supersymmetric leptons can be searched for in $W \to \tilde{\ell}\tilde{\nu}$ and $Z \to \tilde{\ell}^+\tilde{\ell}^-$ decays. In each case $\tilde{\ell} \to \ell\tilde{\gamma}$ is assumed to dominate, and one expects $\ell + \not\!\!E_T$ final states from

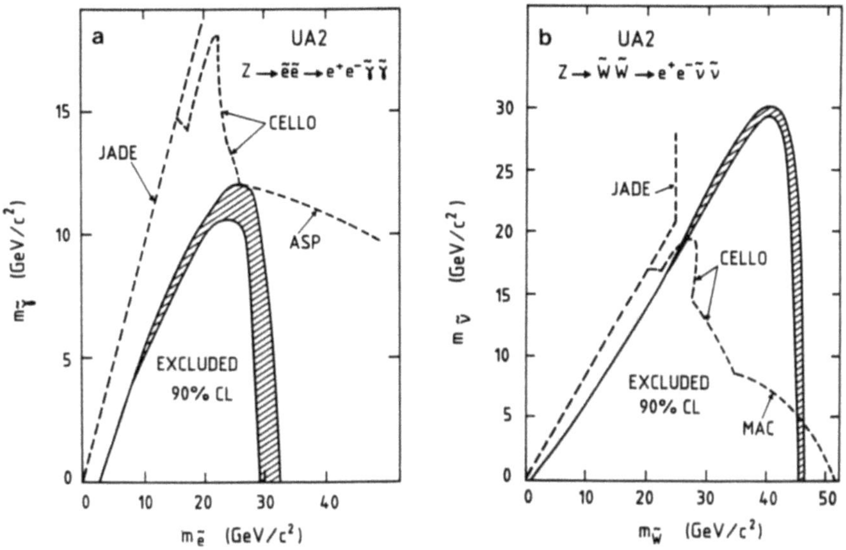

Fig. 19. Limits (90% CL) on (a) $\tilde{e}$ and $\tilde{\gamma}$ masses and (b) $\tilde{W}$ and $\tilde{\nu}$ masses obtained by UA2 [44]. The hatched area illustrates the systematic uncertainty. Also shown are limits from e^+e^- experiments

W decays and $\ell^+\ell^- + \not{E}_T$ final states from Z decays. The former may be distinguished from the conventional $W \to \ell\nu$ decays by its softer lepton spectrum and different angular distribution. The $Z \to \tilde{\ell}^+\tilde{\ell}^-$ decay can be distinguished from conventional Drell-Yan background by the additional $\not{E}_T$ signature.

The Z data sample can also be used to extract a limit for the mass of the wino under the assumption that the decay $Z \to \tilde{W}\tilde{W}$ is kinematically possible. In this case one expects to observe again $\ell^+\ell^-$ pairs in association with large $\not{E}_T$, under the assumption that $\tilde{W} \to e\tilde{\nu}$, where $\tilde{\nu}$ escapes detection (stable $\tilde{\nu}$ or $\tilde{\nu} \to \nu\tilde{\gamma}$).

Figures 19a,b show the 90% CL for $m_{\tilde{e}}$ and $m_{\tilde{W}}$ as a function of $m_{\tilde{\gamma}}$ and $m_{\tilde{\nu}}$, respectively, as obtained by UA2 [44]. In the $\tilde{W}$ case the UA2 result significantly improves the values reached in e^+e^- experiments so far. Using the $W \to e\nu$ data samples, UA1 [9] obtained a selectron mass limit of $\tilde{m} > 32$ GeV/c^2, for $m_{\tilde{W}}$, $m_{\tilde{Z}} \gg m_{\tilde{e}}$, and $m_{\tilde{e}} = m_{\tilde{\nu}}$.

5.3 Searches for Additional Vector Bosons

Additional vector bosons – presumably heavy – are predicted in many possible extensions of the SM [46]. These heavy particles (W′,Z′) would be detected in UA1 and UA2, depending on their mass, their coupling to quarks, and their branching ratio to leptons.

In the UA1 data sample, no $W \to e\nu$ and no $Z \to e^+e^-$ candidates have been observed with electron-neutrino transverse mass or e^+e^- mass in excess of the expected distribution for Standard Model IVB decays [9]. These results have been used to set limits on the production and decay of more massive W(Z)-like objects. Normalizing the cross-section times branching ratio $(\sigma' \cdot BR')_{\mathrm{IVB}'}$ to those of the

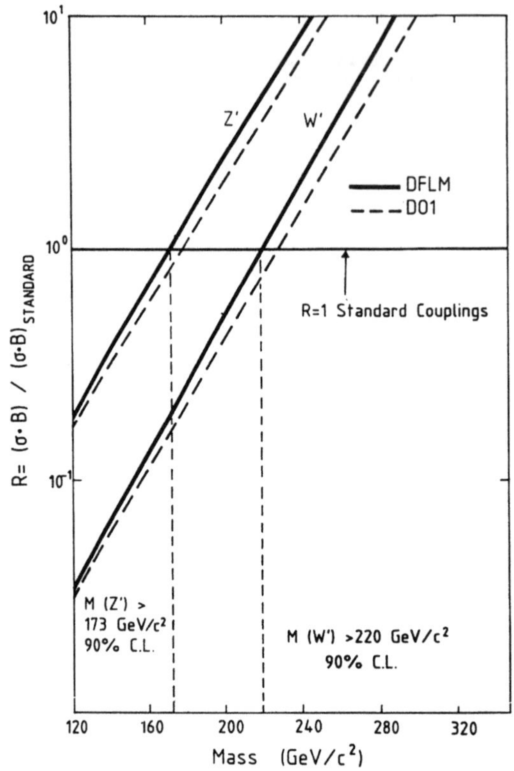

Fig. 20. The W′ (Z′) lower mass limits (90% CL) shown as a function of the W′ (Z′) coupling strength obtained by UA1 [9]

SM vector bosons results in limits on $m_{W'}$ and $m_{Z'}$ as a function of this ratio, as shown in Fig. 20. For standard coupling ($R = 1$), UA1 obtained the limits [9]

$$m_{W'} > 220\ \mathrm{GeV/c^2}, \qquad m_{Z'} > 173\ \mathrm{GeV/c^2} \qquad (90\%\ \mathrm{CL}).$$

From the analysis of the final UA2 data sample, the general limits for $m_{W'}$ and $m_{Z'}$ as a function of $\lambda_q^2 \cdot BR_e$ (coupling times branching ratio) are derived, as shown in Fig. 21 [44]. For standard couplings, i.e. $\lambda_q^2 \cdot BR_e = 1$, UA2 obtains

$$m_{W'} > 209\ \mathrm{GeV/c^2}, \qquad m_{Z'} > 180\ \mathrm{GeV/c^2} \qquad (90\%\ \mathrm{CL}).$$

Most models derived from superstring theories predict at least one extra Z′ with $\lambda_q^2 \approx 0.2$ and BR_e varying from 0.3 to 1.0. For the resulting values of $\lambda_q^2 \cdot BR_e$ between 0.06 and 0.2, UA2 does not exclude any significant mass values.

5.4 Axigluon Search

Axigluons are predicted by gauge theories based on the chiral colour group $SU(3)_L \times SU(3)_R$ [47]. In these theories, the $SU(3)_L \times SU(3)_R$ symmetry is broken down into the QCD $SU(3)_c$ symmetry at some scale that is comparable to the weak scale. A model-independent prediction of all chiral colour theories is the existence of a new massive colour octet of gauge bosons, the axigluons. The dominant decay channel

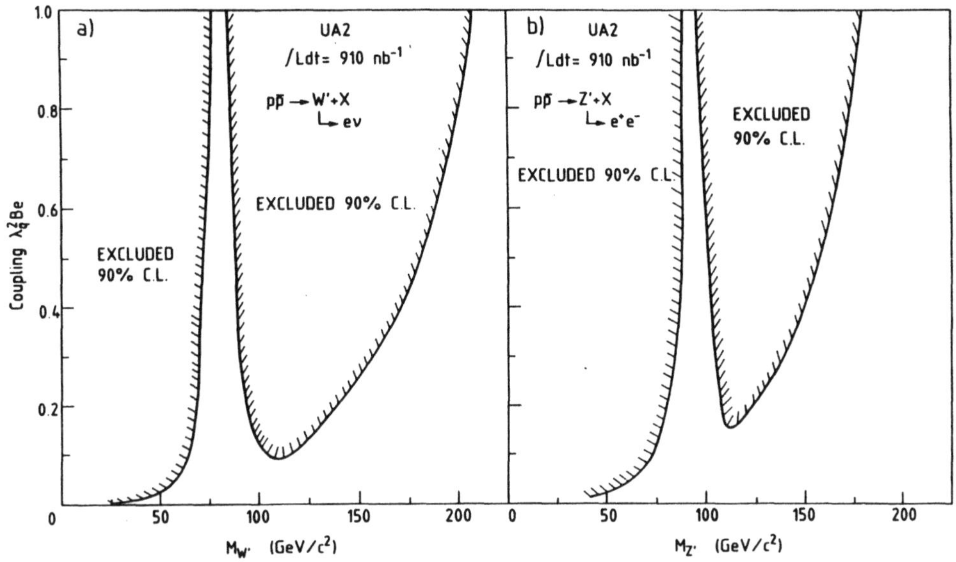

Fig. 21a,b. Limits (90% CL) on additional vector bosons W′ and Z′ depending on $\lambda_q^2 \cdot BR_e$ obtained by UA2 [44]: **(a)** for the W′ obtained from an analysis of single-electron candidates and **(b)** for the Z′ obtained from an analysis of electron pairs

of the axigluon is expected to be into two jets. The width of the axigluon depends on the number of particles into which it can decay. Within the model, the width is expected to be in the range $0.1\ m_A < \Gamma < 0.4\ m_A$.

A search for axigluons in the two-jet channel has been performed by UA1 [48]. Assuming that there is no interference between the normal QCD jet-production mechanism and the axigluon, the QCD and axigluon contributions have been added to obtain the total theoretical prediction in the presence of the axigluon. The data are fitted to the prediction using the maximum likelihood method. Figure 22 displays the limit, at 95% CL, on the production rate of axigluons in pp̄ collisions at $\sqrt{s}$ = 630 GeV. It can be seen that the axigluon is excluded at the 95% CL for

$$150\ \mathrm{GeV/c^2} < m_A < 315\ \mathrm{GeV/c^2}.$$

This result has been generalized to obtain limits on the production rate of a heavy-state X decaying into two jets, for very small and very large widths, and also for the two spin cases (0 and 1). These limits are also shown in Fig. 22.

5.5 Quark Substructure

As already mentioned, there are several theoretical motivations for compositeness. The proliferation of flavour may be a hint that quarks and leptons are composite. A possible manifestation of quark and lepton compositeness could be a new contact interaction having the new strong interaction scale Λ_c [49]. At the CERN Collider, the effect of a four-quark interaction term ($\sim 1/\Lambda_c$) would manifest itself in observed deviations from the QCD jet-production cross-section. Such deviations would be

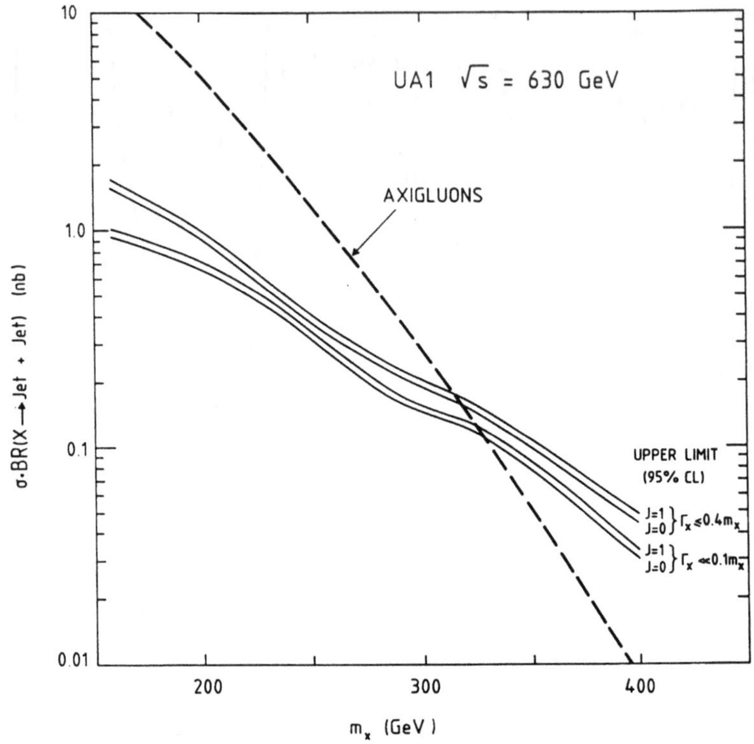

Fig. 22. Cross-section times branching ratio limits (95% CL) for the spin-0 and spin-1 particle X as a function of the mass m_X [48]

observable at scattering energies below Λ_c, and have been searched for by UA1, UA2 and CDF.

UA2 has concentrated on the overall normalization of the large-p_T jet cross-section [50]. Finite values of Λ_c would produce an excess of events at large p_T compared with the standard QCD prediction ($\Lambda_c = \infty$). Taking into account the theoretical and experimental uncertainties, UA2 obtained the limit

$$\Lambda_c > 370 \text{ GeV} \qquad (95\% \text{ CL}).$$

UA1 has used the angular distribution of high-mass jet pairs to set a limit on Λ_c [51]. Figure 23 shows the measured angular distribution, where χ is defined as $(1 + \cos\theta)/(1 - \cos\theta)$, θ being the angle between the axis of the jet pair and the beam direction in the jet-jet rest frame [52]. A finite quark size would modify this angular distribution in such a way that more events are expected at small χ values. By varying the value of the parameter Λ_c in the fit to the measured angular distribution, and taking into account the systematic uncertainty on the jet-energy scale, UA1 obtained

$$\Lambda_c > 415 \text{ GeV} \qquad (95\% \text{ CL}).$$

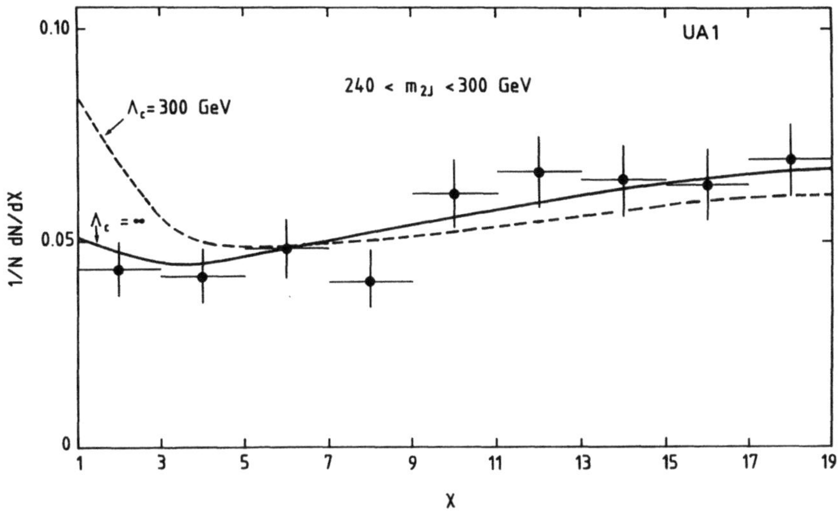

Fig. 23. Normalized angular distribution for very high mass jet-pairs as a function of χ. The solid curve is a QCD prediction ($\Lambda_c = \infty$); the dotted curve corresponds to $\Lambda_c = 300$ GeV, which is clearly excluded by the UA1 data [51]

To search for possible quark substructure, CDF has compared the measured inclusive jet cross-section at $\sqrt{s} = 1.8$ TeV with the predictions of leading-order QCD modified by the addition of a contact interaction with scale Λ_c [53]. Jet transverse energies up to 250 GeV have been observed, and the data with $E_T^{\text{jet}} < 130$ GeV were used to normalize the prediction. Taking into account Poisson fluctuations in the highest E_T-bins, the absence of jets with $E_T > 255$ GeV, and uncertainties in structure functions and process-scales, CDF obtained the limit

$$\Lambda_c > 700 \text{ GeV} \qquad (95\% \text{ CL}).$$

5.6 Outlook

Searches for new physics have so far been fruitless at both CERN and Fermilab. However, in several cases, significant lower mass limits could be placed on hypothetical new particles.

What is the discovery potential of hadron colliders in the near and the distant future?

Searches for new particles using the new data samples of UA1, UA2, and CDF can be divided into two classes: a) where $\sqrt{s}$ does not play an important role, and b) where the production cross-section is strongly dependent on $\sqrt{s}$.

Examples for class (a) are

– fourth-generation of charged leptons searched for in W decays,

– scalar leptons ($\tilde{e}$, $\tilde{\mu}$, $\tilde{\nu}$) and $\tilde{W}$ searched for in W and Z decays.

One expects sensitivity up to the kinematic limit using present data. Figure 24 shows, as an example, the expected limits obtainable by UA2 for 10 pb^{-1} from the decay $Z \to \tilde{e}\tilde{e} \to ee\tilde{\gamma}\tilde{\gamma}$ [54].

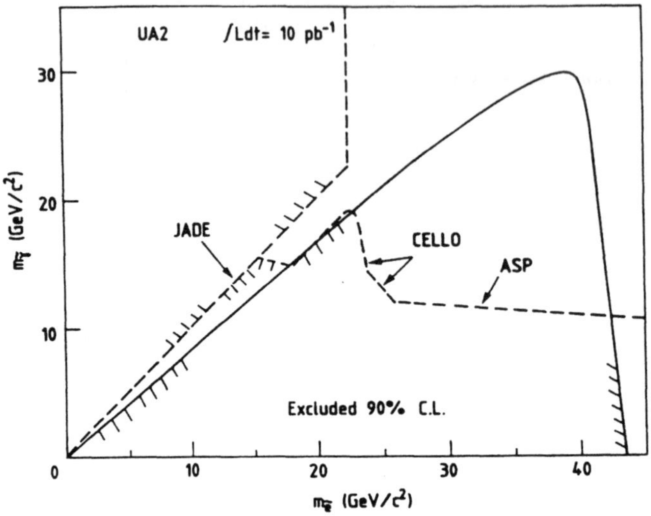

Fig. 24. Expected limits (for 10 pb^{-1}) on $\tilde{e}$ and $\tilde{\gamma}$ at 90% CL from a search for $Z \to \tilde{e}\tilde{e} \to ee\tilde{\gamma}\tilde{\gamma}$ decays in UA2 [54]

Examples for class (b) are
- gluino and squark,
- additional vector bosons,
- compositeness scale Λ_c.

We have already seen the obvious benefit of having higher $\sqrt{s}$ in the search for gluinos and squarks. The mass reach at the Tevatron was about 20 GeV higher for about a factor of 30 less integrated luminosity. From the present data samples, one expects limits of the order of

$$\tilde{m} \approx 120 \text{ GeV/c}^2 \qquad \text{(UA2)},$$
$$\tilde{m} \approx 200 \text{ GeV/c}^2 \qquad \text{(CDF)}.$$

Because of the absence of background in the decay of additional vector bosons (W$'$, Z$'$) into leptons, hadron colliders make a large mass range accessible to experiments. Figure 25 shows the production cross-section for W$'$ and Z$'$ at $\sqrt{s}$ = 0.63 and 1.8 TeV, assuming standard couplings to leptons and quarks, and $W' \to e\nu$ and $Z' \to ee$ decays. The corresponding sensitivity from the present data samples is expected to be

for $m_{W'}$: ~ 300 GeV (UA2) and ~ 600 GeV (CDF),
for $m_{Z'}$: ~ 250 GeV (UA2) and ~ 520 GeV (CDF).

An improvement in the limit on Λ_c of $\geq$ 650 GeV from CERN data will be achievable. CDF expects to reach $\Lambda_c >$ 1.6 TeV from present data samples, again demonstrating the clear benefit of higher $\sqrt{s}$.

Finally, we discuss the possible discovery potential of LHC/SSC with respect to two topics: supersymmetry and Z$'$.

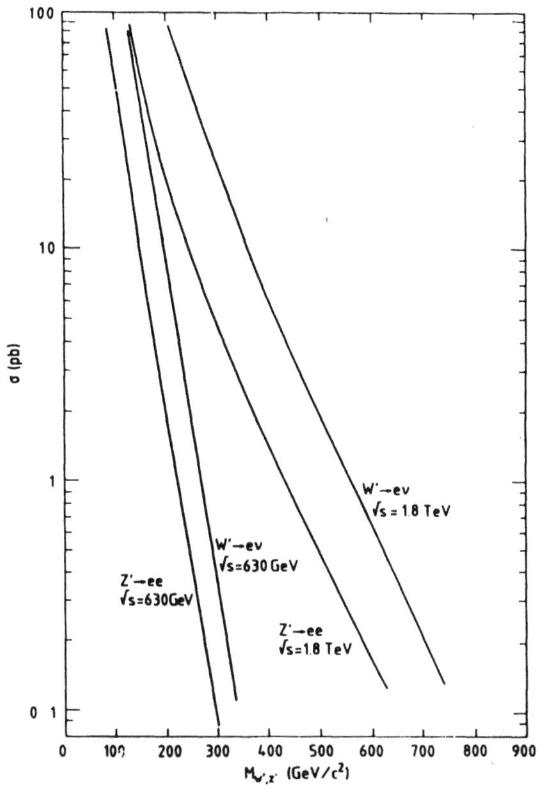

Fig. 25. Production of W′ and Z′ at ACOL and Tevatron as a function of their mass, assuming standard couplings to leptons and quarks [54]

The detection of supersymmetric particles has received a lot of attention in various physics studies dealing with future colliders [55]. Present limits on sparticles are not yet problematic for low-energy supersymmetry. Sparticle masses may still be too high to be produced at the CERN $p\bar{p}$ Collider, at the Tevatron, at LEP, and at the SLC. If this is the case, it will be necessary to await the next generation of high-energy colliders, probing parton-parton collisions at energies of the order of 1 TeV. However, we have to keep in mind that even if the sparticle masses are larger than a few hundred GeV, nearly all low-energy SUSY models imply the existence of at least a light Higgs scalar with mass less than $O(m_Z)$. Therefore, the non-observation of a neutral Higgs boson up to LEP 200 energies would put rather tight constraints on possible supersymmetric parameters [56].

Squarks and gluinos should be copiously produced at the LHC/SSC. For example, for a total integrated luminosity of 10^{40} cm^{-2} = 10 fb^{-1} (corresponding to one year of running at $L = 10^{33}$ cm^{-2} s^{-1}), we expect $\sim$ 104 gluino pairs to be produced at the SSC for $m_{\tilde{g}}$ = 1.5 TeV, or at the LHC for $m_{\tilde{g}}$ = 1 TeV. The sparticle signatures depend on the details of the assumed masses and decay patterns. The 'classical' signature of large $\not{E}_T$ plus two or more high-E_T jets results from the assumption that $\tilde{q}$ and $\tilde{g}$ will decay directly to the LSP. This assumption has been used in the analysis of UA1, UA2 and CDF data. Although these signatures should remain valid at LHC and SSC energies, it has been emphasized that for large squark and gluino masses, direct decays to the LSP become less important [57]. Instead, the decays to heavier

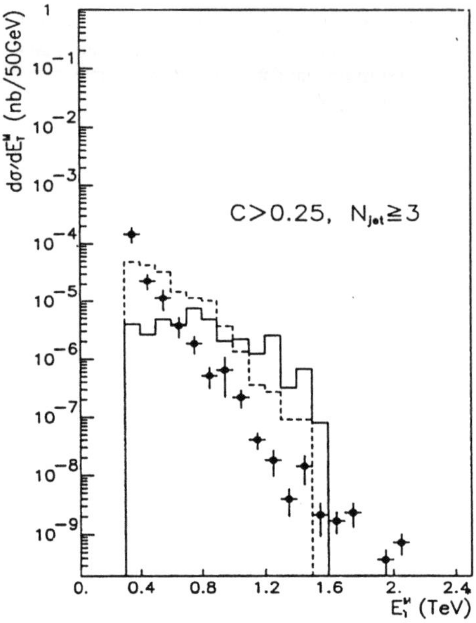

Fig. 26. Missing transverse energy distribution at $\sqrt{s}$ = 17 TeV after possible selection cuts for squark (*solid histogram*) and gluino (*dashed histogram*) production and for the total background (*points with error bars*). A squark and gluino mass of 1 TeV/c^2 is assumed [58]

neutralinos and charginos become dominant. These cascade decays generally occur for both the produced gluinos and squarks, and result in a large variety of possible final states, often containing W and Z bosons in a complex event topology.

Detailed studies of signal and backgrounds have been performed for the direct decays of sparticles into the LSP [58]. Figure 26 shows an example of the $\not{E}_T$ spectrum using one of the possible event selections at LHC energies. The resulting signal-to-background ratios are typically ~ 10:1 at large $\not{E}_T$. Taking into account uncertainties involved in these calculations, it was concluded that squark and gluino masses, up to ~ 1 TeV/c^2 are potentially accessible for 10 fb^{-1} integrated luminosity. Extrapolating the LHC study to SSC energies, a discovery limit of ~ 1.5–2 TeV/c^2 can be expected for the gluino and squark mass, because the signal cross-section increases much faster than the background cross-sections for higher $\sqrt{s}$. The analysis discussed above has been extended to study the feasibility of searching for $\tilde{g}$ and $\tilde{q}$ if the LHC runs at a luminosity of $\sim 5 \times 10^{34}$ cm^{-2} s^{-1} [59]. It was concluded that, provided the experimental problems associated with running at high luminosity can be overcome, an extension of the discovery limit for $\tilde{g}$ and $\tilde{q}$ of masses ~ 2 TeV/c^2 should be feasible at the LHC.

A study of gluino final states containing Z's from cascade decays was carried out for SSC energies [57]. It was concluded that it is possible to find event signatures that give promising results for signal-to-background ratios even for these complex decay modes.

Additional vector bosons are predicted in many possible extensions of the Standard Model. In order to study the discovery potential for an additional neutral gauge boson Z′ at the LHC/SSC, a superstring-inspired model was used [60]. This model predicts one (or two) new Z′ with masses in the range of ≤ 1 TeV/c^2. Figure 27 shows the predicted $\sigma \cdot BR(Z' \to \ell^+\ell^-)$ for the minimal superstring model at LHC

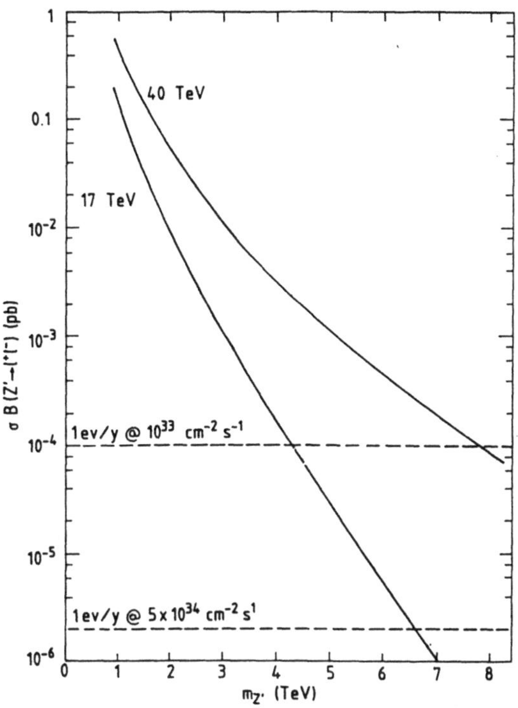

Fig. 27. Cross-section times maximal $\ell^+\ell^-$ branching ratio for Z′ in a minimal rank-5 superstring-inspired model at $\sqrt{s}$ = 17 and 40 TeV [60]

and SSC energies. Since no significant background is expected for the leptonic decay modes, it was concluded that a search can be made for Z′ with masses up to 4 TeV (6 TeV) using the $Z' \to \ell^+\ell^-$ decay channel at the LHC for 10 fb^{-1} (500 fb^{-1}) integrated luminosity. The corresponding discovery limit at the SSC is 7 TeV for 10 fb^{-1} integrated luminosity. This shows that the discovery potential for additional vector bosons is very high in hadron-hadron machines, and also demonstrates the benefit of having higher available centre-of-mass energies.

6. Conclusions

The data collected between 1982 and 1985 by the UA1 and UA2 experiments at the CERN p$\bar{p}$ Collider, corresponding to integrated luminosities of about 1 pb^{-1}, have provided many important physics results. Excellent agreement with the Standard Model has been observed. Searches for new physics have so far been fruitless; however, in several cases, significant lower mass limits could be placed on hypothetical new particles.

First results from CDF at the Fermilab Collider are in agreement with the expectations owing to the increase in $\sqrt{s}$, and no deviations from the Standard Model have been observed. The benefit of the larger $\sqrt{s}$ has been best demonstrated in searches for new physics ($m_{\tilde{g}}$, $m_{\tilde{q}}$, Λ_c).

From the data taken with ACOL and at Fermilab, it is expected that the Standard Model predictions can be tested more precisely, and that hopefully at least one of its missing elements – the top quark – will be found. The preliminary results on searches for the t-quark using a subsample of available data, lead to the conclusion that the t-quark must be very heavy compared with other known quarks and leptons, provided the Standard Model is corrrect and that no new physics influences the production and decay properties of the t-quark. The goal of discovering new physics still remains, but will in some cases be more easily accessible at the Tevatron and at LEP.

References

1 S.L. Glashow et al.: Phys. Rev. **D18** (1978) 1724;
E. Eichten et al.: Rev. Mod. Phys. **56** (1984) 579

2 M.E. Peskin: In *Proc. Int. Symp. on Lepton and Photon Interactions at High Energies, Kyoto, 1985*, eds. M. Konuma and C. Takahashi (Kyoto Univ., 1985), p. 714;
H. Harari: In *Proc. 5th Topical Workshop on Proton-Antiproton Collider Physics, Saint-Vincent, 1985*, ed. M. Greco (World Scientific, Singapore, 1985), p. 429

3 P. Fayet and S. Ferrara: Phys. Rep. **32** (1977) 249;
H.P. Nilles: Phys. Rep. **110** (1984) 1;
J. Ellis et al.: Nucl. Phys. **B238** (1984) 453;
H.E. Haber and G.L. Kane: Phys. Rep. **117** (1985) 75

4 J. Dowell (UA1 Collab.): In *Proc. 6th Topical Workshop on Proton-Antiproton Collider Physics, Aachen 1986*, eds. K. Eggert et al. (World Scientific, Singapore,1987), p. 419

5 C. Booth (UA2 Collab.): In *Proc. 6th Topical Workshop on Proton-Antiproton Collider Physics, Aachen 1986*, eds. K. Eggert et al. (World Scientific, Singapore,1987), p. 381

6 F. Abe et al.: Nucl. Instrum. Methods Phys. Res. **A271** (1988) 387

7 UA1 Collaboration papers:
G. Arnison et al.: Phys. Lett. **122B** (1983) 103; **126B** (1983) 398; **129B** (1983) 273; **134B** (1984) 469; **147B** (1984) 241; **166B** (1986) 484; Europhys. Lett. **1** (1986) 327;
C. Albajar et al.: Phys. Lett. **185B** (1987) 233

8 UA2 Collaboration papers:
M. Banner et al.: Phys. Lett. **122B** (1983) 476;
P. Bagnaia et al.: Phys. Lett. **129B** (1983) 130; Z. Phys. **C24** (1984) 1;
J.A. Appel et al.: Z. Phys. **C30** (1986) 1;
R. Ansari et al.: Phys. Lett. **186B** (1987) 440

9 C. Albajar et al. (UA1 Collab.): preprint CERN-EP/88-168 (1988), submitted to Z. Phys

10 F. Abe et al. (CDF Collab.): Phys. Rev. Lett. **62** (1989) 1005

11 H. Abramowicz et al.: Phys. Rev. Lett. **57** (1986) 298;
J.V. Allaby et al.: Z. Phys. **C36** (1987) 611;
D. Bogert et al.: Phys. Rev. Lett. **55** (1985) 1969;
P. Reutens et al.: Phys. Lett. **152B** (1985) 404

12 F. Jegerlehner: Z. Phys. **C32** (1986) 425;
W.J. Marciano and A. Sirlin: Phys. Rev. **D22** (1980) 2695; **D29** (1984) 945

13 U. Amaldi et al.: Phys. Rev. **D36** (1987) 1385;
J.V. Allaby et al.: preprint CERN-EP/87-140 (1987);
G. Costa et al.: Nucl. Phys. **B297** (1988) 244;
G. Gounaris and D. Schildknecht: preprint CERN-TH.4940 (1987)

14 G. Altarelli et al.: Nucl. Phys. **B157** (1979) 461; **B246** (1984) 12;
G. Altarelli et al.: Z. Phys. **C27** (1985) 617

15 R. Voss: In *Proc. Int. Symp. on Lepton and Photon Interactions at High Energies, Hamburg, 1987*, eds. W. Bartel and R. Rückl (North-Holland Publ. Co., Amsterdam, 1988), p. 525

16 P. Jenni: In *Proc. Int. Symp. on Lepton and Photon Interactions at High Energies, Hamburg, 1987*, eds. W. Bartel and R. Rückl (North-Holland Publ. Co., Amsterdam, 1988), p. 341

17 C. Albajar et al. (UA1 Collab.): Phys. Lett. **193B** (1987) 389
18 G. Arnison et al. (UA1 Collab.): Phys. Lett. **172B** (1986) 461
19 C. Albajar et al. (UA1 Collab.): Phys. Lett. **209B** (1988) 385
20 C. Albajar et al. (UA1 Collab.): Phys. Lett. **209B** (1988) 397
21 W.J. Stirling: In *Proc. 6th Topical Workshop on Proton-Antiproton Collider Physics, Aachen, 1986*, eds. K. Eggert, H. Faissner and E. Radermacher (World Scientific, Singapore, 1987), p. 301
22 P. Aurenche et al.: Nucl. Phys. **B297** (1988) 661; Phys. Lett. **140B** (1984) 87
23 R. Ansari et al. (UA2 Collab.): Phys. Lett. **215B** (1988) 175
24 C. Albajar et al. (UA1 Collab.): Phys. Lett. **198B** (1987) 271
25 L. Di Lella: In *Proc. 7th Topical Workshop on Proton-Antiproton Collider Physics, Fermilab, 1988*, eds. R. Raja et al. (World Scientific, Singapore, 1989), p. 849
26 F.E. Paige: In *Proc. Workshop on Experiments, Detectors, and Experimental Areas for the Supercollider, Berkeley, USA, 1987*, eds. R. Donaldson and M. Gilchriese (World Scientific, Singapore, 1989), p. 750
27 C. Albajar et al. (UA1 Collab.): Z. Phys. **C37** (1988) 489; Phys. Lett. **186B** (1987) 237; **200B** (1988) 380; **213B** (1988) 405
28 F.E. Paige and S.D. Protopopescu: Brookhaven National Laboratory report, BNL-38034 (1986)
29 P. Nason, S. Dawson and R.K. Ellis: Nucl. Phys. **B303** (1988) 607
30 G. Altarelli et al.: preprint CERN-TH.4978/88 (1988)
31 C. Albajar et al. (UA1 Collab.): Phys. Lett. **186B** (1987) 247
32 H. Albrecht et al.: Phys. Lett. **192B** (1987) 246;
H. Schröder: In *Proc. 24th Int. Conf. on High-Energy Physics, Munich, 1988*, eds. R. Kotthaus and J.H. Kühn (Springer-Verlag, Berlin, Heidelberg, 1989), p. 73
33 A. Breakstone et al.: Phys. Lett. **135B** (1984) 510
34 K. Eggert and H.G. Moser: In *Proc. 7th Topical Workshop on Proton-Antiproton Collider Physics, Fermilab, 1988*, eds. R. Raja et al. (World Scientific, Singapore, 1989), p. 599
35 T. Kamae: In *Proc. 24th Int. Conf. on High-Energy Physics, Munich 1988*, eds. R. Kotthaus and J.H. Kühn (Springer-Verlag, Berlin, Heidelberg, 1989), p. 156
36 J. Ellis, J.S. Hagelin and S. Rudaz: Phys. Lett. **192B** (1987) 201;
V. Barger, T. Hau and D.V. Nanopoulos: Phys. Lett. **194B** (1987) 312;
I.I. Bigi and A.I. Sanda: Phys. Lett. **194B** (1987) 307;
L.L. Chau and W.Y. Keung: UC Davis preprint, UCD-87-02 (1987);
H. Harari and Y. Nir: Stanford preprint SLAC-PUB-4341 (1987);
G. Altarelli and P.J. Franzini: preprint CERN-TH-4745/87 (1987);
H. Albrecht et al. (ARGUS Collab.): Phys. Lett. **192B** (1987) 245
37 C. Albajar et al. (UA1 Collab.): Z. Phys. **C37** (1988) 504
38 J. Kroll: In *Proc. 24th Int. Conf. on High-Energy Physics, Munich, 1988*, eds. R. Kotthaus and J.H. Kühn (Springer-Verlag, Berlin, Heidelberg, 1989), p. 1416
39 Seminar at CERN: *Search for the top quark*, 18 April 1989:
M. Della Negra (UA1),
L. Fayard (UA2),
G. Bellettini (CDF)
40 D. Froidevaux: In *Proc. Workshop on Physics at Future Accelerators, La Thuile, Italy and CERN, Geneva, Switzerland*, ed. J. Mulvey, CERN 87-07 (1987), Vol. I, p. 61
41 C. Albajar et al. (UA1 Collab.): Phys. Lett. **185B** (1987) 241
42 S. Dawson et al.: Phys. Rev. **D31** (1985) 1581
43 C. Albajar et al. (UA1 Collab.): Phys. Lett. **198B** (1987) 261
44 R. Ansari et al. (UA2 Collab.): Phys. Lett. **195B** (1987) 613
45 F. Abe et al. (CDF Collab.): Phys. Rev. Lett. **62** (1989) 1825
46 P. Langacker et al.: Phys. Rev. **D30** (1984) 1470;
E. Cohen et al.: Phys. Lett. **165B** (1985) 76;
M.B. Green, J.H. Schwarz and E. Witten: *Superstring theory* (Cambridge Univ. Press, Cambridge, 1987);
F. del Aguila et al.: Nucl. Phys. **B284** (1987) 530; **B287** (1987) 419
47 P.H. Frampton and S.L. Glashow: Phys. Lett. **190B** (1987) 157
48 P.A. Sphicas: In *Proc. 7th Topical Workshop on Proton-Antiproton Collider Physics, Fermilab, 1988*, eds. R. Raja et al. (World Scientific, Singapore, 1989), p. 239
49 E. Eichten et al.: Phys. Rev. Lett. **50** (1983) 811
50 J.A. Appel et al. (UA2 Collab.): Phys. Lett. **160B** (1985) 349
51 G. Arnison et al. (UA1 Collab.): Phys. Lett. **177B** (1986) 244
52 J.C. Collins and D.E. Soper: Phys. Rev. **D16** (1977) 2219

53 F. Abe et al. (CDF Collab.): Phys. Rev. Lett. **62** (1989) 613
54 D. Froidevaux and P. Jenni: In *Proton-Antiproton Collider Physics*, eds. G. Altarelli and L. Di Lella (World Scientific, Singapore, 1989), p. 323
55 For a recent review, see F. Pauss: In *Proc. 24th Int. Conf. on High-Energy Physics, Munich 1988*, eds. R. Kotthaus and J.H. Kühn (Springer-Verlag, Berlin, Heidelberg, 1989), p. 1275; and references therein
56 R. Barbieri: In *Proc. Rencontres de Physique de la Vallée d'Aoste, La Thuile, Italy, 1987*, ed. M. Greco (Editions Frontières, Gif-sur-Yvette, 1987), p. 523;
J. Gunion et al.: Univ. Calif. Davis report UCD-88-11 and LBL-25033 (1988);
G.F. Giudice: Phys. Lett. **B208** (1988) 315
57 R.M. Barnett et al.: In *Proc. Workshop on Experiments, Detectors, and Experimental Areas for the Super Collider, Berkeley, 1987*, eds. R. Donaldson and H. Gilchriese (World Scientific, Singapore, 1989), p. 178;
H. Baer and E. Berger: Phys. Rev. **D34** (1986) 1361;
H. Baer et al.: Phys. Rev. **D36** (1987) 96
58 R. Batley: In *Proc. Workshop on Physics at Future Accelerators, La Thuile, Italy, and CERN, Geneva, Switzerland*, ed. J. Mulvey, CERN 87-07 (1987), Vol. II, p. 109
59 F. Pauss et al.: In *The Feasibility of Experiments at High Luminosity at the Large Hadron Collider*, ed. J. Mulvey, CERN 88-02 (1988), p. 79
60 J. Ellis and F. Pauss: In *Proc. Workshop on Physics at Future Accelerators, La Thuile, Italy, and CERN, Geneva, Switzerland*, ed. J. Mulvey, CERN 87-07 (1987), Vol. I, p. 80;
J. Ellis: In *Proc. of the 1987 ICFA Seminar on Future Perspectives in High-Energy Physics, BNL, Upton, NY, USA, October 1987*, ed. P.F. Dahl, BNL 52114, p. 117

Weak Mixing, CP Violation and the Standard Model

Konrad Kleinknecht

Institut für Physik, Johannes-Gutenberg-Universität Mainz
Postfach 3980, D-6500 Mainz 1, Fed. Rep. of Germany

These lectures summarize experimental information on weak couplings of quark flavours from weak decays and weak production of heavy flavours and the relation of the observed direct CP violation to a non-trivial phase in the quark mixing matrix. The interest in this field is twofold: on the one side, we can measure precisely the parameters of the standard model in its own framework; on the other side, any inconsistency amongst these measurements or evidence for any process forbidden in the standard model will lead beyond that model. The outline of these lectures is the following:

1. Weak Mixing of Quarks
   1.1 Kobayashi-Maskawa-Matrix
   1.2 Experimental Constraints on the KM Matrix
   1.3 Allowed Ranges of Mixing Angles for 3 or 4 Generations
2. CP Violation
   2.1 Phenomenology and Models
   2.2 New Experiments on ϵ'/ϵ
      - E731 at Fermilab
      - NA31 at CERN: First Evidence for Direct CP Violation
3. Implications of Experimental Results on the Standard Model Parameters
4. Beyond the Standard Model: Searches for Rare Kaon Decays
5. Conclusions – Outlook

1. Weak Mixing of Quarks

While all leptons have the same weak coupling constant, this is not so for different quark flavours. When this became experimentally clear in weak hyperon decays, Cabbibo [1] realized that quark mass eigenstates are not the same as weak eigenstates, and that a linear combination of quark states with charge $-1/3$

$$d_c = d\cos\theta + s\sin\theta \tag{1}$$

was the weak eigenstate coupling to the u quark. The orthogonal combination

$$s_c = -d\sin\theta + s\cos\theta \tag{2}$$

Physics for a New Generation Editors: H. Latal · H. Mitter

was then invoked by Glashow, Iliopoulos and Maiani [2] as a partner for the newly postulated charm quark, in order to cancel the first-order diagrams for the decay $K_L \rightarrow \mu\,\mu$. For four quarks, the weak mixing matrix connecting the weak eigenstates and the mass eigenstates then reads

$$\begin{pmatrix} d_c \\ s_c \end{pmatrix} = \begin{pmatrix} \cos\theta & \sin\theta \\ -\sin\theta & \cos\theta \end{pmatrix} \begin{pmatrix} d \\ s \end{pmatrix} . \tag{3}$$

This unitary 2×2 matrix has only one rotation angle θ and no (non-trivial) complex phase. There is therefore *no* possibility of incorporating a direct CP violation in weak $\Delta S = 1$ decays in this model.

1.1 Kobayashi–Maskawa Matrix

For 6 quarks, the generalization of this quark mixing matrix introduced by Kobayashi and Maskawa [3] is a 3×3 unitary matrix V operating on the charge $-1/3$ quarks (d, s, b). By convention the quarks with +2/3 charge remain unmixed. The matrix between weak eigenstates (d′, s′, b′) and mass eigenstates (d, s, b) reads:

$$\begin{pmatrix} d' \\ s' \\ b' \end{pmatrix} = \begin{pmatrix} V_{ud} & V_{us} & V_{ub} \\ V_{cd} & V_{cs} & V_{cb} \\ V_{td} & V_{ts} & V_{tb} \end{pmatrix} \begin{pmatrix} d \\ s \\ b \end{pmatrix} . \tag{4}$$

This matrix can be parametrized with four free parameters, three angles and one non-trivial phase. The important point remarked by Kobayashi and Maskawa was the presence of this complex phase in the matrix. They suggested that this could be the origin of CP violation.

There are several parametrizations of the K-M matrix. We adopt here the standard form advocated in the 1988 Review of Particle Properties [4]:

$$V = \begin{pmatrix} c_{12}c_{13} & s_{12}c_{13} & s_{13}e^{-i\delta_{13}} \\ -s_{12}c_{23} - c_{12}s_{23}s_{13}e^{i\delta_{13}} & c_{12}c_{23} - s_{12}s_{23}s_{13}e^{i\delta_{13}} & s_{23}c_{13} \\ s_{12}s_{23} - c_{12}c_{23}s_{13}e^{i\delta_{13}} & -c_{12}s_{23} - s_{12}c_{23}s_{13}e^{i\delta_{13}} & c_{23}c_{13} \end{pmatrix} . \tag{5}$$

This notation was proposed for three generations by Chau and Keung [5], and for an arbitrary number of generations by Harari and Leurer [6] and by Fritzsch and Plankl [7]. Here $c_{ij} = \cos\theta_{ij}$ and $s_{ij} = \sin\theta_{ij}$, with the indices i and j numbering the generation in increasing order of mass. Related parametrizations have been given by Maiani [8] and Wolfenstein [9]. The advantage of this parametrization is that each angle θ_{ij} is related to the mixing between generation i and j, and decoupled from the other angles. In the limit $\theta_{23} = \theta_{13} = 0$, the third generation is completely decoupled, and we have two-generation mixing with one angle θ_{12} identical to the original Cabibbo angle [1]. If we anticipate the experimental fact that $s_{13}^2 \ll s_{23}^2 \ll 1$ and call $\delta = \delta_{13}$, then an approximate form of Eq. (5) is the following:

$$V = \begin{pmatrix} c_{12} & s_{12} & s_{13}e^{-i\delta} \\ -s_{12} & c_{12} & s_{23} \\ s_{12}s_{23} - s_{13}c_{12}e^{i\delta} & -s_{23} & 1 \end{pmatrix} . \tag{6}$$

CP violating amplitudes are then all proportional to the product of four sines, the quantity [10]

$$J = s_{12}\, s_{13}\, s_{23}\, \sin\delta\ . \tag{7}$$

If any of these angles vanishes, CP violation cannot be described within the standard model.

1.2 Experimental Constraints on the KM Matrix

1.2.1 Light Quark Couplings

Coupling V_{ud}. This parameter is obtained by comparing decay rates of nuclear beta decays and muon decays. A new analysis of radiative corrections in the nuclear beta decays has been presented by Marciano and Sirlin [11], and more recently, the inconsistencies between ft-values of low-Z Fermi transitions have been revised and removed by Sirlin and Zucchini [12]. This result has been confirmed by Jaus and Rasche [13] and by Sirlin [14] who, as a final result, quotes:

$$|V_{ud}| = 0.9744 \pm 0.0010\ . \tag{8}$$

An additional correction caused by isospin breaking due to $\rho - \omega -$ mixing is advocated by Lopez-Castro and Pestieau [15]. However, the size of this correction, estimated at $-(2 \pm 1)\,10^{-3}$, seems to be at variance [16] with the Ademollo-Gatto theorem [17] which ensures that first-order symmetry-breaking effects in vector currents vanish.

Coupling V_{us}. Two kinds of experimental information on the coupling of strange and up-quarks exist; one comes from the analysis [18] of K_{e3} decays, $K_L \rightarrow \pi e \nu$ and $K^+ \rightarrow \pi e \nu$. This result, $|V_{us}| = 0.2196 \pm 0.0023$, is unaffected by SU(3) breaking effects [17] because only vector currents are involved. The other input comes from the selfconsistent analysis of weak semileptonic hyperon and neutron decays, as done by the WA2 collaboration [19] with the result $|V_{us}| = 0.231 \pm 0.003$. Here, SU(3) breaking effects come in for the axial vector part. The procedure proposed by Donoghue et al. [20] applied to the WA2 data [21] gives a corrected value $|V_{us}| = 0.222 \pm 0.003$. The average of these two consistent results is

$$|V_{us}| = 0.2205 \pm 0.0018\ . \tag{9}$$

1.2.2 Charm-Quark Couplings

Coupling V_{cd}. This coupling has been determined from measurements of single charm production in neutrino and antineutrino reactions. The coupling parameter is obtained from the measured ratios of dimuon to single muon production cross-sections in neutrino reactions ($R^\nu = \sigma^\nu_{\mu^-\mu^+}/\sigma^\nu_{\mu^-}$) and antineutrino reactions ($R^{\bar\nu} = \sigma^{\bar\nu}_{\mu^-\mu^+}/\sigma^{\bar\nu}_{\mu^-}$):

$$\beta\, V_{cd}^2 = \frac{R^\nu - R R^{\bar\nu}}{1 - R} \cdot \frac{2}{3}$$

where β is the semileptonic branching ratio of the mixture of charmed particles produced in the neutrino reactions, and $R = \sigma^{\bar{\nu}}/\sigma^{\nu}$ is the ratio of total charged current neutrino cross-sections.

The result of the CDHS collaboration [22],

$$\beta V_{cd}^2 = (0.41 \pm 0.07)\, 10^{-2}$$

has been used to extract V_{cd}. The effective semileptonic branching ratio is obtained [23] from e^+e^- data on charmed particle decays and from emulsion data on neutrino production of different charmed hadron species to be $\beta = (9.3 \pm 1.0)\%$. This gives:

$$|V_{cd}| = 0.21 \pm 0.03\,. \tag{10}$$

Coupling V_{cs}. In the past, this coupling was obtained from neutrino and antineutrino dimuon production data. Since here the charm production from strange sea quarks is considered, the quantity measured is the product $|V_{cs}|^2 \cdot 2S$, where $S = \int xs(x)dx$ is the integral of the strange sea structure function. In the absence of an independent determination of S, one obtains from SU(3) as an upper limit $2S \leqq \bar{U} + \bar{D}$, where $\bar{U}$ and $\bar{D}$ are the momentum fractions of non-strange sea antiquarks. We call $\alpha = 2S/(\bar{U} + \bar{D})$ the ratio of these momentum fractions, and

$$\alpha^* = 2S\left(|V_{us}|^2 + |V_{cs}|^2/r_s\right)/(\bar{U} + \bar{D})$$

the same ratio modified by the threshold suppression factor r_s (= 1.53 [22]) for the charm quark mass. Then from the x-distribution of neutrino dimuons [22] one obtains

$$\frac{|V_{cs}|^2}{|V_{cd}|^2} = (6.26 \pm 0.73)\frac{1 + \alpha^*}{\alpha}\,. \tag{11}$$

With the conservative assumption $\alpha \leqq 1$, this leads to the bound $|V_{cs}| > 0.59$ at 90% C.L. [22].

An alternative method of obtaining V_{cs} is based on the measurement and calculation of the decay rate for $D^+ \to \overline{K^0} e^+ \nu_e$ [24, 25]. This rate can be expressed as

$$\Gamma(D^+ \to \overline{K} e^+ \nu_e) = |f_+^D(0)|^2\, |V_{cs}|^2 (1.54 \cdot 10^{11}\, s^{-1})$$

where $|f_+^D(0)|$ is the form factor for D_{l3} decay for zero momentum transfer. Combining data on the branching ratios of the decays $D^+ \to \overline{K^0} e^+ \nu$ and $D^0 \to K^- e^+ \nu$ and with the world average values of D^+ and D^0 lifetimes, the experimental decay width is $\Gamma(D \to \overline{K} e^+ \nu) = (0.79 \pm 0.11) \times 10^{11}\, s^{-1}$. This leads to

$$|f_+^D(0)|^2\, |V_{cs}|^2 = 0.51 \pm 0.07\,. \tag{12}$$

In three theoretical calculations using constituent quark models [26], three-point function QCD sum rules [27] and two-point function QCD sum rules [28], the values $0.75 - 0.82$, 0.6 ± 0.1, and 0.75 ± 0.05 were obtained for the transition form factor $|f_+^D(0)|$. We use here an intermediate value

$$|f_+^D(0)| = 0.68 \pm 0.08 \tag{13}$$

which, together with Eq. (12), leads to

$$|V_{cs}| = 0.96 \pm 0.10\ . \tag{14}$$

If, on the other hand, the calculation of the form factor is not relied upon, at least the limit

$$|f_+^D(0)| \leqq 1 \tag{15}$$

can be used, together with Eq. (12), to obtain a lower limit on $|V_{cs}|$.

1.2.3 Bottom-Quark Couplings

Ratio $|V_{ub}/V_{cb}|$. There are two sources of information on the matrix element V_{ub}: one is the inclusive semileptonic decay $B \to e\nu X_u$ with an unidentified charmless state X_u, the other is the observation of an exclusive charmless hadronic B decay.

The measurements of the inclusive electron momentum spectrum from B decays is a very sensitive method for obtaining a limit on the ratio $|V_{ub}/V_{cb}|$. The energy of electrons from semileptonic decays of the kind $B \to e\ \nu X_u$ extends beyond the maximum energy possible in decays with a charmed hadronic state, $B \to e\ \nu X_c$. The limit on $|V_{ub}/V_{cb}|$ deduced from experimental electron spectra depends on the calculated spectral shapes for these two decay modes.

There are two places where models for these spectral shapes enter: First, when the ratio of decay rates $\Gamma^u/\Gamma^c = \Gamma(b \to u)/\Gamma(b \to c)$ is extracted from experimental data, the predicted shape for the channel $b \to u$ above the end-point of the $b \to c$ decay enters; second, when the ratio of coupling strengths $|V_{ub}/V_{cb}|$ is deduced from Γ^u/Γ^c, the phase-space integrals f_u and f_c are used:

$$\frac{\Gamma(b \to u)}{\Gamma(b \to c)} = \frac{f_u}{f_c}\frac{|V_{ub}|^2}{|V_{cb}|^2}\ . \tag{16}$$

The parton spectator model of Altarelli et al. [29] gives a ratio of phase space integrals of $f_c/f_u \sim 0.5$ (see Eq. (20)) and relatively soft electron momentum spectrum. Alternatively, form-factor models for exclusive channels can be used to obtain the inclusive spectral shape as a sum of many such exclusive channels [26, 30, 31]. In these latter models, the electron momentum spectra for both decays are harder than in [29]. Measurements of this inclusive electron spectrum at the Υ(4S) resonance in e^+e^- storage rings have not found evidence for the $b \to u$ decay. Therefore, only upper limits for $|V_{ub}/V_{cb}|$ are derived from these experiments. A new result of the ARGUS collaboration [32], is shown in Fig. 1. Upper limits from different experiments are given in Table 1.

If one uses the conservative approach of extracting Γ^u/Γ^c with the model giving the least sensitivity for the upper end of the $b \to u$ momentum spectrum, a limit around the latest ARGUS value is obtained. However, for deducing $|V_{ub}/V_{cb}|$, the phase space ratio of the ACM [29] model seems to be more reliable than the one of GIW [30]. The result is then, at 90% C.L.:

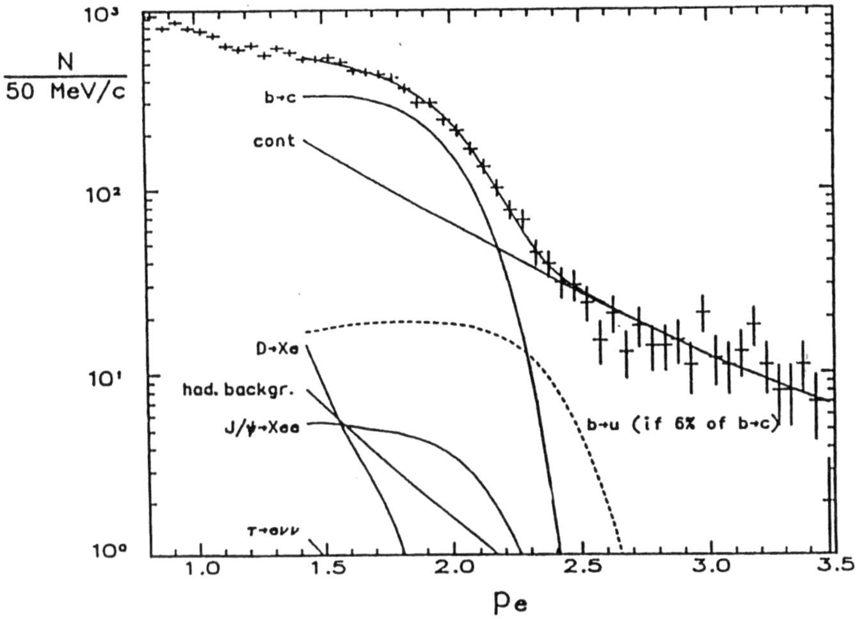

Fig. 1. Electron momentum spectrum from B decays produced at the Y(4S) resonance from ARGUS [32]

Table 1. Upper limits on $|V_{ub}/V_{cb}|$ from inclusive semileptonic B decays (90% C.L.)

Experiment	Ref.	Γ^u/Γ^c	Model used
ARGUS 88	32	< 0.03	GIW [30]
CLEO 87	33	< 0.04	GIW [30]
X BALL 87	34	< 0.046	GIW [30]
CUSB 83	35	< 0.055	ACM [29]

$$|V_{ub}/V_{cb}| < 0.13 \,. \tag{17}$$

While in this search the b → u transition was not seen, the observation of exclusive charmless decays has been reported by the ARGUS collaboration [36]. The decay channels $B^+ \to \bar{p}p\pi^+$ and $B^0 \to \bar{p}p\pi^+\pi^-$ were seen with branching ratios of $(5.2 \pm 1.4 \pm 1.9)\,10^{-4}$ and $(6.0 \pm 2.0 \pm 2.2)\,10^{-4}$, respectively. This observation has not been confirmed by the CLEO collaboration [37], as discussed by Kreinick [38]. The discrepancy between these results, at the 3 s.d. level, has not been cleared up, as shown by Schröder [39]. If the ARGUS result is taken at face value, by using rough estimates for the matrix element of this decay, one can obtain a lower limit [40]:

$$|V_{ub}/V_{cb}| > 0.07 \,. \tag{18}$$

We will see below that a *much* smaller value for V_{ub} would lead to difficulties with the explanation of CP violation in the Kobayashi-Maskawa model. Of course, $V_{ub} = 0$ would exclude such an explanation.

Coupling V_{cb}. This can be determined from the measured lifetime τ_B of bottom quark hadrons and their semileptonic branching ratio R_ℓ or, less accurately, from

Table 2. Measurements of semileptonic branching ratios R_ℓ of B mesons produced on the Υ(4S) (%)

Group	Ref.	$R(\mathrm{B} \to \mathrm{X}e\nu)$	$R(\mathrm{B} \to \mathrm{X}\mu\nu)$	$R_\ell(\mathrm{B} \to \mathrm{X}\ell\nu)$
CLEO 88	43			$10.0 \pm 0.2 \pm 1.2$
CLEO 87	33	$11.0 \pm 0.4 \pm 0.7$	$11.0 \pm 0.3 \pm 1.1$	
ARGUS 88	32	$9.9 \pm 0.8 \pm 0.2$	$10.5 \pm 1.1 \pm 0.2$	
X BALL 88	34	$11.7 \pm 0.4 \pm 0.7$		
CUSB 84	35	$13.2 \pm 0.8 \pm 1.4$	$11.2 \pm 0.9 \pm 1.0$	
average 88				10.9 ± 0.8

the semileptonic decay width of B decays into exclusive final states and models for these decays.

The semileptonic inclusive decay rate is

$$\Gamma(\mathrm{b} \to x\ell\nu) = \frac{R_\ell}{\tau_\mathrm{B}} = \frac{m_\mathrm{b}^5}{m_\mu^5}\frac{1}{\tau_\mu}\left(f_\mathrm{u}|V_\mathrm{ub}|^2 + f_\mathrm{c}|V_\mathrm{cb}|^2\right) . \tag{19}$$

Here the coefficients f_q (q = u or c) are the product of a phase space factor and a QCD correction in lowest order [41,29]:

$$\begin{aligned} f_\mathrm{q} = &\left(1 - 8r^2 - 12r^4 \ln r^2\right) \\ &\times \left[1 - 2\alpha_s \frac{m_\mathrm{b}^2}{3\pi} \cdot \left(\pi^2 - \frac{25}{4}\right) \cdot \left(1 - 1.23r + 0.7r^2\right)\right] \end{aligned} \tag{20}$$

with $r = m_\mathrm{q}/m_\mathrm{b}$. Using the new B lifetime average of $\tau_\mathrm{B} = (1.15 \pm 0.14)$ ps [42], and the new contributions of the ARGUS [32], CLEO [43] and Crystal Ball [34] collaborations on this semileptonic branching ratio of B mesons produced on the Υ(4S) resonance, one obtains the semileptonic decay rate. Table 2 summarizes data on R_ℓ from the four experiments, and gives the new average $R_\ell = (10.9 \pm 0.8)\%$.

V_cb is then obtained using Eq.(19). Here an important part of the error comes from the uncertainty in m_b, usually taken to be (5.0 ± 0.2) GeV [4]. It was pointed out by Rückl [44] that the experimentally measured shape of the inclusive lepton spectrum from $\mathrm{b} \to \mathrm{c}$ decay determines very precisely the mass difference $m_\mathrm{b} - m_\mathrm{c}$. Indeed, the ARGUS collaboration obtains [32] $m_\mathrm{b} - m_\mathrm{c} = (3.37 \pm 0.05)$ GeV. If m_b is varied keeping this value of $m_\mathrm{b} - m_\mathrm{c}$ fixed, the variation of the coefficient f_c according to Eq.(20) compensates to a large part the m_b^5 variation in Eq. (19). With $\alpha_s = 0.23$ and $m_\mathrm{b} = (5.0 \pm 0.2)$ GeV, $f_\mathrm{c} = 0.43 \pm 0.03$ is obtained. The product $f_\mathrm{c} \cdot m_\mathrm{b}^5$ varies only by $\pm 8\%$ within the range of uncertainty in m_b. Using this procedure, the error on V_cb is reduced substantially:

$$\begin{aligned} |V_\mathrm{cb}| &= 0.045 \pm 0.003 \pm 0.002 \\ &= 0.045 \pm 0.004 . \end{aligned} \tag{21}$$

The first error in the upper line is the one from τ_B and R_ℓ, and the second the one from the variation of $f_\mathrm{c} \cdot m_\mathrm{b}^5$.

1.3 Allowed Ranges of Mixing Angles for 3 or 4 Generations

Three Generations. Using the constraints of Eqs. (8–12, 15, 17, 21) we first do a fit using the six-quark parametrization of Eq. (5), along the lines of Ref. [45]. We

Table 3. Elements of 3 × 3 quark mixing matrix $|V_{ik}|$ from fit of experimental constraints (90% C.L. allowed ranges)

	d	s	b
u	0.9746 to 0.9758	0.218 to 0.224	(0.003) to 0.006
c	0.218 to 0.224	0.9736 to 0.9749	0.039 to 0.051
t	0.004 to 0.016	0.038 to 0.050	0.9987 to 0.9993

obtain, for a $\chi^2/D.F. = 1.01/1$, the values for the mixing angles $\sin\theta_{12} = 0.2205 \pm 0.0018$, $\sin\theta_{23} = 0.045 \pm 0.004$, and $\sin\theta_{13} < 0.006$, and $2S/(\bar{U}+\bar{D}) = 0.42 \pm 0.06$. The corresponding 90% C.L. allowed ranges for the mixing matrix elements are given in Table 3. A lower limit of $|V_{ub}| > 0.003$ would follow from Eq. (18).

Four Generations. We use the standard parametrization [4–7]. The allowed range of parameters was examined by a numerical scan of the parameter space. The free parameters of the scan were the six angles $\theta_{12}, \theta_{23}, \theta_{13}, \theta_{14}, \theta_{24}$ and θ_{34}, the three phases and the values of α and $f_+^D(0)$. The parameters were varied in a region around the solution with the minimal $\chi^2 = \chi_0^2$, the phases were varied between 0 and 2π. As experimental constraints, Eqs. (8–13, 17, 21) were used with the value of Eq. (18) as an additional constraint. All combinations of parameters leading to a χ^2 value below $\chi_0^2 + 1.6^2$ were accepted as allowed values. Table 4 gives the resulting bounds for the elements of the 4 × 4 unitary matrix.

Also, in Fig. 2 contour plots for allowed regions of each pair of angles $\theta_{ik}(i < k)$ are given. In this analysis, several experimental constraints which have implications on some of the elements of the mixing matrix, were not used because their interpretation in terms of the KM elements needs more theoretical input. These experiments include: observation of B - $\bar{\text{B}}$ - mixing, of direct CP violation, and of direct b → u transitions. This will be discussed in Sect.3.

Discussion. It is evident from this analysis that for the first three generations of quarks, we observe a pattern of decreasing mixing angles: the angle connecting the first two generations, θ_{12}, is larger than the one connecting generation 2 and 3, θ_{23}, and this one again is larger than θ_{13}. This pattern remains to be explained by theories beyond the standard model which should explain the pattern of quark masses as well. At the moment, only a few phenomenological models exist which connect the values

Table 4. Elements of 4 × 4 quark mixing matrix $|V_{ik}|$ from fit of experimental constraints (90% C.L. allowed ranges)

	d	s	b	b′
u	0.9730 – 0.9757	0.218 – 0.224	0. – 0.006	0. – 0.07
c	0.181 – 0.230	0.84 – 0.98	0.039 – 0.051	0. – 0.49
t	0. – 0.13	0. – 0.48	0.002 – 0.999	0. – 0.999
t′	0. – 0.13	0. – 0.47	0. – 0.999	0. – 0.999

allowed ranges for angles:

$0.218 < s_{12} < 0.223$	$0.039 < s_{23} < 0.054$
$0. < s_{13} < 0.006$	$0. < s_{24} < 0.49$
$0. < s_{14} < 0.072$	$0. < s_{34} < 1.0$

other parameters:

$0.27 < 2S/(\bar{U}+\bar{D}) < 0.57$

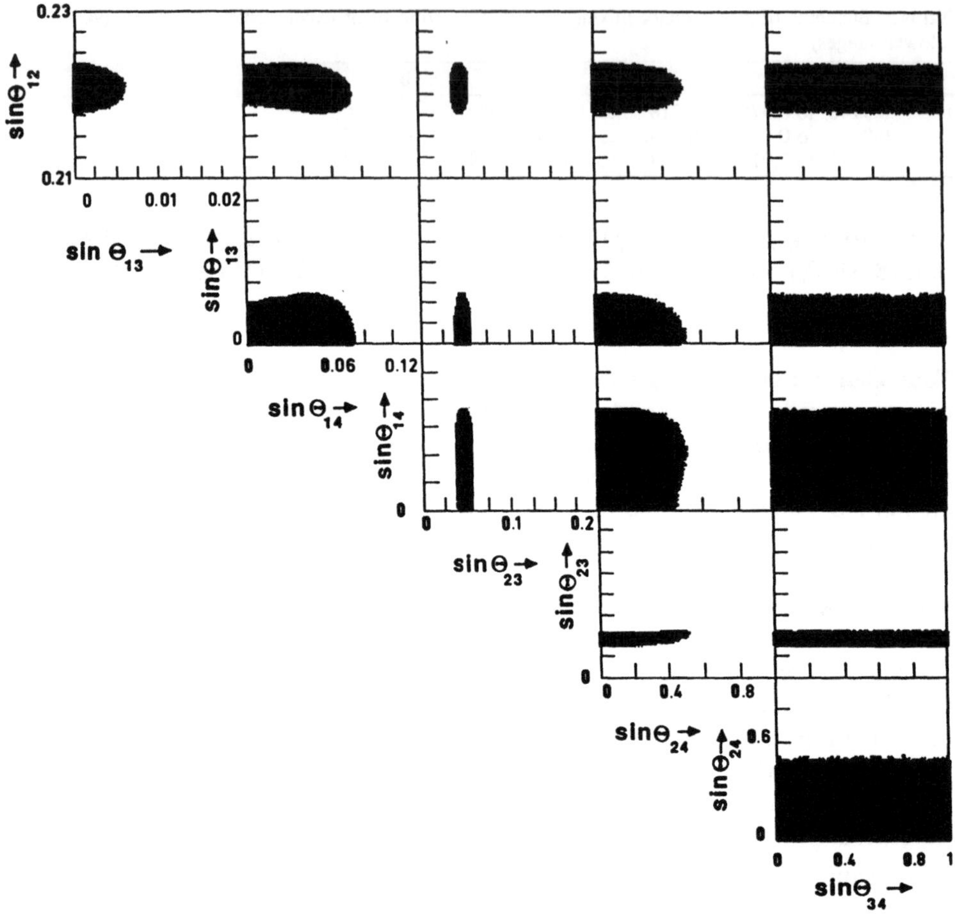

Fig. 2. Allowed ranges (90% C.L.) for the six mixing angles in the standard parametrization for 4 quark generations

of the mixing angles θ_{ik} to the ratio of quark masses in the generations i and k, m_i/m_k [46, 47].

For the angles connected to the fourth generation, only θ_{14} is constrained significantly. The upper limits obtained here follow the sequence $s_{34}^{\max} > s_{24}^{\max} > s_{14}^{\max}$. It is therefore possible that s_{34} and s_{24} are larger than s_{12}, contrary to the expectation of the simple models. The experimental situation can be improved significantly if the top quark is observed and if the observation of W decays into heavy quark pairs at LEP 200 or at $\bar{p}p$ Colliders allows a determination of their weak couplings.

Outlook. An analysis of the precision with which some of these weak coupling parameters could be obtained with the LEP machine at 200 GeV c.m. energy was done at the LEP 200 workshop at Aachen by Longo et al. [48]. The WW production cross-section is around 15 pb, significant measurements can be done with an integrated luminosity of 500 pb^{-1}. Using decays of the type WW $\rightarrow \ell\nu$ q $\bar{\text{q}}$ an estimate of the precision gives results of the sort:

Table 5. Allowed ranges of elements of 4 × 4 quark mixing matrix $|V_{ik}|$ from experimental data including anticipated results from experiments at LEP 200 (90% C.L. limits)

	d	s	b	b′
u	0.9729 – 0.9757	0.218 – 0.223	0. – 0.006	0. – 0.07
c	0.196 – 0.228	0.93 – 0.98	0.039 – 0.051	0. – 0.3
t	0. – 0.03	0. – 0.1	0.948 – 0.999	0. – 0.3
t′	0. – 0.12	0. – 0.29	0. – 0.3	0.93 – 1.

$$|V_{cs}|^2 = 1. \pm 0.1$$
$$|V_{tb}|^2 = 1. \pm 0.07$$
$$|V_{td}|^2 + |V_{ts}|^2 = 0. \pm 0.08$$

if the angles of the fourth generation are small, as anticipated here. With this information, it is possible to constrain significantly 8 of the 16 elements in the 4 × 4 matrix. In particular, it will be possible with this information to find out whether the 2 × 2 submatrix in the right hand lower corner is diagonal or antidiagonal. Table 5 shows allowed ranges for the matrix elements of the 4 × 4 mixing matrix if these anticipated results of experiments at LEP are included in the fit. The results can be improved if one assumes (optimistically) that also the decays WW → $q\bar{q}$ $q\bar{q}$ can be analyzed.

2. CP Violation

2.1 Phenomenology and Models

The phenomenon of CP violation in decays of neutral K^0 mesons is now with us for 24 years [49]. The first ten years of intense experimentation after the discovery of the decay $K_L \to \pi^+\pi^-$ were devoted to the observation of other manifestations of the phenomenon, like the decay $K_L \to \pi^0\pi^0$ [50] and the charge asymmetry in the decays $K_L \to \pi^\pm e^\mp \nu$ and $K_L \to \pi^\pm \mu^\mp \nu$ [51], and to precision experiments on the moduli and phases of the CP violating amplitudes [52]. These experimental results excluded a large number of theoretical models proposed to explain CP violation, such that at the time of the London 1974 conference [53] essentially two classes of models survived: The superweak model postulating a new, very weak, CP violating interaction with $\Delta S = 2$ [54], and milliweak models invoking a small (10^{-3} part of the normal $\Delta S = 1$ weak interaction as the source of CP violation. The key question then became: which of these models is describing the phenomenon? Can one devise experiments distinguishing between those models?

In this context it was very important that a specific milliweak model within the standard model was proposed by Kobayashi and Maskawa [3] in 1973. At the time of the discovery of CP violation, only 3 quarks were known, and there was no possibility of explaining CP violation as a genuine phenomenon of weak interactions. This situation remained unchanged with the fourth quark because the 2 × 2 mixing matrix has only one free parameter, the Cabibbo angle, and no non-trivial complex

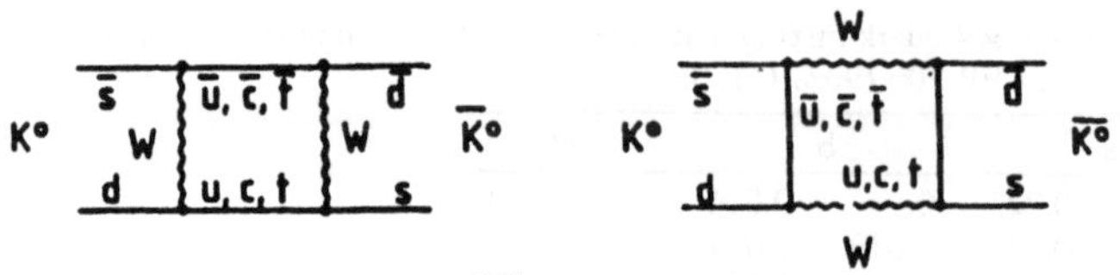

Fig. 3. Box diagram for $K^0 - \overline{K^0}$ mixing connected to CP violating parameter ϵ

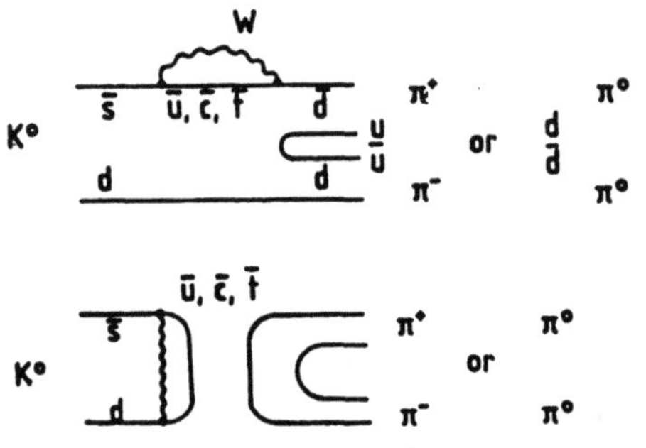

Fig. 4. Penguin diagrams for $K^0 \to 2\pi$ decay with direct CP violation (amplitude ϵ')

phase. However, as remarked by Kobayashi and Maskawa, the picture changes if six quarks are present. Then the 3×3 mixing matrix naturally contains a phase, apart from three mixing angles (see Sect. 1.1). It is then possible to construct CP violating weak amplitudes from "box-diagrams" of the form shown in Fig. 3.

A necessary consequence of this model of CP violation is the non-equality of the relative decay rates for $K_L \to \pi^+\pi^-$ and $K_L \to \pi^0\pi^0$. This "direct CP violation" is due to "penguin diagrams" of the form given in Fig. 4.

For a quantitative discussion, we use the conventional notations. Let $\overline{K^0}$ = CP K^0, then the eigenstates of CP are:

$$\begin{aligned} K_1 &= \left(K^0 + \overline{K^0}\right)/\sqrt{2} = +\mathrm{CP}\,K_1 \\ K_2 &= \left(K^0 - \overline{K^0}\right)/\sqrt{2} = -\mathrm{CP}\,K_2\,. \end{aligned} \tag{22}$$

The physical long-lived (K_L) and short-lived (K_S) states are then

$$\begin{aligned} K_S &= (K_1 + \epsilon_S K_2)\,/\,\left(1 + |\epsilon_S|^2\right)^{1/2} \\ K_L &= (K_2 + \epsilon_L K_1)\,/\,\left(1 + |\epsilon_L|^2\right)^{1/2}. \end{aligned} \tag{23}$$

With CPT invariance, $\epsilon_S = \epsilon_L = \epsilon$. The admixture parameter

$$\epsilon = \frac{\mathrm{Im}\Gamma_{12}/2 + i\,\mathrm{Im}M_{12}}{i\,(\Gamma_S - \Gamma_L)/2 - (m_S - m_L)} \tag{24}$$

is given in terms of the $K^0 - \overline{K^0}$ mass matrix M and decay matrix Γ. The phase of ϵ, assuming CPT, is

$$\mathrm{Arg}(\epsilon) = \arctan(2\Delta m/\Gamma_S) = 43.7^\circ \pm 0.2^\circ\,. \tag{25}$$

The experimentally observable quantities are

$$|\eta_{+-}|\,e^{i\phi_{+-}} = \eta_{+-} = < \pi^+\pi^-|T|K_L > / < \pi^+\pi^-|T|K_S >,$$
$$|\eta_{00}|\,e^{i\phi_{00}} = \eta_{00} = < \pi^0\pi^0|T|K_L > / < \pi^0\pi^0|T|K_S >. \tag{26}$$

It can be shown that these amplitude ratios consist of the contribution from CP violation in the $K^0-\overline{K^0}$ mixing (box diagrams in the KM model), called ϵ above, and another one from CP violation in the weak $K \to 2\pi$ amplitudes (penguin diagrams in the KM model), called ϵ':

$$\eta_{+-} = \epsilon + \epsilon'$$
$$\eta_{00} = \epsilon - 2\epsilon'. \tag{27}$$

In this way η_{+-}, η_{00} and $3\epsilon'$ form a triangle in the complex plane, the Wu-Yang triangle. The CP violating decay amplitude ϵ' is due to interference of $\Delta I = 1/2$ (A_0) and $3/2$ (A_2) amplitudes:

$$\epsilon' = \frac{i}{2}\frac{\mathrm{Im}A_2}{A_0}\exp[i(\delta_2 - \delta_0)] \tag{28}$$

and its phase is given by the $\pi\pi$ phase shifts in the $I = 0$ and $I = 2$ states, δ_0 and δ_2 (CPT assumed):

$$\mathrm{Arg}(\epsilon') = (\delta_2 - \delta_0) + \pi/2 \tag{29}$$

which experimentally is $(61 \pm 9)°$ [55] or $(45 \pm 9)°$ [56]. The two models discussed above differ significantly: the superweak model predicts vanishing direct CP violation in weak decays, $\epsilon' = 0$, and therefore $\eta_{+-} = \eta_{00} = \epsilon$, while in milliweak models one expects $\epsilon' \neq 0$. All data obtained up to 1987 are consistent with the superweak model, as shown in Table 6.

The prediction for ϵ' within the KM model can be estimated [62, 63] if one infers the magnitude of the mixing angles from other experiments (Sect. 1) and if the hadronic matrix elements for box graphs and penguin graphs are calculated (see Sect. 3). Typical values [63] for $|\epsilon'/\epsilon|$ are in the range $+(1-5) \times 10^{-3}$ for three generations of quarks and between -1% and $+1\%$ for four generations [64]. A measurement of this quantity to this level of precision therefore becomes the "experimentum crucis" for our understanding of CP violation. Since the phase of ϵ' (Eq. (29)) is close to the one of ϵ (Eq. (25)), to a good approximation we get:

$$\mathrm{Re}(\epsilon'/\epsilon) = \left(1 - |\eta_{00}/\eta_{+-}|^2\right)/6\,. \tag{30}$$

Table 6. Experimental results up to 1987 and predictions of the superweak model for CP violation parameters

Quantity	Superweak Model	Exp. Value	Ref.
$\|\eta_{+-}\|$	–	$(2.27 \pm 0.02)10^{-3}$	[57]
$\|\eta_{00}/\eta_{+-}\|$	1.0	1.01 ± 0.03	[58, 60]
ϕ_{+-}	43.7°	$44.6° \pm 1.2°$	[59]
ϕ_{00}	43.7°	$55° \pm 5°$	[60]
$\mathrm{Re}\,\epsilon$	1.65×10^{-3}	$(1.62 \pm 0.09)10^{-3}$	[61]

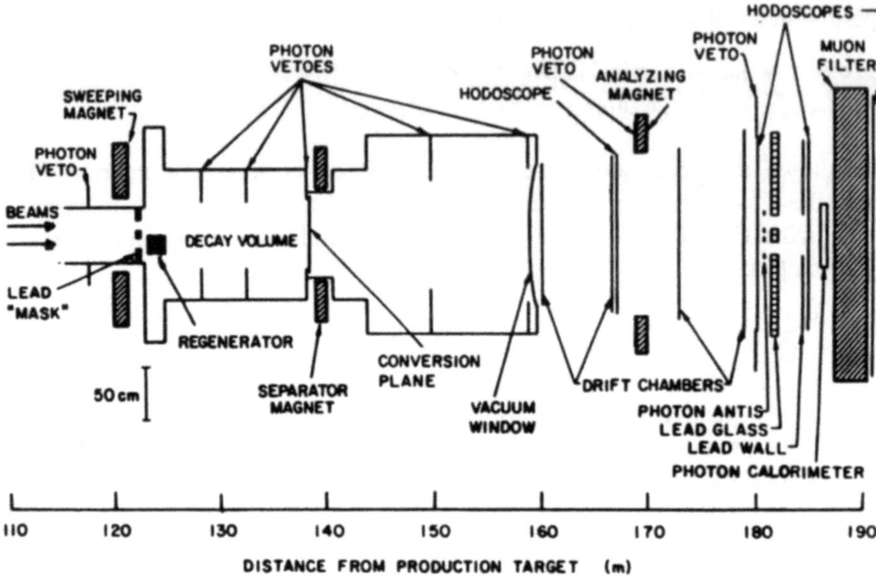

Fig. 5. Plan view of experiment E731 at Fermilab on $|\eta_{00}/\eta_{+-}|^2$

A measurement of the double ratio

$$R = \frac{|\eta_{00}|^2}{|\eta_{+-}|^2} = \frac{\Gamma\left(\mathrm{K_L} \to 2\pi^0\right)/\Gamma\left(\mathrm{K_L} \to \pi^+\pi^-\right)}{\Gamma\left(\mathrm{K_S} \to 2\pi^0\right)/\Gamma\left(\mathrm{K_S} \to \pi^+\pi^-\right)} \tag{31}$$

to a precision of 0.5% is therefore required to distinguish between the two remaining models and test the Standard Model prediction. Two such experiments are on the floor at Fermilab and CERN.

2.2 New Experiments on ϵ'/ϵ

Experiment E731 at Fermilab. The Chicago–Elmhurst–Fermilab–Princeton–Saclay collaboration [65] measures the double ratio R by recording $\mathrm{K_L}$ and $\mathrm{K_S}$ decays simultaneously in a split neutral beam with half the area covered by a regenerator producing coherently and incoherently $\mathrm{K_S}$ mesons. Only one decay mode (neutral or charged) at a time is recorded. For the detection of the neutral mode, a thin ($0.1\,X_0$) Pb converter sheet is introduced, and one of the four photons from $\mathrm{K_L} \to 2\pi^0$ decay is required to convert into an e^+e^- pair in this foil (see Fig. 5). The purpose of this conversion is twofold: the charged (double) track emerging from the sheet is used for a reduction of the trigger rate, and the reconstructed direction of this track helps to disentangle events originating from $\mathrm{K_L}$ and $\mathrm{K_S}$ beams. In this way, the transverse position of the Kaon trajectory at the regenerator, $\sim$ 50 meters upstream from the analyzing magnet, was measured with a precision $\sigma_x \sim 1$ cm, while the two beams are separated by 6 cm.

Neutral $2\pi^0$ events were constructed from the converted photon and three other electromagnetic energy clusters in the lead-glass array. The best pairing of the four

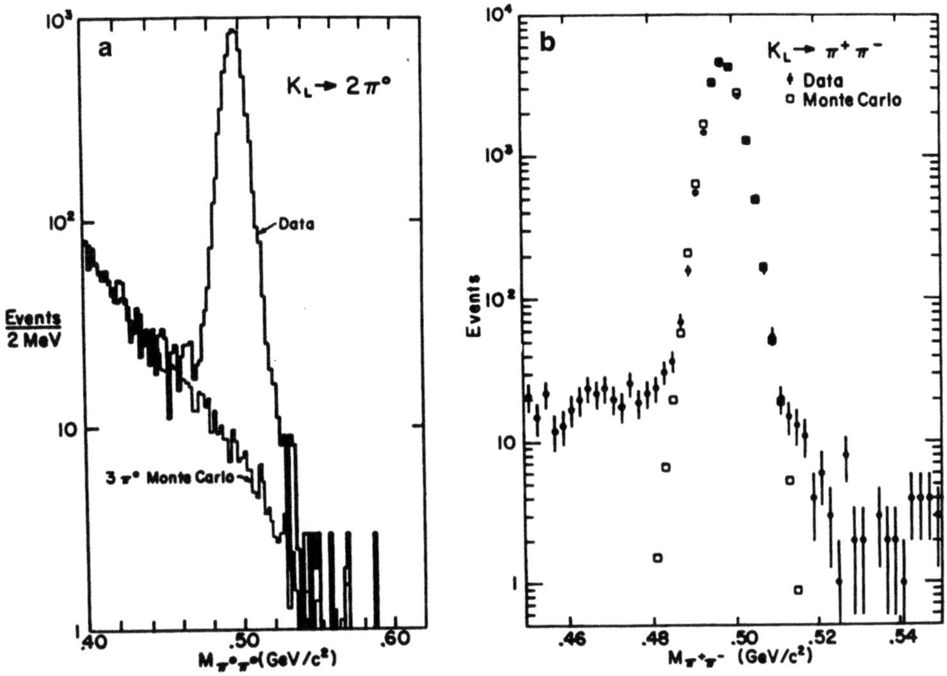

Fig. 6. (a) Invariant $2\pi^0$ mass from K_L decays. (b) Invariant $\pi^+\pi^-$ mass from K_L decays (E731)

gamma rays was selected on the basis of consistency between the two z-coordinates (along the beam), of the vertices from the two π^0's, obtained from $z^2 = E_1 E_2 r_{12}^2/m_\pi^2$. Here E_1 and E_2 are the energies of two photons, and r_{12} the transverse distance of their impact points at the lead-glass array. From the combined vertex of the two neutral pions, the invariant $2\pi^0$ mass is reconstructed, as shown in Fig. 6. The background below the peak comes from $3\pi^0$ decays where two photons have escaped detection. Monte Carlo calculations of the shape of this background below the peak are used and yield a background of $(1.56 \pm 0.30)\%$. The measured event numbers for the four decay modes are listed in Table 7. The main *systematic* errors come from the differences in the geometrical apparatus acceptance for K_S and K_L decays due to their different decay z vertex distribution along the beam (see Fig. 7). For the charged decay mode, with a vertex resolution of $\sigma_z \sim 15$ cm, the systematic uncertainty is estimated at 0.25%, while for the neutral mode the poorer vertex resolution, $\sigma_z \sim 110$ cm, and the uncertainty in the neutral energy scale increase this error to 0.5%. Combining all systematic errors in quadrature, the total uncertainty is 0.75%. The result of this experiment (called E731A) is:

$$\mathrm{Re}(\epsilon'/\epsilon) = (3.2 \pm 2.8 \pm 1.2)\,10^{-3}\,. \tag{32}$$

In order to decrease the statistical error, the collaboration has modified the experiment recently and taken new data [66]. In addition to a run with about $3 \times 10^4\, K_L \to 2\pi^0$ events taken in the previous apparatus, there were data taken without the requirement of one photon conversion in the lead sheet (2×10^5 events) and with the converter physically removed (8×10^4 events). This increase in event

Table 7. Recent experiments on ϵ'/ϵ

Collaboration	Chicago-Saclay	Yale-BNL	E731 Chicago-Elmhurst-FNAL-Princeton-Saclay	NA31 CERN-Dortmund-Edinburgh-Mainz-Orsay Pisa-Siegen
	Bernstein et al. [58]	Black et al. [58]	Woods et al. [65]	Burkhardt et al. [67]
$K_L \to 2\pi^0$	3152	1122	6.747	109.000
$K_S \to 2\pi^0$	5663	3317	21.788	932.000
$K_L \to \pi^+\pi^-$	10638	8506	35.838	295.000
$K_S \to \pi^+\pi^-$	25751	20960	130.025	2300.000
Data taken	1983-84	1983-84	1985	1986
Result	1985	1985	1988	1988
R	1.028 $\pm$0.032 $\pm$0.014	0.990 $\pm$0.043 $\pm$0.026	0.981 $\pm$0.017 $\pm$0.007	0.980 $\pm$0.004 $\pm$0.005
$\epsilon'/\epsilon \times 10^3$	−4.6 $\pm$5.3 $\pm$2.4	1.7 $\pm$8.2	3.2 $\pm$2.8 $\pm$1.2	3.3 $\pm$0.66 $\pm$0.8

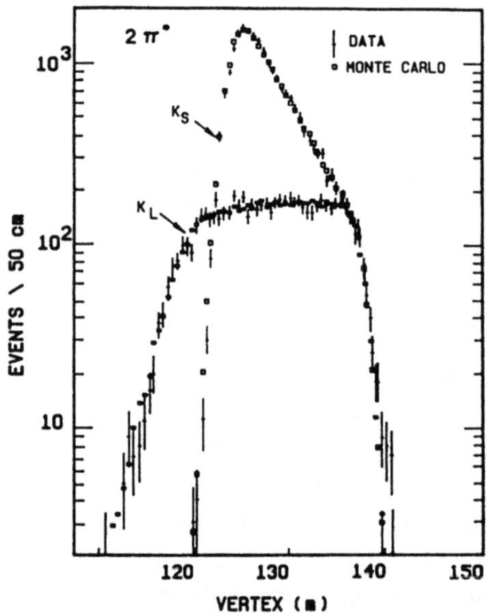

Fig. 7. Reconstructed longitudinal decay vertex for $K_{S,L} \to 2\pi^0$ from the E731 experiment

numbers is accompanied by a larger error in the transverse position of the decay point of neutral events, and therefore by a confusion of events from the K_S– and K_L–part of the beam. Figure 8 shows the transverse beam profile of $2\pi^0$ events at the place of the regenerator [66]. The regenerated K_S and the K_L beams are separated by 6 cm. The cross-over of events from the K_S to the K_L beam by reconstruction errors is 4%, leading to an additional systematic uncertainty. This collaboration, E731B, is planning to get a preliminary result by early 1989, and a final one in summer 1989.

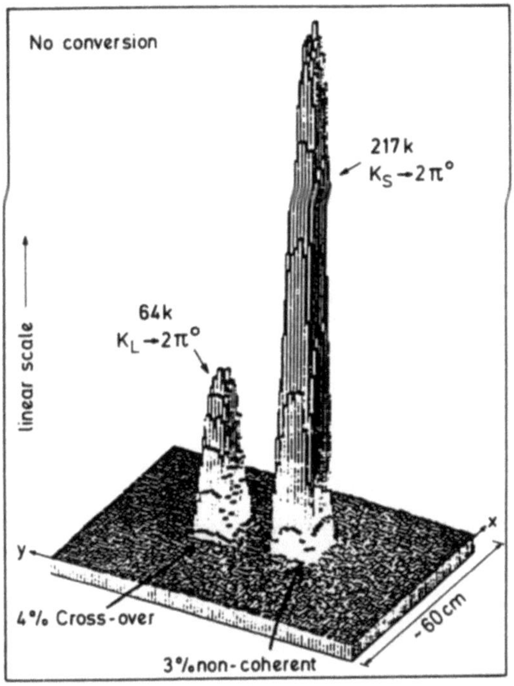

Fig. 8. Transverse beam profile of $K^0 \to 2\pi^0$ events, obtained from the center of energy in leadglass, E731B [66]

Experiment NA31 at CERN: First Evidence for Direct CP Violation [67]. In this experiment [68], both neutral and charged decay modes are recorded concurrently. K_S mesons are not produced by regeneration in a K_L beam, but directly from an external proton target near the detector. Collinear K_S and K_L beams are used alternately, changing frequently from one to the other. In order to have the maximum acceptance also for charged decays, no magnet is used, but hadron calorimetry instead. The systematic difference between the vertex distributions of short-lived (6 m decay length) and long-lived Kaon decays is overcome by taking data with the K_S target displaced in 1.2 m steps over the entire K_L decay region of 50 m length. Figure 9 shows a side view of the NA31 detector. The decay products and the neutral beam enter from the left side. On their way, they encounter two drift chambers, the electromagnetic calorimeter with Pb plates in liquid argon, and the hadron calorimeter made of iron and scintillator.

The 4 photons from $K \to \pi^0\pi^0$ decay are detected in the electromagnetic calorimeter with an energy resolution of $\Delta E_\gamma / E_\gamma = 7\% / \sqrt{E_\gamma(\mathrm{GeV})}$. Two of them are combined to give a neutral pion. Of the three possible combinations, the one fitting best the $2\pi^0$ hypothesis is chosen. The distance of the decay vertex from the calorimeter is calculated, using the K^0 mass as a constraint, with about 1% precision. Constraints on the masses of two-photon pairs are used to reduce the background, which is primarily due to $K_L \to 3\pi^0 \to 6\gamma$ decays with undetected photons. This background is uniformly distributed in a two-dimensional scatter plot of photon-pair masses. The π^0 mass resolution is ~ 2 MeV. Signal and background events are counted in equal-area χ^2 contours around the region defined for accepted events (see Fig. 10). The signal region is taken as $\chi^2 < 9$. Background is subtracted by linear

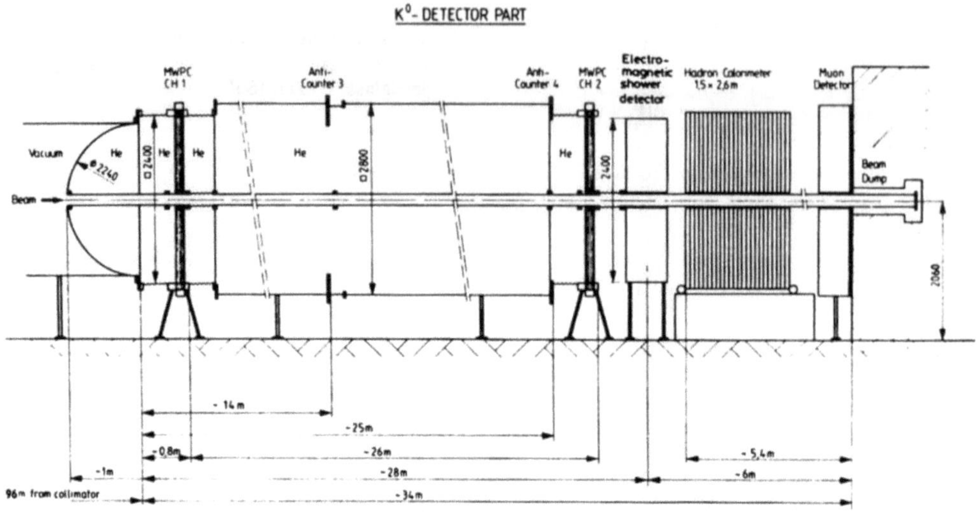

Fig. 9. Detector of the NA31 collaboration for the measurement of ϵ'/ϵ in Kaon decays [68]

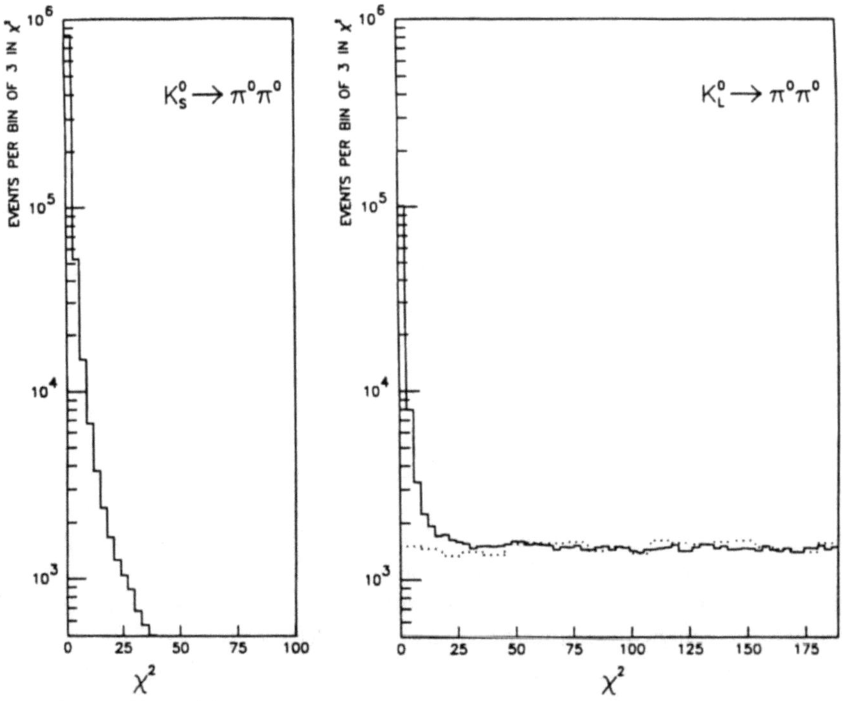

Fig. 10. χ^2 distribution of 4γ events in K_S and K_L beams, and Monte Carlo calculation for background from $K_L \to 3\pi^0$ from NA31 [67]

extrapolation into the signal region. It is about 4% in the K_L beam, and negligible in the K_S beam.

Charged pions from $K^0 \to \pi^+\pi^-$ decays are detected in the chambers and the calorimeters, the invariant K mass is calculated from the vector momenta of both

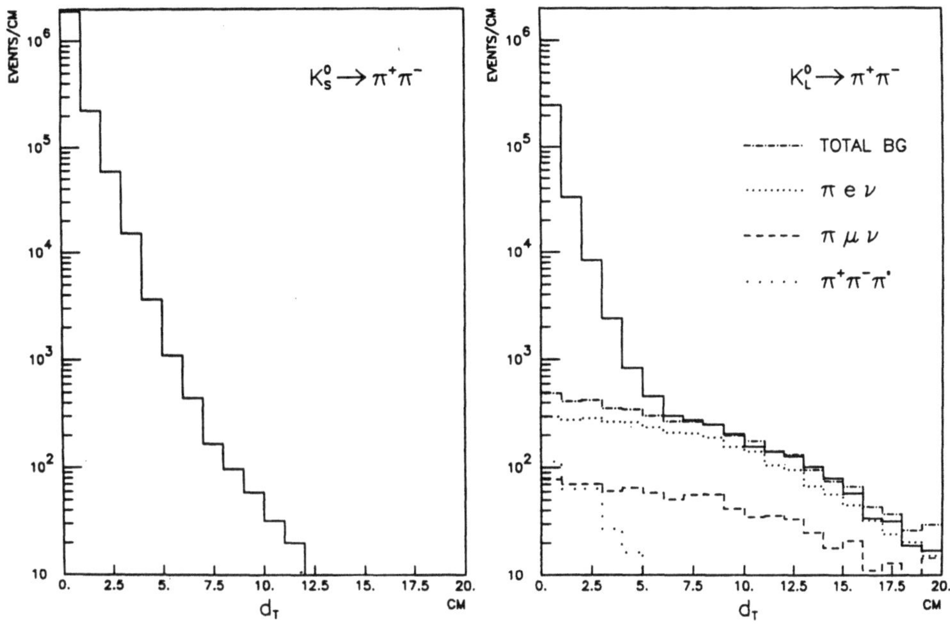

Fig. 11. Event distribution for charged decays in the distance between decay plane and production target, from NA31 [67]

pions. The longitudinal vertex resolution is better than 1 m. The K^0 energy is calculated with 1% precision from the Kaon mass, the opening angle between the two tracks, and the ratio of track energies as measured in the calorimeter. Events with isolated photons, such as $K^0 \to \pi^+\pi^-\pi^0$ decays, $K^0 \to \pi^+\pi^-\gamma$ decays and events with accidental photons, are rejected. The $K^0 \to \pi e \nu$ events are identified and rejected by comparing, for each track, the energy deposited in the front half of the electromagnetic calorimeter with the energy deposited in the hadron calorimeter. After cuts on the $\pi^+\pi^-$ mass and on the reconstructed center of gravity with respect to the beam axis, a residual background of the three-body decays must be subtracted. From the measured vector momenta of the two pions, the distance d_T of the decay plane from the production target is calculated. While two-body decays come from the target, $d_T \sim 0$, three-body decays show a broad distribution. This is illustrated in Fig. 11, where the d_T distribution of events in the K_S and K_L beams are shown. For $K_S \to \pi^+\pi^-$, the background is below the 10^{-3} level. For $K_L \to \pi^+\pi^-$, three-body decays constitute the background.

The signal is taken to be $d_T < 5$ cm, and the control region for the background extrapolation is taken as 7 cm $< d_T <$ 12 cm. The fraction of events in the background region is $(3.6 \pm 0.1) \times 10^{-3}$ of the signal. This background consists mainly of $K_L \to \pi e \nu$ decays and has contributions from $K \to \pi \mu \nu$, from $K \to \pi^+\pi^-\pi^0$ where one photon overlaps the shower of one of the charged pions, and a small amount of K_S production in the final K_L collimator. The $K^0 \to \pi^+\pi^-\pi^0$ background is subtracted directly by counting events with identified photons as a function of the distance between the photon and the nearest track. The background below the peak around $d_T = 0$ is then obtained from an extrapolation of the observed events

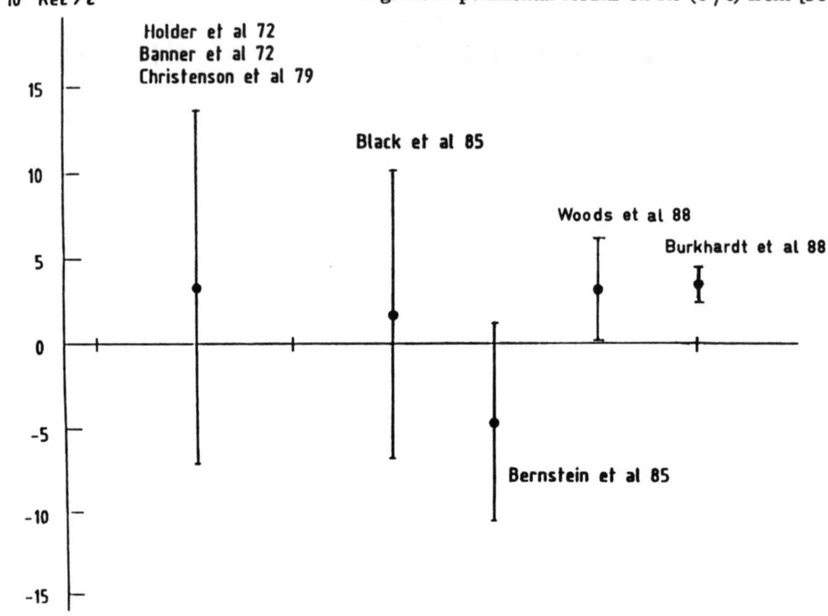

Fig. 12. Experimental results on Re (ϵ'/ϵ) from [58, 60, 65, 67]

in the control region below the peak using Monte Carlo calculations (see Fig. 11). The resulting event numbers in the four decay channels are listed in Table 7. The main systematic error sources are the uncertainties in the background subtraction for $K_L \to 2\pi^0$ and $K_L \to \pi^+\pi^-$ decays (each 0.2%), the uncertainty in the difference between the two energy scales in the neutral and charged mode (0.3%), and the error in the gains and losses by accidentals in the double ratio (0.2%). All systematic errors add up (in quadrature) to 0.5%. The result of this experiment is then $R = 0.980 \pm 0.004 \pm 0.005$, or

$$\mathrm{Re}(\epsilon'/\epsilon) = (3.3 \pm 1.1)\, 10^{-3}. \tag{33}$$

This result is the first evidence for $\epsilon' \neq 0$ i.e. the first observation of direct CP violation, and therefore against the superweak model. The value of ϵ'/ϵ is, however, in good agreement with the numbers calculated in the standard model for three generations of quarks. This result is compared to previous experimental measurements of this quantity in Fig. 12.

A confirmation of the NA31 result in Eq. (33) can be expected from three sources:

i) the new Fermilab experiment E731B is expected to produce a result with sufficient sensitivity by 1989
ii) the NA31 collaboration has improved their detector by adding a transition radiation detector for electron rejection and a photon veto hodoscope in the beam, and by running at different proton energy for K_S and K_L beams in order to reduce the uncertainty in ϵ'/ϵ from the energy scales; with this new set-up, new data have been taken in 1988

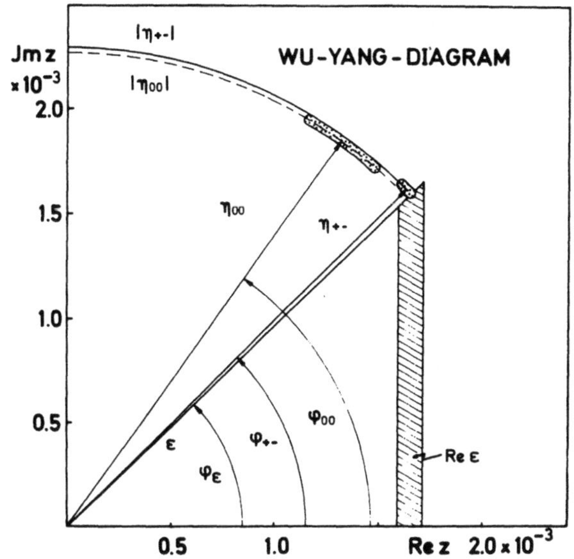

Fig. 13. Wu-Yang triangle, present experimental status

iii) at the low energy antiproton ring LEAR, an experiment producing pure strangeness K^0 states is being prepared [69].

If all available information on the CP parameters in the Kaon system is used, the present status of the Wu-Yang triangle (Eq. (27)) can be obtained [70]. The result is shown in Fig. 13.

The shaded quasi-ellipsoidal areas give the experimental 90% C.L. domains for the amplitude vectors η_{+-} and η_{00}. Direct CP violation shows up as a difference in length of these vectors, or a component of ϵ' parallel to ϵ. However, a finite phase difference $\Delta\phi = \phi_{00} - \phi_{+-}$ would violate CPT. In fact, using the value of $\mathrm{Re}(\epsilon'/\epsilon)$ from (33) and the phases of ϵ and ϵ' from (25, 29), we obtain a bound on the transverse component of ϵ'

$$|\epsilon'_T| < 3 \times 10^{-6} \qquad 90\%\ \mathrm{C.L.} \tag{34}$$

and from the Wu-Yang relations, Eq. (27), follows

$$\begin{aligned} &\Delta\phi = \arctan(3\epsilon'_T/|\epsilon|) \\ &|\Delta\phi| < 0.2^\circ . \end{aligned} \tag{35}$$

The present experimental value of ϕ_{00} is about 2 st.dev. away from the CPT prediction. Both the NA31 and the E731 collaborations are in the process of measuring $\Delta\phi$ with improved precision.

3. Implications of Experimental Results on the Standard Model Parameters

If we consider the quark sector of the standard model, assuming three generations, there are six quark masses, three mixing angles and one CP phase δ. Of these ten parameters, seven are determined with some accuracy, the angles θ_{12} and θ_{23} (Sect. 1) and the masses of the five known quarks [71]. However, little is known about the top mass, m_t, the angle θ_{13} and the phase δ. The relevance of the new results on $|\epsilon'/\epsilon|$ and the B–$\bar{\text{B}}$ mixing has been widely discussed [72].

i) Implications of the ϵ'/ϵ Measurement. CP violating amplitudes are proportional to the product J (Eq. (7)) of four sines of mixing angles. This is one source of uncertainty. On the other hand, an additional uncertainty comes from the calculation of the hadronic matrix elements for the box graph (ϵ) and the penguin graph (Sect. 2). In particular, the K meson decay constant f_K (160 – 170 MeV) and the B_K factor in the K system,

$$B_K = <\mathrm{K}^0|T|\overline{\mathrm{K}^0}> / <\mathrm{K}^0|T|\overline{\mathrm{K}^0}>_{\text{vac.sat.}}$$

i.e. the $\mathrm{K}^0-\overline{\mathrm{K}^0}$ transition matrix element, normalized to its vacuum saturation value, are needed. Considerable effort has been devoted to calculations of this B_K factor, using different approximation methods in low energy QCD, including Hadronic Sum Rules, 1/N expansion, lattice calculation, and QCD sum rules. Table 8 lists some results of these calculations. The conclusion to be drawn at the moment is probably that $B_K = 2/3 \pm 1/3$. Using this value for B_K (which coincides with the value from the 1/N expansion in second order) and the hadronic penguin matrix element (also from 1/N expansion) Buras and Gerard [63] obtain a prediction for ϵ'/ϵ as shown in Fig. 14. The two groups of boundaries are given for two values of the mixing angle θ_{13}, i.e. for $\overline{R} = \Gamma(\mathrm{b} \to \mathrm{u})/\Gamma(\mathrm{b} \to \mathrm{c})$ taking the values 0.01 or 0.1. If the upper limit from Sect. 1 is used instead, $R < 0.03$, the predicted values are narrowed down to the interval $0.002 < \mathrm{Re}\ (\epsilon'/\epsilon) < 0.005$ for $m_t \sim 60$ GeV.

Table 8. B_K Parameter for K System in QCD calculations

Hadronic sum rules	
PICH, DE RAFAEL [73]	0.33 ± 0.09
1/N expansion (2nd order)	
BARDEEN, BURAS, GERARD [74]	0.66 ± 0.10
Lattice	
GAVELA et al. [75]	0.87 ± 0.20
QCD sum rules	
REINDERS, YAZAKI [76]	0.84 ± 0.08
CHETYRKIN et al. [77]	1.2 ± 0.1
DECKER [78]	0.58 ± 0.16
BILIC, DOMINGUEZ, GUBERINA [79]	0.50 ± 0.22
Present estimate:	2/3 ± 1/3

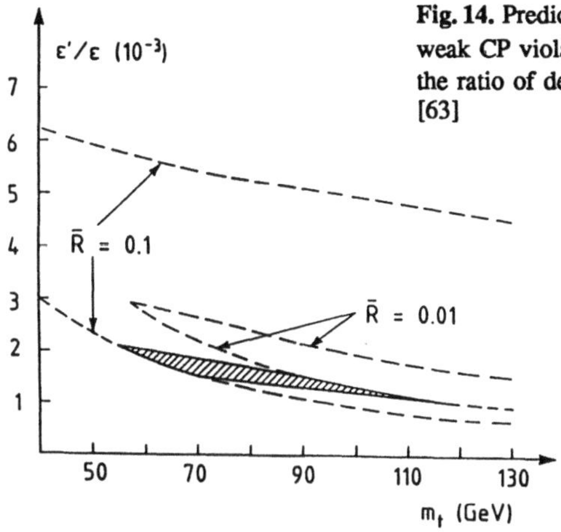

Fig. 14. Predicted values for ϵ'/ϵ in the six-quark model of milliweak CP violation, as a function of the top-quark mass m_t. $\overline{R}$ is the ratio of decay rates of b quarks, $\overline{R} = \Gamma(b \to u)/\Gamma(b \to c)$ [63]

There is perfect agreement of this value with experiment in sign and magnitude of the effect. The prediction of KM models with four generations by Türke et al. [64] can be positive or negative, in the range $-0.01 < \mathrm{Re}(\epsilon'/\epsilon) < +0.01$.

If we use the ϵ' calculation for penguin diagrams and the experimental value of $|\epsilon|$, instead of the calculated one, the predicition for ϵ'/ϵ depends on the mixing angles [63]:

$$\mathrm{Re}(\epsilon'/\epsilon) \sim 16\, s_{23} s_{13} \sin\delta \,.$$

Using Eqs. (21, 33) we obtain

$$|s_{13} \cdot \sin\delta| \gtrsim 0.002\,. \tag{36}$$

This implies that s_{13} is non-zero, $|s_{13}| > 0.002$, and together with the upper limit on s_{13} from Sect. 1, that

$$|\sin\delta| > 0.3\,. \tag{37}$$

The CP phase is consistent with being large, $\delta = (90 \pm 70)°$.

ii) Implications of the Large $B^0 - \overline{B^0}$ Mixing. $B - \bar{B}$ mixing has been observed by the ARGUS collaboration [94] and confirmed by the CLEO collaboration [43]. First indications had been obtained by the UA1 experiment [80]. The mixing parameter $x_d = \Delta M/\Gamma$ is the ratio of the mass difference ΔM of the weak eigenstates (B_1 and B_2) and the decay width Γ. Experimentally, the two values of x_d agree very well. We use their average

$$x_d = 0.70 \pm 0.13\,. \tag{38}$$

In the standard model, this mixing is due to a box graph similar to the one in the $K^0 - \overline{K^0}$ system. The dominant process is the one with an exchange of a virtual top quark, and therefore:

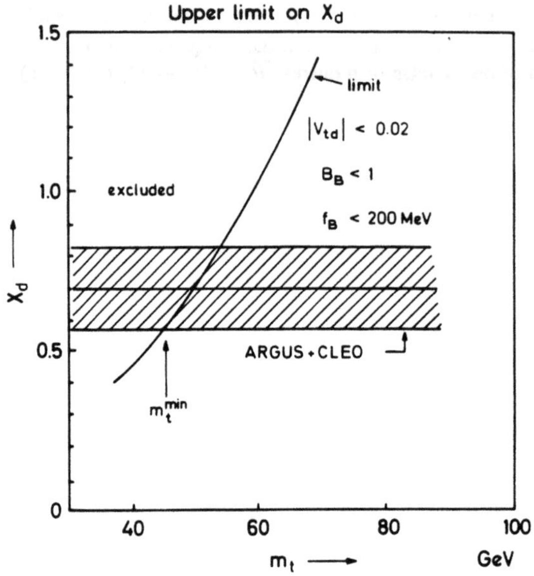

Fig. 15. Upper limit on the B $\bar{B}$ mixing parameter x_d, as a function of the top quark mass, m_t

$$x_d \sim B_B\, f_B^2\, m_t^2\, |V_{td}|^2 \tag{39}$$

Here B_B is the B factor for the B system and f_B the B decay constant. This relation can be used to derive a lower limit on m_t. Following Altarelli and Franzini [72] and using the conservative bounds $B_B < 1$, $f_B < 200$ MeV, and $|V_{td}| < 0.02$, the upper limit on x_d in Fig. 15 is obtained.

From the measured x_d range one can then infer the conservative lower limit $m_t > 45$ GeV. A number of limits have been obtained [72] in the range between 45 GeV and 55 GeV. They depend on the choice of f_B and B_B. If $f_B < 140$ MeV was assumed, the lower limit on m_t would increase to ~ 60 GeV. The directly measured lower limit from UA1 is $m_t > 41$ GeV [81].

Mixing in the B_s meson system is expected to be maximal. Since the B_B and f_B in Eq. (39) are expected to be similar for the B_d and B_s mesons, the mixing parameter x_s is expected to be larger than x_d:

$$x_s/x_d \sim |V_{ts}|^2/|V_{td}|^2\,. \tag{40}$$

From Sect. 1 one infers that $x_s/x_d > 5.6$ with 90% confidence. The results of the UA1 collaboration [80] together with the ones of the ARGUS [36] and CLEO [43] collaborations then imply that mixing in the B_s system is maximal. Direct evidence for B_s mixing is still missing, however.

iii) The CP parameters ϵ and ϵ' and the B – $\bar{B}$ mixing parameter x_d can be used, together with the knowledge on KM matrix elements from Sect. 1 to determine the poorly known remaining parameters, m_t, θ_{13} and δ. One of these analyses [72] is the one of Harari and Nir. Their result is presented in Fig. 16 as a two-dimensional diagram in the plane with $q = s_{13}/s_{23}$ and δ as variables, for a fixed value of m_t, and $B_K = 0.7$. Constraints from the CP violation parameter are given as dashed contour

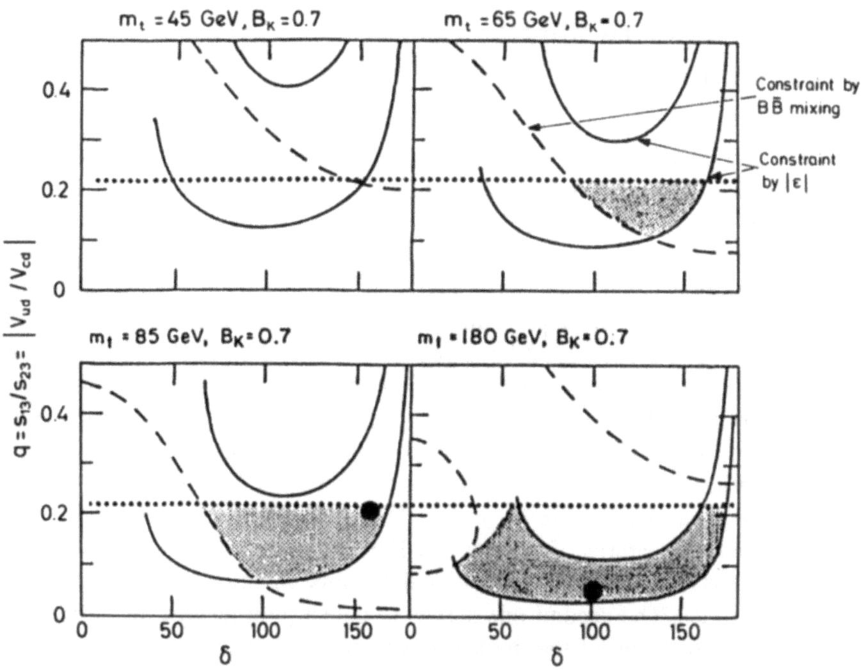

Fig. 16. Allowed range for $q = s_{13}/s_{23}$ and δ, for fixed values of m_t and $B_K = 0.7$ [72]. Solid lines give band allowed by value of CP parameter $|\epsilon|$, dashed lines are constraints from B–B̄ mixing parameter x_d. Allowed region is shaded

lines, the ones from the B – B̄ mixing parameter as solid lines. Areas allowed by both constraints and by the experimental limit $|q| < 0.2$ are shaded. From this analysis again it appears that no solution exists for $m_t \lesseqq 50$ GeV, and that the allowed range for δ always includes the value 90°. This range increases with increasing m_t. An upper limit on m_t has been derived from the comparison of precision measurements of the Weinberg angle in neutrino-quark scattering with the measured values of the intermediate boson masses, M_Z and M_W, by Amaldi et al. [82], with the result $m_t < 180$ GeV.

The origin of the mixing angles and quark masses is not explained in the Standard Model. There are, however, some attempts of relating the mixing angles to the quark masses, one based on a model of Fritzsch [46], the other devised by Stech [47]. These models lead to relations between $\sin\theta_{ik}$ and the ratios m_i/m_k, where m_i are the quark masses in generation i. With the present experimental data, and using the physical quark masses at an energy scale of 1 GeV from Gasser and Leutwyler [71], Harari and Nir conclude that the Stech scheme is excluded, while for the Fritzsch scheme a unique solution remains, with $m_t \sim 85$ GeV, $s_{13} \sim 0.0035$ and $\delta \sim 100°$.

4. Beyond the Standard Model: Searches for Rare Kaon Decays

While the direct search for phenomena beyond the Standard Model is done at the highest energies to be reached at the future colliders LEP, HERA and SSC, indirect searches for such phenomena can be performed at medium energy machines. These investigations search for rare decay modes, absolutely forbidden or highly suppressed in the Standard Model. For Kaons, particularly attractive decay modes for such a search are the ones mediated by lepton-number violating neutral currents. Contributions to the Munich conference 1988 have been summarized by Zeller [83]. The decay modes studied were $K^+ \rightarrow \pi^+\mu^+e^-$, $K_L \rightarrow \mu e$, $K_L \rightarrow e\,e$, and $K_L \rightarrow \pi^0 e^+ e^-$, at Brookhaven, Fermilab, CERN and KEK. As an example, we discuss here the experiment E791, done by a UCLA-LANL-Pennsylvania-UCI-Stanford-Temple-William + Mary-collaboration at Brookhaven [84], searching for the decay $K_L \rightarrow \mu e$, and the search for the decay $K_L \rightarrow \pi^0 e^+ e^-$ by the NA31 collaboration [85].

The experimental apparatus of experiment E791 is shown in Fig.17.

Compared to earlier experiments in this field, this detector has two magnetic spectrometers in series. The second one serves to detect pion decays inside the first magnet which constitute one of the main backgrounds, originating from $K_L \rightarrow \pi e \nu$ decay. The apparatus is able to work in a beam with 4×10^{12} protons per pulse on target. Its mass resolution is demonstrated in Fig. 18, with 75 events of the rare decay $K_L \rightarrow \mu\mu$. This exposure corresponds to a sensitivity of $\sim 10^{-10}$ per event.

Figure 19 then shows candidates with the (μe)-signature. The background from three-body decays has invariant (μe)-masses below 492 MeV, and an upper limit of 3×10^{-10} can be derived for the branching ratio of $K_L \rightarrow \mu e$.

The mass scale of new phenomena tested by such small upper limits on the branching ratio R can be estimated with a simple comparison of two diagrams

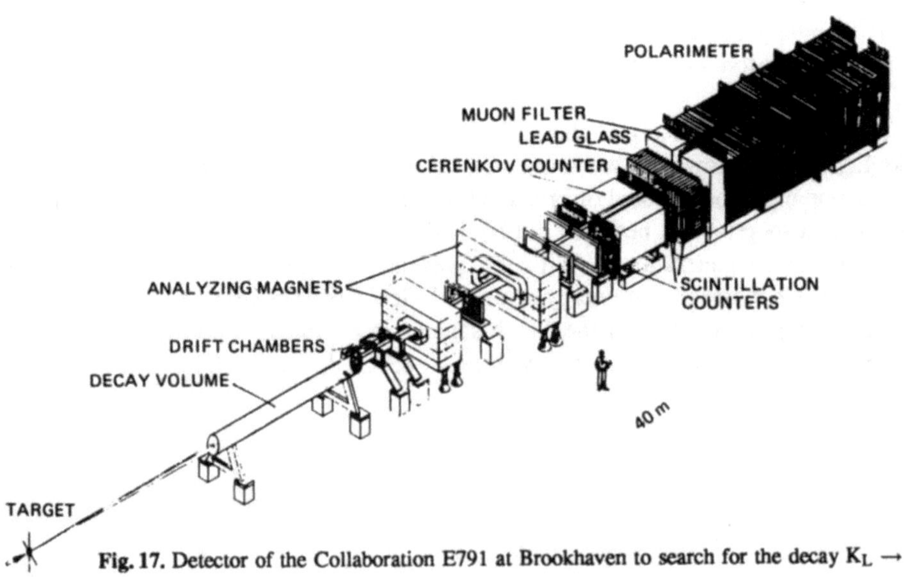

Fig. 17. Detector of the Collaboration E791 at Brookhaven to search for the decay $K_L \rightarrow \mu e$

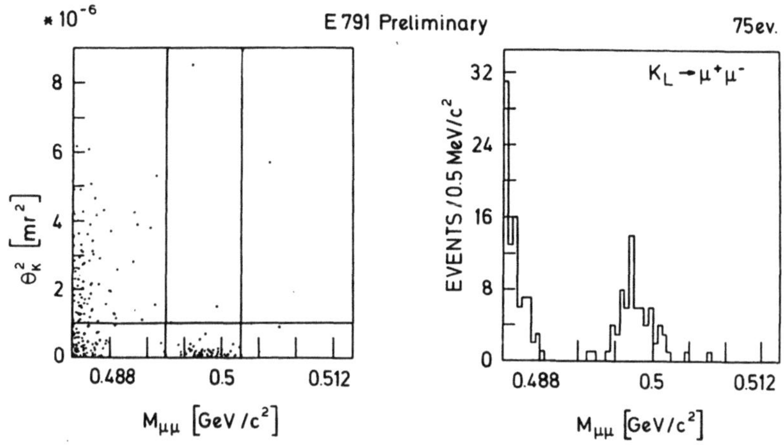

Fig. 18. Invariant $\mu\mu$ mass $M_{\mu\mu}$ vs. squared angle of reconstructed Kaon direction relative to beam axis, θ_K^2, and projection on $M_{\mu\mu}$ axis, for decay $K_L \to \mu\mu$

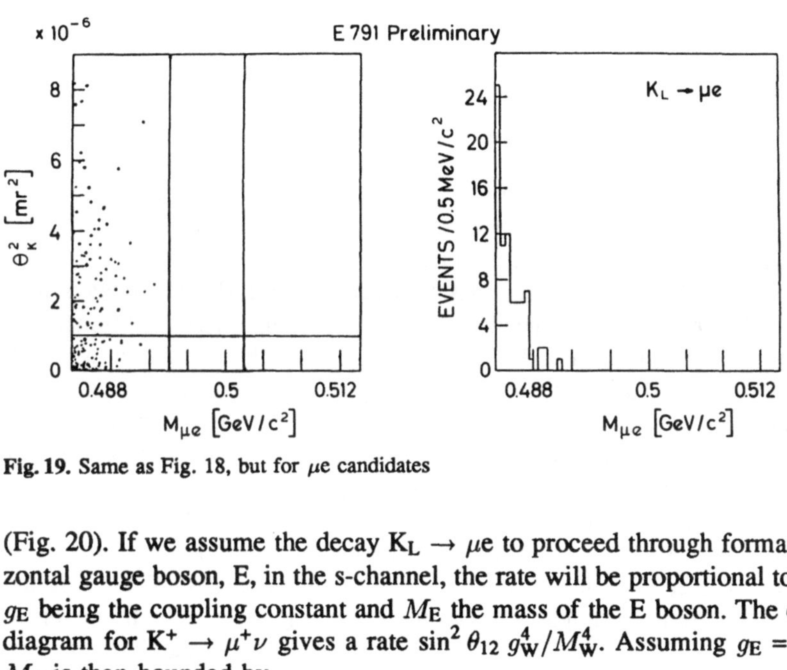

Fig. 19. Same as Fig. 18, but for μe candidates

(Fig. 20). If we assume the decay $K_L \to \mu e$ to proceed through formation of a horizontal gauge boson, E, in the s-channel, the rate will be proportional to g_E^4/M_E^4, with g_E being the coupling constant and M_E the mass of the E boson. The corresponding diagram for $K^+ \to \mu^+\nu$ gives a rate $\sin^2\theta_{12}\, g_W^4/M_W^4$. Assuming $g_E = g_W$, the mass M_E is then bounded by

$$M_E > 70\,\mathrm{TeV}\left(\frac{10^{-10}}{R}\right)^{1/4} \tag{41}$$

The mass values derived are of course different for different specific models (SUSY, Multi-Higgs, Left-Right-Symmetric,...).

As a second example, we discuss the search for the decay $K_L \to \pi^0 e^+e^-$ and search for a light Higgs particle in the decay chain $K_L \to H^0\pi^0$ with H^0 decaying to e^+e^-, performed by the NA31 collaboration at CERN [85].

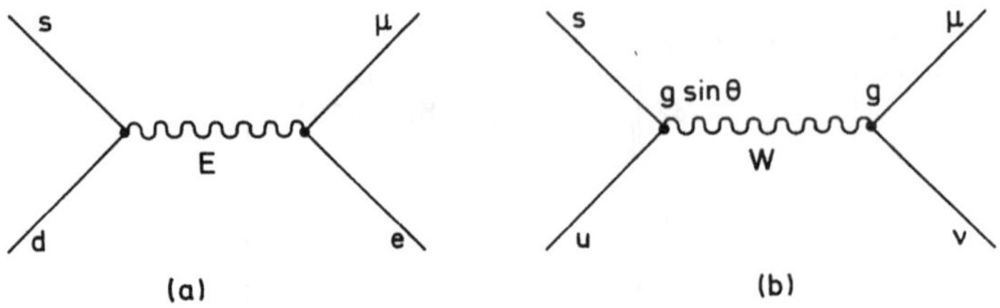

Fig. 20. Diagrams contributing to the decays (a) $K_L \to \mu e$ via exchange of horizontal gauge boson E, (b) $K^+ \to \mu^+\nu$ via weak interaction

This decay mode is of special interest because the one-photon exchange process is a CP violating transition. In the Standard Model, CP violating processes through state mixing ("indirect CP violation") and through the decay Hamiltonian in $K_2 \to \pi^0 e^+e^-$ ("direct CP violation") give comparable contributions at the 10^{-11} level in branching ratio [86]. In addition, CP conserving processes via two-photon exchange may be present at the same level.

The data were collected while performing the measurement of ϵ'/ϵ described above [67, 68]. The $K_L \to \pi^0 e^+e^-$ events fulfil the same trigger conditions as the $K_L \to \pi^0\pi^0$ events with four identified photons. 10^5 events with two (electron) tracks in the drift chambers pointing to electromagnetic showers in the liquid Argon calorimeter and at least two additional photon showers were collected. A common vertex is calculated from the energies and lateral positions of the showers in the calorimeter, using the K^0 mass as a constraint. The invariant mass of the photon pair and a four-particle invariant mass is calculated. The $\pi^0 e^+e^-$ candidates and the $\pi^0\pi^0$ events used as normalization were subject to the following cuts: total energy between 50 and 190 GeV; shower energies between 5 and 100 GeV; decays in a fiducial volume of 50 m length. For the electron candidates, there were additional cuts in order to reject events with charged pion tracks, based on the shower information from electromagnetic and hadronic calorimeters. Figure 21 shows the distribution of two-photon mass versus four-particle mass for $K_L \to \pi^0 e^+e^-$ events generated by Monte Carlo. The events cluster around the π^0 and K^0 masses with resolutions of 2 MeV and 9 MeV, respectively. The corresponding plot for real events is shown in Fig. 22. One event is observed in the 3σ signal region of Fig. 21. This event has an invariant mass of the e^+e^- pair of 130 MeV, consistent with the π^0 mass. This event is therefore also consistent with the expected "double-conversion" background where two photons from one single π^0 from the decay $K_L \to \pi^0\pi^0$ both convert to an e^+e^- pair in the 0.004 radiation length of material upstream of the first chamber. The pairs do not open up and are recorded as one electron. We expect 1.5 events of this kind in the exposure. The upper limit on a possible signal for the decay $K_L \to \pi^0 e^+e^-$ is therefore < 3.1 events at 90% C.L.. From the $\pi^0\pi^0$ sample used for normalization, we obtain a flux of 6×10^8 K_L decays, and an upper limit to the branching ratio of

$$\Gamma\,(K_L \to \pi^0 e^+e^-)\,/\,\Gamma\,(K_L \to \text{all}) < 4 \times 10^{-8}$$

with 90% confidence.

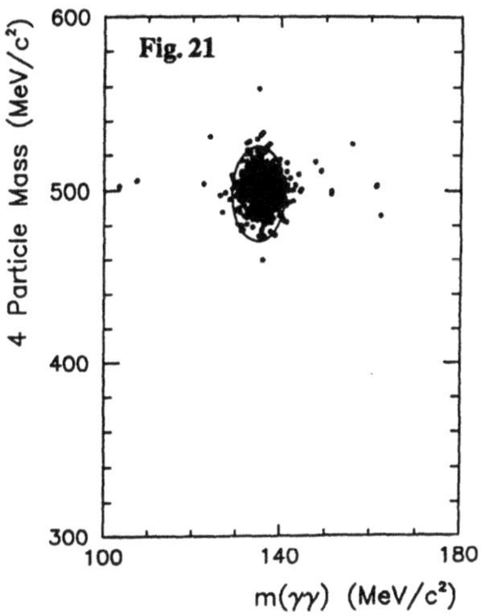

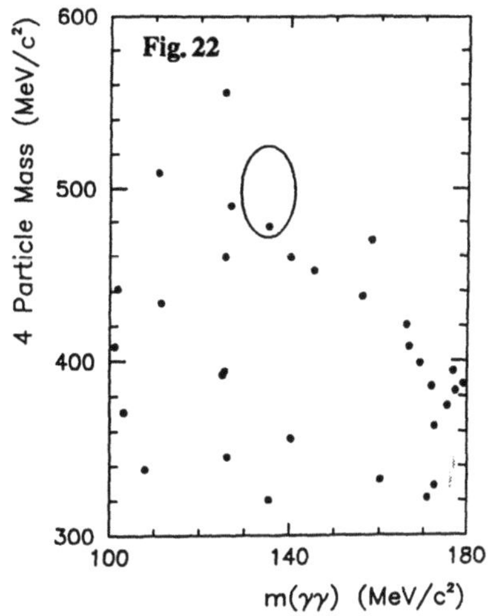

Fig. 21. Invariant two-photon mass versus four-particle mass for Monte-Carlo events. Ellipse is three-standard deviation contour around the signal [85]

Fig. 22. Same as Fig. 21 for real events. Ellipse indicates signal region [85]

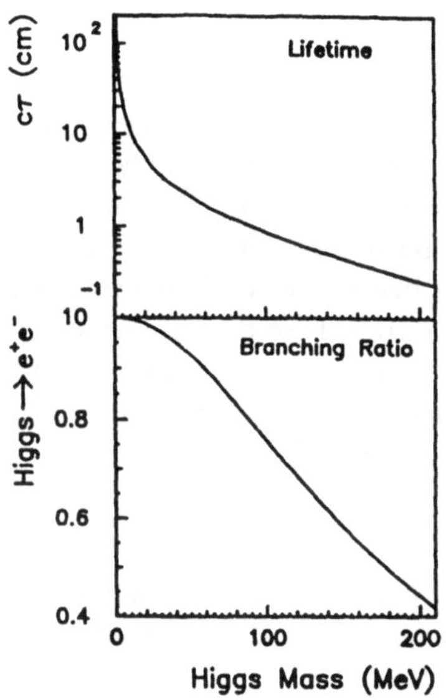

Fig. 23. Lifetime (decay path) of standard Higgs particle and branching ratio for decay $H^0 \rightarrow e^+e^-$

The same data sample can also be used to search for a light Higgs particle in the decay chain $K_L \rightarrow \pi^0 H^0$, $H^0 \rightarrow e^+e^-$. If the Higgs particle has a mass between 1 MeV and 211 MeV, its lifetime and the branching ratio into electron pairs are well predicted in the Standard Model. Figure 23 shows the lifetime and branching ratio as a function of the Higgs mass. The branching ratio for $K_L \rightarrow \pi^0 H^0$ has been calculated [87, 88, 89]. Figure 24 gives the result as a function of the top quark mass. These branching ratios are well above the sensitivity of this experiment.

In the analysis searching for the Higgs particle, the constraint of a common vertex of the e^+e^- pair and the $\gamma\gamma$ pair has to be abandoned (see Fig. 25). There are

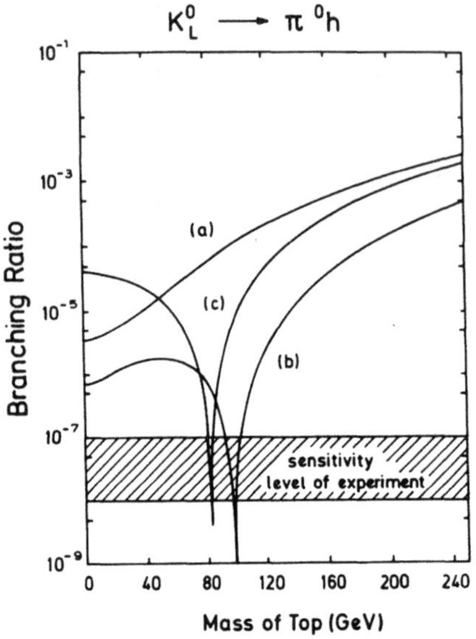

Fig. 24. Theoretical predictions of the branching ratio for the decay $K_L \to \pi^0 H^0$: (*a*) [87], (*b*) [88], (*c*) [89]

still four electromagnetic showers required, two of them linked to charged tracks. The K^0 vertex is formed from the $\gamma\gamma$ pair, using the π^0 mass, and the H^0 vertex from the charged e^+e^- tracks. The trajectories of H^0 and π^0 are required to meet within the resolution of the reconstruction (σ_{CDA} = 1 cm at 100 GeV energy, 2.2 cm at 50 GeV, "CDA cut"). The e^- and e^+ tracks are required to be coplanar, the electrons have to be identified in both the transition radiation detector and the liquid Argon calorimeter. Finally the H^0 vertex must not be more than 10 m upstream of the K^0 vertex ("vertex cut"). Figure 26 displays distributions of 61 candidate events before the CDA and vertex cuts, and 15 events after these cuts. In the signal region, at $m_K \pm$ 25 MeV, there is one event. Indeed this event is the same as in the $\pi^0 e^+e^-$ decay analysis, presumably due to double conversion of photons. Therefore, the upper limit for the product of branching ratios is, at 90% C.L.,

$$BR(K_L \to \pi^0 H^0) \cdot BR(H^0 \to e^+e^-) < 4 \times 10^{-8} .$$

This limit excludes a standard single neutral Higgs particle as long as the decay path of the Higgs is short enough to fit in the decay volume of about 50 m length. This situation is illustrated in Fig. 27. The experiment is sensitive down to Higgs masses of 20 MeV. The experiment therefore excludes a standard light Higgs particle in the mass range from 20 MeV to 211 MeV.

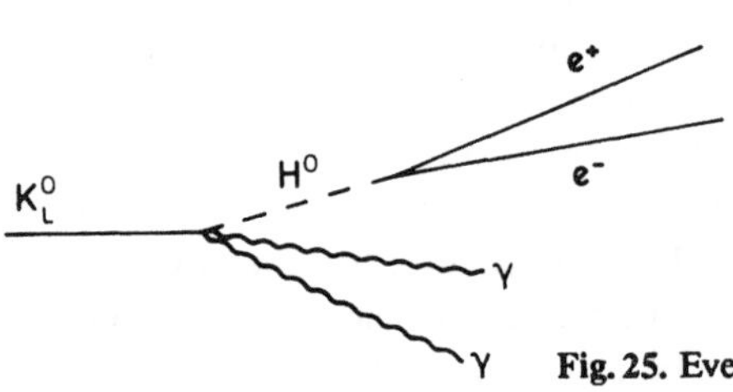

Fig. 25. Event topology for decay chain $K_L \to \pi^0 H^0, H^0 \to e^+e^-$

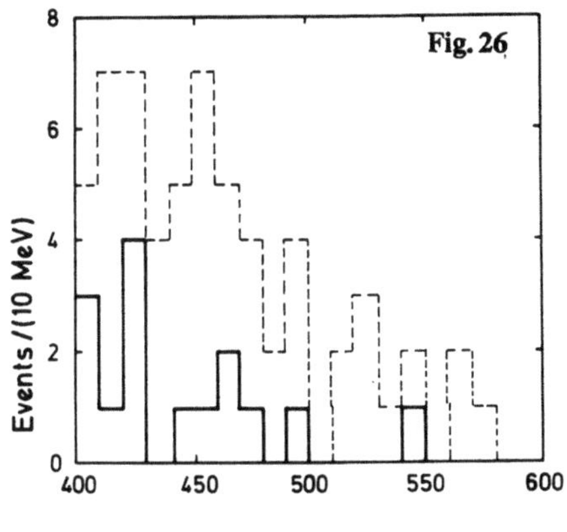

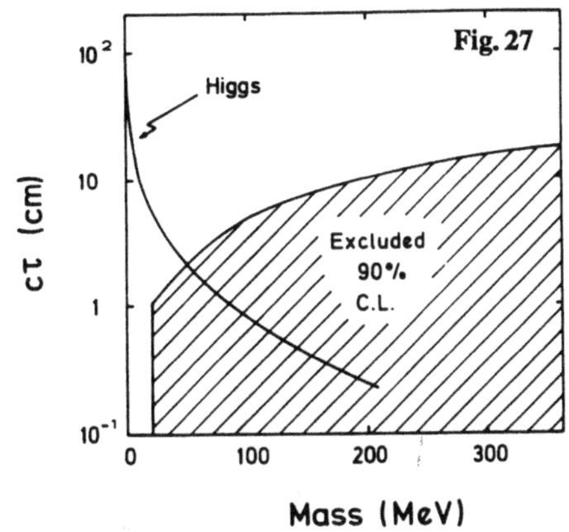

Fig. 26. Distribution of candidates for the Higgs decay chain in the mass of π^0H^0 system

Fig. 27. Decay path vs. mass of Higgs particle. Region excluded by experiment is shaded

Table 9. Experimental results on rare Kaon decays (90% C.L. upper limits on branching ratios) [83]

Decay mode	Previous limit	Munich Conf. (preliminary results)	Experiment
$K^+ \to \pi^+\mu^+e^-$	$< 5 \times 10^{-9}$	$< 1.1 \times 10^{-9}$	BNL - Yale E777 [90]
$K_L \to \mu\ e$	$< 7 \times 10^{-9}$	$< 1.3 \times 10^{-9}$	BNL - Yale E780
		$< 5.1 \times 10^{-9}$	KEK [93]
		$< 3 \times 10^{-10}$	E791 at BNL
$K_L \to e\ e$	$< 2 \times 10^{-7}$	$< 1.2 \times 10^{-9}$	BNL - Yale E780 [91, 92]
		$< 5.8 \times 10^{-9}$	KEK [93]
$K_L \to \pi^0e^+e^-$	$< 2.3 \times 10^{-6}$	$< 3.2 \times 10^{-7}$	BNL - Yale E780
		$< 5.1 \times 10^{-8}$	Fermilab E731
		$< 4.0 \times 10^{-8}$	NA31 [85]

This and other experimental searches for rare Kaon decays are summarized in Table 9.

5. Conclusions – Outlook

This review can be summarized in the following way:

1. Direct CP violation has been observed in $K_L{}^0$ decays. The parameter $\epsilon' \neq 0$, and the superweak model is excluded as sole source of CP violation.
2. Sign and magnitude of ϵ' are in very good agreement with calculations in the Kobayashi-Maskawa model for three generations of quarks.
3. The elements of the KM quark mixing matrix and the mixing angles are well constrained by data on weak decays of b, c, and s quarks, except for the b $\to$ u channel.

4. The weak mixing angles show a hierarchy, $\sin\theta_{12} > \sin\theta_{23} > \sin\theta_{13}$.
5. Experimental constraints imply $m_t > 50$ GeV and the CP phase $\sin\delta > 0.3$.
6. The origin of mixing angles and quark masses is not understood.
7. Rare Kaon decay experiments will probe new physics at the mass scale of 10–100 TeV.

Open problems and possibilities include the following:

1. The experimental discrepancy at the 2–3 st.dev. level between the ARGUS and CLEO results on the existence of the $b \to u$ transition needs clarification.
2. The positive result on ϵ'/ϵ of the NA31 collaboration awaits confirmation by the experiment E731B at Fermilab within less than a year, and also by the improved NA31 experiment at CERN and by the CP-LEAR experiment.
3. In the more distant future, the weak couplings involving top quarks can be studied at LEP 200 through $W^+ W^-$ production if $m_t < m_W - m_b$.
4. A better theoretical understanding of exclusive hadronic weak decays is important; QCD sum rules, the 1/N expansion and lattice calculations require considerable effort.
5. CP violation in the B meson system can be studied through large $B - \bar{B}$ partial decay rate asymmetries; due to small branching ratios of these channels, $\sim 10^8$ decays are needed, which can be achieved in a dedicated B factory with a luminosity of $\sim 5 \times 10^{33}\,cm^{-2}\,s^{-1}$.

References

1 N. Cabibbo: Phys. Rev. Lett. **10**, 531 (1963)
2 S. Glashow, Iliopoulos, L. Maiani: Phys. Rev. **D2**, 1285 (1970)
3 M. Kobayashi, T. Maskawa: Prog. Theor. Phys. **49**, 652 (1973)
4 F. Gilman, K. Kleinknecht, B. Renk: In *Review of Particle Properties*, Phys.Lett. **204B**, 1 (1988)
5 L.L. Chau, W.Y. Keung: Phys. Rev. Lett. **53**, 1802 (1984)
6 H. Harari, M. Leurer: Phys. Lett. **181B**, 123 (1986)
7 H. Fritzsch, J. Plankl: Phys. Rev. **D35**, 1732 (1987)
8 L. Maiani: *Proc.Int. Symp. on Lepton and Proton Interactions, Hamburg* (DESY, Hamburg, 1977) p. 867
9 L. Wolfenstein: Phys. Rev. Lett. **51**, 1945 (1984)
10 C. Jarlskog: Phys. Rev. Lett. **55**, 1039 (1985) and Z. Phys. **C29**, 491 (1985)
11 W.J. Marciano, A. Sirlin: Phys. Rev. Lett. **56**, 22 (1986)
12 A. Sirlin, R. Zucchini: Phys. Rev. Lett. **57**, 1994 (1986)
13 W. Jaus, G. Rasche: Phys. Rev. **D35**, 3420 (1987)
14 A. Sirlin: Phys. Rev. **D35**, 3423 (1987)
15 G. Lopez-Castro, J. Pestieau: Phys. Lett. **203B**, 315 (1988)
16 D. Wyler, P. Langacker: Priv. comm. (1988)
17 M. Ademollo, R. Gatto: Phys. Rev. Lett. **13**, 264 (1964)
18 H. Leutwyler, M. Roos: Z. Phys. **C25**, 91 (1984)
19 M. Bourquin et al.: Z. Phys. **C21**, 27 (1983)
20 J.F. Donoghue, B.R. Holstein, S.W. Klimt: Phys. Rev. **D35**, 934 (1987)
21 J.M. Gaillard, G. Sauvage: Priv. comm. (1988)
22 H. Abramowicz et al.: Z. Phys. **C15**, 19 (1982)
23 K. Kleinknecht, B. Renk: Z. Phys. **C34**, 209 (1987)
24 F.J. Gilman, K. Kleinknecht: In *Review of Particle Properties*, Phys. Lett. **170B**, 74 (1986)
25 F. Bletzacker, H.T. Nieh, A. Soni: Phys. Rev. **D16**, 732 (1977)

26 M. Wirbel et al.: Z. Phys. **C29**, 637 (1985)
27 T.M. Aliev et al.: Sov. J. Nucl. Phys. **40**, 527 (1984)
28 C.A. Dominguez, N. Paver: Phys. Lett. **207B**, 499 (1988)
29 G. Altarelli et al.: Nucl. Phys. **B208**, 365 (1982)
30 B. Grinstein, N. Isgur, M.B. Wise: Phys. Rev. Lett. **56**, 298 (1986);
N. Isgur: Priv. comm.
31 J.G. Körner, G.A. Schuler: Z. Phys. **C38**, 511 (1988)
32 H. Albrecht et al.: Contribution to 24th Int. Conference on High Energy Physics, Munich 1988
33 S. Berends et al.: Phys. Rev. Lett. **59**, 407 (1987)
34 K. Wachs et al.: DESY-report 88-111 (1988), unpublished
35 C. Klopfenstein et al.: Phys. Lett. **130B**, 444 (1983);
G. Levman et al.: Phys. Lett. **141B**, 271 (1984)
36 H. Albrecht et al.: Phys. Lett. **209B**, 119 (1988)
37 C. Bebek et al.: Contribution to 24th Int. Conference on High Energy Physics, Munich 1988
38 D. Kreinick: Talk at 24th Int. Conference on High Energy Physics, Munich 1988
39 H. Schröder: Plenary talk at 24th Int. Conference on High Energy Physics, Munich 1988
40 W. Schmidt-Parzefall: Plenary talk, *Proc. 1987 Int. Symp. on Lepton and Photon Interaction, Hamburg*, ed. by W. Bartel and R. Rückl, Nucl. Phys. **B Suppl.3**, 257 (1988)
41 A. Ali, E. Pietarinen: Nucl. Phys. **B154**, 519 (1979)
42 D. Müller: Talk at 24th Int. Conference on High Energy Physics, Munich 1988
43 R. Fulton et al.: Contribution to 24th Conference on High Energy Physics, Munich 1988
44 R. Rückl: Habilitationsschrift Univ. München 1983;
K.R. Schubert: Priv. comm.
45 K. Kleinknecht, B. Renk: Phys. Lett. **130B**, 459 (1983)
46 H. Fritzsch: Nucl. Phys. **B155**, 189 (1979); Phys. Lett. **70B**, 436 (1977)
47 B. Stech: Phys. Lett. **130B**, 189 (1983)
48 E. Longo, B. Renk et al.: W decay working group, ECFA Workshop LEP 200, Aachen
49 J.H. Christenson, J.W. Cronin, V.L. Fitch, R. Turlay: Phys. Rev. Lett. **13**, 138 (1964)
50 J.M. Gaillard et al.: Phys. Rev. Lett. **18**, 20 (1967);
J.W. Cronin et al.: Phys. Rev. Lett. **18**, 25 (1967)
51 S. Bennett et al.: Phys. Rev. Lett. **19**, 993 (1967);
D. Dorfan et al.: Phys. Rev. Lett. **19**, 987 (1967)
52 Reviewed in K. Kleinknecht, Ann. Rev. Nucl. Sci. **26**, 1 (1976)
53 K. Kleinknecht: *Proc. 17th Int. Conf. High Energy Phys., London 1974*, ed. J.R. Smith (Chilton, Didcot, UK), p. III-23
54 L. Wolfenstein: Phys. Rev. Lett. **13**, 562 (1964)
55 N.N. Biswas et al.: Phys. Rev. Lett. **47**, 1378 (1981); and priv. comm. to W. Ochs
56 W. Ochs: Priv. comm. from analysis of CERN-Munich data
57 C. Geweniger et al.: Phys. Lett. **48B**, 487 (1974);
R. Messner et al.: Phys. Rev. Lett. **30**, 876 (1973);
D.P. Coupal et al.: Phys. Rev. Lett. **55**, 566 (1985)
58 M.Banner et al.: Phys. Rev. Lett. **28**, 1597 (1972);
M. Holder et al.: Phys. Lett. **B40**, 141 (1972);
R.H. Bernstein et al.: Phys. Rev. Lett. **54**, 1631 (1985);
J.K. Black et al.: Phys. Rev. Lett. **54**, 1628 (1985)
59 C. Geweniger et al.: Phys. Lett. **52B**, 119 (1974)
60 J.H. Christenson et al.: Phys. Rev. Lett. **43**, 1209 (1979)
61 C. Geweniger et al.: Phys. Lett. **48B**, 483 (1974)
62 J. Ellis et al.: Nucl. Phys. **B109**, 213 (1976);
F.J. Gilman, M.B. Wise: Phys. Lett. **83B**, 83 (1979), Phys. Rev. **D20**, 2392 (1979);
B. Guberina, R. Peccei: Nucl. Phys. **B163**, 289 (1980);
L. Wolfenstein: Ann. Rev. Nucl. Sci. **36**, 137 (1986)
63 A.J. Buras, J.M. Gérard: Phys. Lett. **203**, 272 (1988)
64 U. Türke et al.: Nucl. Phys. **B285**, 313 (1985)
65 M. Woods et al.: Phys. Rev. Lett. **60**, 1695 (1988)
66 H. Yamamoto: Talk at 24th Int. Conference on High Energy Physics, Munich 1988
67 H. Burkhardt et al.: Phys. Lett. **206B**, 169 (1988)
68 H. Burkhardt et al.: Nucl. Instr. Meth. **A268**, 116 (1988)
69 Athens - Basel - CERN - Fribourg - Liverpool - Saclay - SIN - Stockholm - Thessaloniki - Zürich - Collaboration, CERN Proposal PSCC/85-6/P82

70 K. Kleinknecht: In *CP Violation*, ed. by C. Jarlskog (World Scientific, Singapore 1988)
71 J. Gasser, H. Leutwyler: Phys. Rev. **87**, 77 (1982)
72 J. Ellis et al.: Phys. Lett. **192B**, 201 (1987);
G. Altarelli, P. Franzini: Z. Phys. **C37**, 271 (1988);
H. Harari, Y. Nir: Phys. Lett. **195B**, 586 (1987);
J. Maalampi, M. Roos: U. of Helsinki preprint (1988);
A. Datta, E.A. Paschos, U. Türke: Phys. Lett. **196B**, 376 (1987);
L.L. Chau, W.Y. Keung: Prepr. UCD-87-02 (1987);
C.H. Albright, C. Jarlskog, B.A. Lindholm: Phys. Lett. **199B**, 553 (1987);
V. Barger et al.: Prepr. MAD-PH-341 (1987);
I.I. Bigi et al.: Phys. Lett. **194B**, 307 (1987);
W.S. Hou, A. Soni: Phys. Lett. **196B**, 92 (1987)
73 A. Pich, E. de Rafael: Phys. Lett. **158B**, 477 (1985);
A. Pich, B. Guberina, E. de Rafael: Nucl. Phys. **B277**, 197 (1986)
74 W.A. Bardeen, A.J. Buras, J.M. Gérard: Phys. Lett. **211**, 343 (1988)
75 M.B. Gavela et al.: CERN-TH 4905/87 (1987)
76 L.J. Reinders, S. Yazaki: Nucl. Phys. **B288**, 789 (1987)
77 K.G. Chetyrkin et al.: Phys. Lett. **174B**, 104 (1986)
78 R. Decker: Nucl. Phys. **B277**, 661 (1986)
79 N. Bilic, C.A. Dominguez, B. Guberina: DESY 87-162 (1987)
80 C. Albajar et al.: Phys. Lett. **186B**, 247 (1987);
K. Eggert: talk at la Thuile (1988)
81 C. Albajar et al.: Z. Phys. **C37**, 505 (1988);
G. Altarelli et al.: CERN-TH-4978/88 (1988);
F. Pauss: these proceedings
82 U. Amaldi et al.: Phys. Rev. **D36**, 1385 (1987)
83 M. Zeller: Minirapporteur's talk at 24th Int. Conference on High Energy Physics, Munich 1988
84 See Ref. 83, and priv. comm. by W. Morse
85 G.D. Barr et al.: Phys. Lett. **214B**, 303 (1988)
86 C.O. Dib, I. Dunietz and F.J. Gilman: SLAC-PUB-4762 (Nov. 1988)
87 R.S. Willey: Phys. Lett. **173B**, 480 (1986)
88 T.N. Pham and D.G. Sutherland: Phys. Lett. **151B**, 444 (1985)
89 R.S. Chivukula and A.V. Manohar: Phys. Lett. **207B**, 86 (1988)
90 M.E. Zeller et al.: E777, contr. No. 394 to 24th Int. Conference on High Energy Physics, Munich 1988
91 M. Mannelli et al.: BNL-Yale, contr. No. 484D to 24th Int. Conference on High Energy Physics, Munich 1988
92 E. Jastrzembski et al.: BNL-Yale, contr. No. 655B to 24th Int. Conference on High Energy Physics, Munich 1988
93 T. Inagaki et al.: KEK-Tokyo-Kyoto, contr. to 24th Int. Conference on High Energy Physics, Munich 1988
94 H. Albrecht et al.: Phys. Lett. **192B**, 245 (1987)

Recent Developments in Perturbative QCD

Z. Kunszt

Institute of Theoretical Physics, Eidgenössische Technische Hochschule
CH-8093 Zürich, Switzerland

1. Introduction

In these lectures I review some of the recent theoretical developments of the applications of perturbative QCD to hard scattering processes which can be studied at the present and future high energy e^+e^- , e^-p , $p\bar{p}$ and pp colliders.

In Sect. 2, after a brief summary of the basic concepts, I describe the most important factorization and cancellation theorems. The validity of almost all applications is based on some of these asymptotic theorems. The proofs of the theorems are rather technical and still incomplete.

The region of the validity of the perturbative description depends also on the size of the expansion parameter α_s defined with the help of the renormalization group equation. In the second part of this Section I give a short summary of the renormalization, the running coupling constant and the renormalization group. As a simplest application I consider the total cross section of e^+e^- annihilation into hadrons. The surprisingly large $\mathcal{O}(\alpha_s^3)$ corrections are discussed as well as some new features appearing at the Z-pole.

Sect. 3 is devoted to deep inelastic scattering (DIS). The central topic in this Section is the spin dependent scattering. The recent EMC data triggered some new interest in this field. In principle, the spin dependent DIS offers a measurement of the axial anomaly. The QCD interpretation of the EMC data suggests a large spin dependent gluon number density in the proton which may be tested in the future.

In Sect. 4 I describe the latest theoretical results obtained for the large momentum transfer reactions in $p\bar{p}$ collisions. Important new calculations appeared on the transverse momentum distribution of the W and Z production and on heavy quark production.

Finally, Sect. 5 is devoted to jet production in e^+e^- and $p\bar{p}$ collisions. Our main concern here is the description of jet production in the next-to-leading order accuracy.

Physics for a New Generation Editors: H. Latal · H. Mitter

2. General Properties of Perturbative QCD

2.1 On the Validity of the Perturbative Description

The pointlike nature of the deep inelastic structure functions of electron-proton scattering at large momentum transfer (Bjorken scaling) has given the first indication that strong interactions become weak at short distances, although they are strong at distance scales of the order of the proton size. The suggestion that QCD is the correct theory of the strong interactions [1] has obtained its decisive support when it has been discovered by Gross, Wilczek and Politzer [2] that quantum chromodynamics is an asymptotically free field theory in which the effective coupling constant goes to zero as the relevant distance scale of the phenomena goes to zero. *Asymptotic freedom* gave the clue to resolve the apparent contradiction, namely that i) QCD is the theory of quark and gluon interactions; ii) QCD is the theory of hadrons and their interactions.

The concept of coloured quarks and gluons has been found from the study of the multiplet structure observed in hadron spectroscopy[1]. Quarks come in six different flavours (u, d, s, c, b, t) and three colours (red, blue, green). Hadrons are made from quarks. They are bounds states which are held together by the colour force mediated by the coloured gluons. The conserved quantum numbers of the hadrons (electric charge, isospin, strangeness, charm, etc.) are derived from the corresponding conserved quantum numbers of the quarks. According to the *Confinement Hypothesis* [4] the colourless hadrons are the physical states of the theory in which the colour degrees of freedom are permanently confined. For example

$$|\text{proton}>=|\text{uud}> \quad \text{in color singlet}$$
$$|\pi^->=|\bar{\text{u}}\text{d}> \quad \text{in color singlet.}$$

In general, a hadron state can also have gluons and quark-antiquark pairs.

Since the theory is asymptotically free, the properties of physical phenomena at short distances and at finite time intervals can be described perturbatively in terms of weakly interacting quarks and gluons. At large distances, however, the colour confining force becomes strong, the perturbative treatment fails. Therefore we can classify the various time and space regions naturally into perturbative and non-perturbative regimes.

We refer to the *Confinement Hypothesis*, since a rigorous derivation of the property of colour confinement does not exist. It is generally believed, however, that the hypothesis is correct. Large scale computer simulations of QCD on the lattice also give support to quark confinement [5]. Furthermore, there are rigorous non-perturbative theorems which indicate the consistency of the confinement hypothesis. (For example the vector symmetries of QCD remain unbroken, or that the pion is the lightest hadron etc. [6].) In the meson spectroscopy and in low energy pion and kaon interactions the spontaneously broken chiral symmetry is the most fundamental concept. Although no rigorous derivation of chiral symmetry breaking exists, large scale

[1] See for example [3].

computer simulations indicate that confinement and chiral symmetry breaking might be derived directly from the QCD Lagrangian. There are important non-perturbative theorems on the chiral properties (for example if the quarks are massless the pion mass must vanish, etc. [6]). In particular, chiral perturbation theory [7] is a powerful tool to describe quantitatively some of the basic properties of the low lying hadrons.

Asymptotic freedom is also crucial in the non-perturbative treatment. It gives the consistency condition for the existence of the continuum limit when QCD is defined as the limit of some cutoff theory (for example lattice QCD [4,5]).

It is rather important for collider physics that the perturbative description is valid for large momentum transfer reactions since perturbation theory is the only systematic method for calculating scattering cross sections directly from the QCD Lagrangians. The application of perturbative QCD to scattering phenomena with large momentum transfer, however, is not entirely straightforward. In all the scattering processes, some long distance effects also take place. Fortunately for many quantities these long distance effects can be neglected or they can be decomposed into some universal process-independent factors.

In the the numerous applications of perturbative QCD we can find reliable results where factorization is proved rigorously, as well as approximate models which are only inspired by QCD. In the first category we have, for example, the total cross section of e^+e^- annihilation into hadrons, while in the second category we find the string or cluster model of jet fragmentation. Since the QCD Lagrangian, as well as the Feynman rules, are described in great detail in many textbooks and lecture notes [8,9,10] I summarize here only the basic theorems which provide the foundations for the application of perturbative QCD. [2]

2.2 Theorems of Perturbative QCD

The most fundamental properties needed for applications of perturbative QCD are given by the following theorems:

i) QCD is a renormalizable non-abelian gauge theory.
ii) The S-matrix is renormalization invariant.
iii) QCD is asymptotically free, and the QCD coupling constant can be specified in terms of a fundamental dimensional parameter with the help of the renormalization group equation.
iv) In some simple processes like e^+e^- annihilation into hadrons all infrared sensitivity cancels.
v) In simple hard scattering processes involving hadrons in the initial state only the soft gluon effects cancel. The infrared sensitivity due to collinear long distance effects, however, is universal. It factors into parton distribution functions for incoming hadrons and decay functions for outgoing partons. This theorem gives the required consistency condition for the validity of the QCD improved parton model in which the cross sections of the large momentum transfer reactions are calculated according the diagram in Fig. 1 which leads to the formula

[2] For an excellent exposition with many references see the review by Collins and Soper [11].

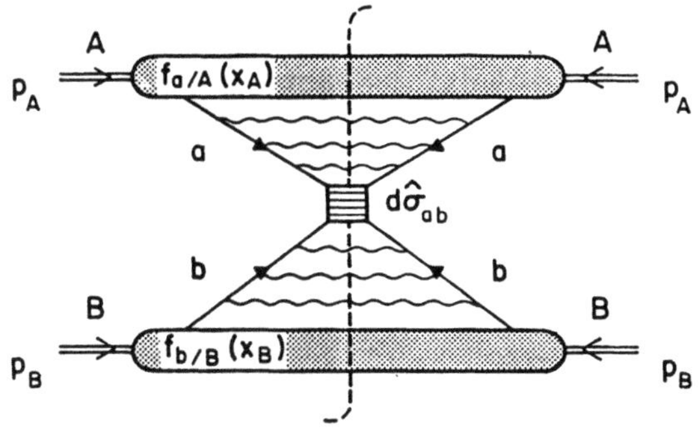

Fig. 1. Cut diagram for the physical cross section (2.1) in the QCD-improved parton model

$$d\sigma_{AB}(p_A, p_B, ..) = \sum_{A,B} \int dx_A dx_B f_{a/A}(x_A) f_{b/B}(x_B) d\hat{\sigma}_{ab}(x_a p_A, x_b p_B, ...) \ . \tag{2.1}$$

vi) Theorems at the boundary of the kinematical limits of the hard processes, like the p_T distribution of the production of the W-boson produced in p$\bar{\text{p}}$ collisions at small p_T values, or like the behaviour of the structure functions at x near one or at x near zero (small x region).

The cancellation of the soft gluon contribution is highly non-trivial since the soft gluon structure of QCD is different from the soft photon structure of QED as a consequence of the non-abelian nature of gluons [12]. The factorization theorem is proven when in the initial state we have only one hadron [13]. When both initial particles are hadrons the proof is more complicated. Nevertheless factorization is more or less rigorously proven for the Drell-Yan process. A further improvement of some loose points of the earlier proof has been recently given [14]. No explicit proof appeared for other processes like inclusive jet production or heavy flavour production in hadron collisions (see however [11]). But the main strategy of the proof of the Drell-Yan case should be applicable also in these latter cases.

At the edge of the kinematical limits the cancellation of the soft and collinear singularities is not perfect. Certain large logarithmic factors appear which destroy the validity of the naive perturbative expansion [15]. These logarithms are not controlled by the the renormalization group behaviour. For example, in the case of the p_T distribution of the W-boson produced in p$\bar{\text{p}}$ collisions these logarithmic factors are the so called Sudakov logarithms. They can be resummed and resummation gives rise to the so called Sudakov form factors [15, 16, 17].

In these applications we assume that we can extend the validity of perturbative QCD if we are able to resum the logarithmic factors appearing due to the incomplete cancellation of the soft and collinear gluon configurations. Clearly, it is important to test experimentally the predictions of perturbative QCD in the most general situations. This topic requires further theoretical studies, and it has large potential for further interesting theoretical and experimental results [18].

In many applications of perturbative QCD the short distance scale relevant for a given experiment is not small enough for the reliable leading order application of perturbation theory. The difficulty is signalled by a large K-factor and a sensitivity to the choice of the "reasonable" scale in the expansion parameter α_s . A reliable discussion on the validity of the perturbative results in general requires the evalution of the next-to-leading corrections.

2.3 Renormalization

2.3.1 Lagrangian

QCD is defined as the quantum field theory given by the Lagrangian

$$\begin{aligned}\mathcal{L} = &-\frac{1}{4}G^i_{\mu\nu}G^{\mu\nu}_i + \sum_a \bar{q}_a(i\hat{D} - m)_{ab}q_b \\ &- \frac{1}{2\lambda}(\partial^\mu G^i_\mu)^2 + \partial_\mu \bar{c}^i D^\mu_{ij} c^j \ ,\end{aligned} \tag{2.2}$$

where $G^i_{\mu\nu}$ is the gluon field strength, q is the quark field, c^i is the ghost field, λ is the gauge fixing parameter and

$$\begin{aligned}(D_\mu)_{ab} &= \partial_\mu \delta_{ab} - igt^i_{ab}G^i_\mu, \quad a,b = 1,2,3 \\ (D_\mu)_{ij} &= \partial_\mu \delta_{ij} - igT^k_{ij}G^k_\mu, \quad i,j = 1,2,...,8 \\ G^i_{\mu\nu} &= \partial_\mu G^i_\nu - \partial_\nu G^i_\mu + gf^{ijk}G^j_\mu G^k_\nu \ .\end{aligned} \tag{2.3}$$

The t^i, T^i denote the color SU(3) matrices in the fundamental and in the adjoint representations, respectively. It is convenient to introduce the colour factors as

$$Tr(t^a t^b) = T_R \delta_{ab}, \quad Tr(T^i T^i) = N_C \delta^{ij}, \quad \sum_a (t^a t^a)_{\alpha\beta} = C_F \delta_{\alpha\beta} \ .$$

For SU(3) we have

$$C_F = \frac{N_C^2 - 1}{2N_C}, \quad N_C = 3, \quad T_R = \frac{1}{2} \ . \tag{2.4}$$

From this Lagrangian we obtain the Feyman rules of QCD (see for example [8, 9, 10]). The corresponding Feynman diagrams uniquely define the matrix elements of any physical amplitude or Green function. In higher order, however, ultraviolet divergent integrals appear. They can be eliminated by renormalization which requires the introduction of some convenient regularization and subtraction prescription. The most commonly used regularization is the dimensional regularization, when the integrals which are divergent in four dimensions are regularized by means of analytic continuation into $d = 4 - 2\epsilon$ dimensions. The typical integrals which appear in one-loop corrections have the form

$$\int \frac{d^d k}{(2\pi)^d} \frac{1}{[-k^2 + C - i\epsilon]^m} = i(4\pi)^{-2+\epsilon}[C - i\epsilon]^{2-m-\epsilon} \frac{\Gamma(m - 2 + \epsilon)}{\Gamma(m)} .$$

The dimensional regularization has the advantage that it preserves gauge and Lorentz invariance, and it can also be used to regulate the soft and collinear singularities. With dimensional regularization we can have the mass independent minimal subtraction scheme (MS) [19] or its slightly modified version, the so-called $\overline{\text{MS}}$ scheme [20]. Renormalizability assures that the ultraviolet divergences which appear in higher order in perturbation theory can be removed by renormalization of the wave functions, coupling constants and quark masses while preserving gauge invariance and the decoupling of the unphysical degrees of freedom from the S-matrix elements.

2.3.2 Renormalization Factors

As part of the renormalization program one specifies the regularization scheme and the renormalization prescription which tells us uniquely the form of Z-factors. It is instructive to recall the definition of the renormalization factors and their form in the MS scheme. The renormalized fields (G, q, c) and the renormalized parameters (g, m_i, $\alpha_s = g^2/(4\pi)$) of the theory are defined as

$$\begin{aligned} G^i_\mu &= Z_3^{-1/2} G^{(0)}_\mu, \quad q_a = Z_2^{-1/2} q^{(0)}_a, \\ c_a &= \tilde{Z}_3^{-1/2} c^{(0)}_a, \qquad \lambda = Z_3^{-1} \lambda^{(0)}, \\ g(\mu) &= \mu^{-\epsilon} Z_1^{-1} Z_3^{3/2} g^{(0)}(\epsilon) = \mu^{-\epsilon} \tilde{Z}_1^{-1} Z_3^{1/2} g^{(0)}(\epsilon) \\ &= \mu^{-\epsilon} (Z_1^F)^{-1} Z_2 Z_3^{1/2} g^{(0)}(\epsilon) = \mu^{-\epsilon} Z_4^{-1/2} Z_3 g^{(0)}(\epsilon) \\ m &= Z_m^{-1} m^{(0)}. \end{aligned} \tag{2.5}$$

The gauge invariance is preserved since for the Z_i's we have the Ward identities

$$Z_1/Z_3 = \tilde{Z}_1/\tilde{Z}_3 = Z_1^F/Z_2 = Z_4/Z_1 \ . \tag{2.6}$$

In the MS scheme the renormalization factors have only the pole terms in $1/\epsilon$ (see [9])

$$\begin{aligned} Z_1 &= 1 + \frac{g^2}{16\pi^2}\frac{1}{\epsilon}\Big[N_C\Big(\frac{17}{12} - \frac{3\lambda}{4}\Big) - \frac{4}{3} n_f T_R\Big], \\ Z_2 &= 1 - \frac{g^2}{16\pi^2}\frac{1}{\epsilon}\Big[C_F\lambda\Big], \\ Z_3 &= 1 + \frac{g^2}{16\pi^2}\frac{1}{\epsilon}\Big[N_C\Big(\frac{16}{6} - \frac{\lambda}{2}\Big) - \frac{4}{3} n_f T_R\Big], \\ \tilde{Z}_3 &= 1 + \frac{g^2}{16\pi^2}\frac{1}{\epsilon}\Big[N_C\Big(\frac{3}{4} - \frac{\lambda}{4}\Big)\Big] \ . \end{aligned} \tag{2.7}$$

I note that in any fixed n-th order of the perturbation theory some prescription dependence will remain. Its magnitude, however, is of order $\alpha_s^{(n+1)}$.

2.3.3 Comment on the Axial Gauge

The Lagrangian in Eq. (2.2) has been given in covariant gauge where ghost terms also occur. In this gauge we have well established proof of renormalizability. In the axial gauge, the gauge condition is

$$n_\mu G_i^\mu = 0$$

where n_μ is some fixed four-vector. The conditions $n^2 > 0$ or $n^2 = 0$ or $n^2 < 0$ define the temporal, light-cone or pure axial gauge. In this gauge the longitudinal degrees of freedom of the gauge field do not propagate. But in the gluon propagator extra singularities of the type $1/qn$ appear. These local singularities, unfortunately, become too strong to carry out meaningfully the renormalization program. But if it is modified to the non-local expression

$$1/qn \to \frac{qn^*}{(qn)(qn^*) + i\epsilon}, \quad \epsilon > 0, \quad n_\mu^* = (n_0, -\boldsymbol{n})$$

the Wilson loop in the next-to-leading order can be renormalized meaningfully with non-local counter terms [21]. In the axial gauge the proof of the factorization theorems might become simpler. But how to overcome the obstruction due to the presence of the $1/qn$ terms is not clear.

2.3.4 The Running Coupling Constant

In the dimensional regularization scheme the scale of the ultraviolet subtraction appears via the scale parameter μ which assures the correct physical dimensions in d dimensions. The beta function is defined through the differential equation

$$\mu^2 \frac{d\alpha_s}{d\mu^2} = \frac{1}{2}\beta(\alpha_s) = -\alpha_s^2 b_0 - \alpha_s^3 b_1 - \alpha_s^4 b_2 + \mathcal{O}(\alpha_s^5). \tag{2.8}$$

The coefficients of the beta function b_0 and b_1 [2, 22] are renormalization scheme independent[3]

$$b_0 = \frac{33 - 2n_f}{12\pi}, \quad b_1 = \frac{153 - 19n_f}{24\pi^2} . \tag{2.9}$$

The coefficient b_2 is also known [23] in the $\overline{\text{MS}}$ scheme

$$b_2 = \frac{2857 - (5033/18)n_f + (325/54)n_f^2}{(4\pi)^3} . \tag{2.10}$$

The first order differential equation (2.8) can be solved in terms of a single initial parameter. The Λ parameter of QCD is defined by writing the most general solution in the form

[3] It is a simple exercise to derive from Eqs. (2.5) and (2.7) the value of b_0.

$$\alpha_s = \frac{1}{b_0 \log(\mu^2/\Lambda^2)} \left(1 - \frac{b_1}{b_0^2} \frac{\log(\log(\mu^2/\Lambda^2))}{\log(\mu^2/\Lambda^2)} + \left(\frac{b_1}{b_0^2} \frac{\log(\log(\mu^2/\Lambda^2))}{\log(\mu^2/\Lambda^2)} \right)^2 \right.$$

$$\left. - \frac{b_1^2(\log(\log(\mu^2/\Lambda^2)) + 1) - b_0 b_2}{b_0^3 \log^2(\mu^2/\Lambda^2)} \right) + \mathcal{O}\left(\frac{\log^3[\log(\mu^2/\Lambda^2)]}{\log^4(\mu^2/\Lambda^2)} \right) . \qquad (2.11)$$

The Λ parameter may be considered as the fundamental parameter of QCD. Its physical dimension is GeV. Its numerical value depends on the renormalization scheme and on the value of n_f, therefore it is more appropriate to use the notation $\alpha_s(\mu^2)^{(n_f)}$. When a new threshold of a heavy quark is passed in the $\overline{\text{MS}}$ scheme we have to require the matching condition

$$\alpha_s^{(3)}(m_c^2) = \alpha_s^{(4)}(m_c^2), \quad \alpha_s^{(4)}(m_b^2) = \alpha_s^{(5)}(m_b^2) . \qquad (2.12)$$

These equations can be used to relate the numerical value of $\Lambda_{\overline{\text{MS}}}$ defined with different flavour number. A partial compilation [24] of the most significant measurements of α_s is given in Fig. 2. For five flavours in the $\overline{\text{MS}}$ scheme we have

$$\Lambda_{\overline{\text{MS}}}^{(5)} \approx 140\,\text{MeV} \pm 90\,\text{MeV}$$

which at the Z-pole gives for α_s the value

$$\alpha_s(M_Z) = 0.11 \pm 0.01 .$$

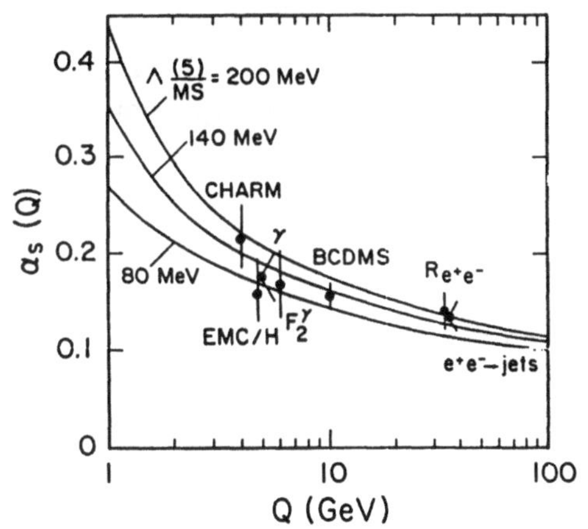

Fig. 2. The various determinations of α_s

2.3.5 The Renormalization Group Equations

In the massless limit the QCD Lagrangian is formally scale invariant. But this formal scale invariance is destroyed by the renormalization prescription, which requires the introduction of some mass parameter μ. As we have seen above, in the dimensional regularization μ appears as the scale of the coupling constant which is not dimensionless if the space-time dimension is different from four.

This scale parameter can be introduced in the MS or in the $\overline{\text{MS}}$ schemes in such a way that the massless limit of the renormalization counter terms exist.

The physical quantities must be invariant with respect to the change of the renormalization scale parameters μ. The parameter μ, however, appears in the physical cross sections also via the renormalized coupling constant and renormalized mass parameters. If μ is changed we obtain a change in α_s and m_j as

$$\mu\frac{d\alpha_s}{d\mu} = \beta(\alpha_s), \quad \mu\frac{dm_j}{d\mu} = \gamma_m(\alpha_s)m_j. \tag{2.13}$$

Therefore the physical quantities must vanish under the linear differential operator

$$\begin{aligned} D_{\mathrm{RG}} &\equiv \mu\frac{\partial}{\partial\mu} + \beta\frac{\partial}{\partial\alpha_s}\gamma_m\sum_i m_i\frac{\partial}{\partial m_i}\,, \\ D_{\mathrm{RG}}S &= 0 \end{aligned} \tag{2.14}$$

In perturbation theory when a physical quantity is calculated to order α_s^n an apparent μ dependence remains which is, however, of the order $\alpha_s^{(n+1)}$. A simple illustration is given in the next Subsection.

2.4 The Total Cross Section of e^+e^- Annihilation into Hadrons

2.4.1 Perturbative Expansion to Order α_s^3

In the limit when the annihilation energy is much larger than the quark and hadron masses

$$\sqrt{s} = Q, \quad Q \gg m_i, \quad i = \mathrm{u,\ d,\ s,} \ldots$$

the ratio of the total cross section of e^+e^- annihilation into hadrons (with one photon exchange) to the Born cross section of $e^+e^- \to \mu^+ + \mu^-$ has the expansion at the renormalization scale μ

$$\begin{aligned} R(Q^2) = (3\sum_i e_i{}^2)\Bigg(1 + c_1\alpha_s(\mu^2) + \left[c_2 + c_1 b_0 \log\left(\frac{\mu^2}{Q^2}\right)\right]\alpha_s(\mu^2)^2 \\ + \left[c_3 + \left\{2c_2 b_0 + c_1 b_1 + c_1 b_0^2 \log\left(\frac{\mu^2}{Q^2}\right)\right\}\log\left(\frac{\mu^2}{Q^2}\right)\right]\alpha_s(\mu^2)^3\Bigg) \end{aligned} \tag{2.15}$$

where the coefficients have the values [25]

$$\begin{aligned} c_1 &= \frac{\alpha_s}{\pi}, \quad c_2 = (1.986 - 0.115 n_f)(\frac{\alpha_s}{\pi})^2, \\ c_3 &= (70.986 - 1.2 n_f - 0.005 n_f{}^2)\left(\frac{\alpha_s}{\pi}\right)^3 . \end{aligned} \tag{2.16}$$

In Eq. (2.15) we used $\alpha_s(\mu^2)$ as expansion parameter. It is easy to see that R satisfies the renormalization group equation (2.14) to order α_s^4. Since the perturbation series have been truncated some apparent scale dependence has remained in the

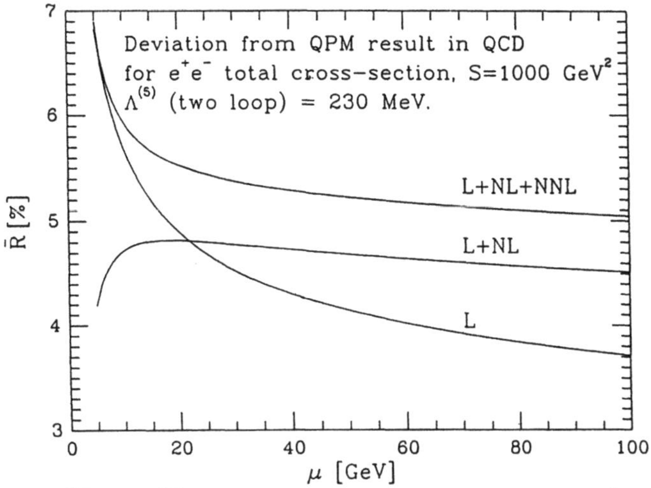

Fig. 3. $\overline{R} = R/R^{(0)} - 1$ **as a function of the scale** μ **to order** α_s **,** α_s^2 **and** α_s^3 **[24], with the leading (LO), next-to-leading (NL) and next-to-next-to-leading (NNL) contributions**

theoretical prediction. It is depicted in Fig. 3. However, in a finite neighborhood of the physical scale $\mu \approx Q$ the cross section shows very weak scale dependence. This variation can be considered as a rough measure of the size of the unknown higher order corrections.

From a precise measurement of R one can extract the value of α_s in second subleading order accuracy. Whether one should use the value of α_s so determined in formulas which are only accurate to next-to-leading order is only a matter of taste.

It is surprising that the second subleading correction is larger than the first one. We should admit that in perturbative QCD, similarly to the case of QED, we have difficulties to get a general idea of the result to be expected. An educated guess can be obtained for the unknown higher order corrections by assuming somewhat arbitrarily that the correct expansion parameter is $\alpha_s(Q^2)$. This is not unreasonable, since the first coefficient is ≈ 0.3, the second is ≈ 0.15, and the coefficient of α_s^3 is around two. Assuming $\alpha_s = 0.12$, the corresponding correction to R would be in this case $\pm 0.1\%$, far beyond the expected experimental accuray. With this assumption the third order correction is large but the second order correction is unusually small. This observation is not inconsistent with other quantities calculated in perturbation theory to next-to-leading order. In most of the cases the first subleading correction is large. It is difficult either to support or to reject this speculation.

According to some wilder speculations the appearance of such a large correction is signalling the breakdown of pertubation theory. It would be very interesting to learn more about the nature of this difficulty. One may use causality, locality, renormalization group behaviour and Borel transforms to constrain the ratios of the perturbative coefficients. At best they can constrain only the large-n behaviour of the perturbative series. While it would be very interesting to know the large-n behaviour of the perturbative expansion of R it is unlikely that it could be tested against the existing result since $n = 3$ is not asymptotically large. Therefore it is of great interest to calculate at least one or two more coefficients. Unfortunately, the higher order calculation seems to be prohibitively difficult with the presently available calculational techniques.

2.4.2 Measurement of R at the Z-Pole

At the Z-pole instead of R one can study the ratio of the hadronic events to the leptonic events. This quantity is directly related to the ratio of the hadronic width to the leptonic width. In leading order in QCD the leptonic and hadronic width is given as

$$\Gamma_i^0 = \frac{G_F M_Z^3}{24\pi\sqrt{2}}(1+\delta_f)C(i)(a_i^2+v_i^2) \quad i = \mathrm{e},\ \mu, \ldots, \mathrm{u},\ \mathrm{d}, \ldots \tag{2.17}$$

where δ_f summarizes the electroweak corrections, $C(i)$ is the colour factor (it is 3 for quarks and 1 for leptons) and a_i, v_i denote the axial and vector coupling of the fermions to the Z-boson. I note that the formula obtained for the perturbative expansion of the QCD corrections is not directly applicable at the Z-pole. Of order α_s^2 and α_s^3 there are diagrams with two (triangle like) quark loops which give vanishing vector contribution but non-vanishing axial contribution. However it is expected that their contribution is negligible [26].[4]

In order to get the value of α_s with a precision of 10%, one needs to measure the ratio of the hadronic width to the leptonic width of the Z-boson with a precision of about 0.5%, which is the expected accuracy of the future LEP experiments.

3. Deep Inelastic Scattering

3.1 Structure Functions and Cross Sections

Deep inelastic scattering of an electron on a proton target played a central role in our understanding of the QCD-improved parton model [2, 8, 11]. To leading order in the electromagnetic interaction the cross section is described by the cut diagram shown in Fig. 4.

The lower blob in Fig. 4 denotes the current correlation function

$$W_{\mu\nu}^{\mathrm{S}}(q,p) + iW_{\mu\nu}^{\mathrm{A}}(q,p,s) = \int dx\ e^{iqx} < p, s|J_\mu(x)J_\nu(0)|p, s > \tag{3.1}$$

where the symmetric part (S) describes the spin independent scattering while the antisymmetric part (A) gives the spin dependent effects. p, q denote the four momenta of the proton and the virtual photon, and s is the spin polarization vector of the proton. Both $W_{\mu\nu}^{\mathrm{S}}$ and $W_{\mu\nu}^{\mathrm{A}}$ are expressed in terms of two dimensionless scalar structure functions $F_1(x,Q^2)$, $F_2(x,Q^2)$ and $g_1(x,Q^2)$, $g_2(x,Q^2)$, respectively. They give the spin independent cross section

[4] At the Z-pole the mass correction due to the finite b-quark mass is less than 0.3%.

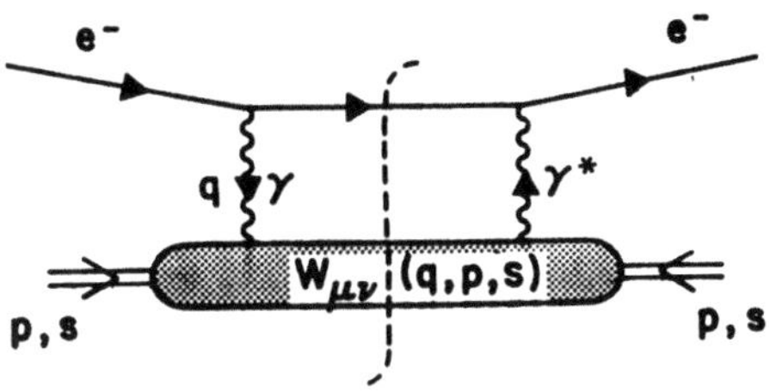

Fig. 4. Cut diagram defining the cross section of deeply inelastic electron proton scattering

$$\frac{d\sigma}{dxdy} = 8\pi\alpha^2\frac{ME}{Q^4}\left[\frac{1+(1-y)^2}{2}2xF_1(x,Q^2) + \{(1-y)(F_2(x,Q^2) - xF_1(x,Q^2))\} - \frac{M}{2E}F_2(x,Q^2)\right] \tag{3.2}$$

and the polarization asymmetry

$$A = \frac{d\sigma^{\downarrow\uparrow} - d\sigma^{\uparrow\uparrow}}{d\sigma^{\downarrow\uparrow} + d\sigma^{\uparrow\uparrow}} \sim [2\frac{E}{M} - (x + \frac{E}{M})(1-y)]g_1(x,Q^2) - 2xg_2(x,Q^2)\ , \tag{3.3}$$

where $x = Q^2/2pq$, $y = pq/M^2$, $Q^2 = -q^2$; E is the electron energy in the rest frame of the proton and M is the proton mass.

Experimentally it has been found that for large Q^2 at any given fixed value of x the structure functions are approximately constant with respect to Q^2 (Bjorken scaling).

In the naive parton model [27] it is assumed that if Q is large in the infinite momentum frame the proton can be treated approximately as a beam of free quarks. The momentum distribution of the quarks is given by the bound state wave function of the proton. It is defined by the long distance properties of the theory. It can not be calculated in weak coupling perturbation theory. In the large $Q^2 \to \infty$ limit with x fixed, the physical cross section is the convolution of the (non-perturbative) number density of the quarks and the (perturbative) cross section of the absorption of the virtual photon on a single quark

$$d\sigma_{\gamma^* p}(p,q) = \sum_a q_{a/P}(x) \otimes d\hat{\sigma}_{\gamma^* a}(xp,q)\ , \tag{3.4}$$

where $q_{a/P}(x)$ denotes the quark number densities $u(x)$, $d(x)$,... etc. in the proton and we introduced the short hand notation for the convolution

$$A(x) \otimes B(x) = \int_0^1 dy \int_0^1 dz A(y)B(z)\delta(x - yz)\ . \tag{3.5}$$

Since the photo-absorption cross section on a single quark is proportional to a delta function

$$d\sigma_{\gamma^* a} \sim e_a^2\delta(Q^2 - 2xpq)\ ,$$

the structure functions become linear combinations of the quark number density functions, e.g.,

$$F_1(x) = \sum_a e_a^2 \left(q_a(x) + \bar{q}_a(x)\right) \tag{3.6}$$

$$g_1(x) = 1/2 \sum_a e_a^2 \left(q_a^{\uparrow}(x) - \bar{q}_a^{\downarrow}(x)\right) , \tag{3.7}$$

where e_a denotes the electric charge of the quark a and $q_a^{\uparrow}$ denotes the number density of positive helicity quarks within the proton. In the parton model the large transverse momentum or highly off-shell regions are separated from the regions where the hadronic final states are formed. It is assumed that this result is valid up to power corrections of $\mathcal{O}(M/Q)$.

It has great practical importance for collider physics that this result remains essentially valid also in QCD [2,11]. The only modification is that the separation of the long distance and short distance regions is not unique since the quark-gluon interaction requires renormalization.

In QCD one can prove that, as in the naive parton model, in the Bjorken limit the cross section for deep inelastic electron proton scattering can be written in the factorized form

$$\begin{aligned} d\sigma_{\gamma^* \mathrm{p}} &= \sum_a f_{a/P}(x,\mu) \otimes_x d\hat{\sigma}_{\gamma^* a}(xp, q, \alpha_s(\mu), \mu) \\ a &= \mathrm{u},\ \mathrm{d},\ \bar{\mathrm{u}}, \ldots, \mathrm{gluon}\ , \end{aligned} \tag{3.8}$$

or projecting out a given structure function we obtain

$$F_i(x, Q^2) = \sum_a \int_x^1 \frac{dy}{y} C_{i,a}\left[x/y, Q/\mu, \alpha_s(\mu)\right] f_{a/P}(y,\mu) + \mathcal{O}(M/Q) \ . \tag{3.9}$$

The long distance and short distance effects are separated by the renormalization scale μ. The physical cross section, of course, is independent of μ. Therefore the scale dependence in the parton density functions must be cancelled by the scale dependence of the short distance cross sections. The short distance cross sections and the structure functions satisfy renormalization group equations with anomalous terms of opposite signs.

The validity of factorization theorems of this kind is a general feature of renormalizable field theories. For example it is a factorization theorem that the coupling constant can be multiplicatively renormalized (see Eq. (2.5)).

The proof of the factorization theorem of Eqs. (3.8–3.9) in the original analysis [2] has been performed with the help of the Wilson operator product expansion [28] in terms of moments. In renormalizable field theories to all orders in perturbation theory the product of two local operators at short distances can be written as

$$J(x)J(0) \sim \sum_n C_n(x^2) x^{\nu_1} x^{\nu_2} \ldots x^{\nu_n} \hat{O}^{(n)}_{\nu_1 \nu_2 \ldots \nu_n} + \ldots \tag{3.10}$$

where all external Lorentz indices and internal quantum numbers are suppressed. In Eq. (3.10) $\hat{O}^{(n)}$ are local operators describing the long distance effects and the coefficients C_n are C-number functions which give the short distance properties.

Since the Green functions with local operator insertions satisfy the renormalization group equation the matrix elements of the product of current operators as well as the matrix elements of the local operators appearing on the right hand side of Eq. (3.10) also satisfy the renormalization group equations (see Eq. (2.14))

$$
\begin{aligned}
&(D_{\mathrm{RG}} + 2d_J + 2d_p) < p|J(x)J(0)|p >= 0 \\
&(D_{\mathrm{RG}} + 2d_p + d_{O^{(n)}}) < p|\hat{O}^{(n)}|p >= 0 \\
&(D_{\mathrm{RG}} + 2d_J - d_{O^{(n)}})C_n(x^2\mu^2, \alpha_s) = 0
\end{aligned}
\tag{3.11}
$$

where d_a denotes the total (physical plus anomalous) dimension of the corresponding operator.

Taking the matrix elements between proton states and performing the appropriate projections on both sides of the Wilson operator product expansion, Eq. (3.10), we obtain for the moments of the structure functions the factorized form

$$
F_i(N, Q) = \sum_a C\left[Q/\mu, \alpha_s(\mu^2)\right] f_{a/P}(N, \mu) \, , \tag{3.12}
$$

where the moments are defined as

$$
F_i(N, Q) = \int_0^1 dx \; x^{N-1} F_i(x, Q) \tag{3.13}
$$

and the coefficients $f_{a/P}(N, \mu)$ are given by the matrix elements of the local operators $O^{(n)}$. Assuming the validity of the result also for non-integer N's, Eq. (3.12) can be inverted by a Mellin transformation leading to the form given by Eq. (3.9).

In more modern treatments the validity of the factorization theorem has been derived with the help of a completely diagrammatic analysis [13, 14, 11]. The diagrammatic analysis has the advantage that it can be generalized from deep inelastic scattering also to the more complicated hadron-hadron reactions. The method requires the investigation of the leading integration regions in all orders of perturbation theory. A carefully defined subtraction procedure had to be used to identify the hard scattering part of the diagrams without double counting. This led to a well defined algorithm for the calculation of the infrared and ultraviolet finite hard scattering cross section $d\hat{\sigma}$ appearing in Eq. (3.8), and to a process-independent definition of the parton distribution functions.

The hard scattering cross section can be extracted from the parton cross section ($d\sigma^B$) defined in general n dimensions in terms of Feynman diagrams with ultraviolet renormalization counter terms, which cancel the ultraviolet divergences appearing in the loop integrals. This parton cross section is not finite since it still contains infrared collinear singularities. These singularities, however, are also regularized in n dimensions and $d\sigma^B$ can be decomposed into the convolution of a finite short distance cross section ($d\hat{\sigma}$) times a singular perturbative "parton in a parton" splitting function (Γ) as

$$
d\sigma_a^{(B)}(q, p, n) = \sum_b \int_0^1 dx \Gamma_{b/a}(x, \epsilon) d\hat{\sigma}_{\gamma^* b}(q, xp, \alpha_s(\mu), \mu) \tag{3.14}
$$

where μ is the ultraviolet renormalization scale. In the $\overline{\mathrm{MS}}$ renormalization scheme the "parton in a parton" splitting function has the perturbative expression

$$\Gamma_{a/b}(x,\epsilon) = \delta_{ab}\delta(1-x) - \frac{\alpha_s}{2\pi}P_{a/b}(x)\left(\frac{1}{\epsilon} + \log(4\pi) - \gamma_E\right) + \dots \,, \tag{3.15}$$

where $P_{a/b}(x)$ denotes the Altarelli-Parisi splitting function. For example for deep inelastic scattering we obtain the structure

$$\hat{\sigma}_{\gamma^* \mathrm{q}} \sim \frac{4\pi\alpha}{Q^2}\left[\delta(x - \frac{Q^2}{2pq}) + \frac{\alpha_s}{2\pi}C_{\mathrm{q}}(x,\mu) + \mathcal{O}(\alpha_s^2)\right] \tag{3.16a}$$

and

$$\hat{\sigma}_{\gamma^* \mathrm{g}} \sim \frac{4\pi\alpha}{Q^2}\left[\frac{\alpha_s}{2\pi}C_{\mathrm{g}}(x,\mu) + \mathcal{O}(\alpha_s^2)\right] . \tag{3.16b}$$

The explicit form of the perturbative coefficients, C_{q} and C_{g}, depends on the renormalization scheme [29]. A virtue of the diagrammatic analysis is that it has been manifestly shown that the soft and final collinear singularities cancel and that the initial collinear singularities can be factored into the process-independent common factors joined to the common hard scattering part of cut Feynman diagrams (see Fig. 5).

Finally, the diagrammatic analysis also helped to find a process-independent definition of the parton distribution functions in terms of the matrix elements of quark and gluon operators. For example for the u-quark number density in the proton we have the definition

$$f_{u/P}(\xi) = \frac{1}{4\pi}\int dx^- \int e^{-i\xi P^+ x^-} < P\left|\overline{u}(0,x^-,0_\perp)\gamma^+ \mathcal{G} u(0,0,0_\perp)\right| P > \tag{3.17a}$$

where the operator $\mathcal{G}$ denotes the path ordered phase factor

$$\mathcal{G} = \mathcal{P}\exp\left\{ig\int_0^{x^-} dy^- A_i^+(0,y^-,0_\perp)T^i\right\} . \tag{3.17b}$$

The labels +, −, $\perp$ refer to the plus, minus and transverse components of a four vector ($A_\pm = (A_0 \pm A_3)/\sqrt{2}$).

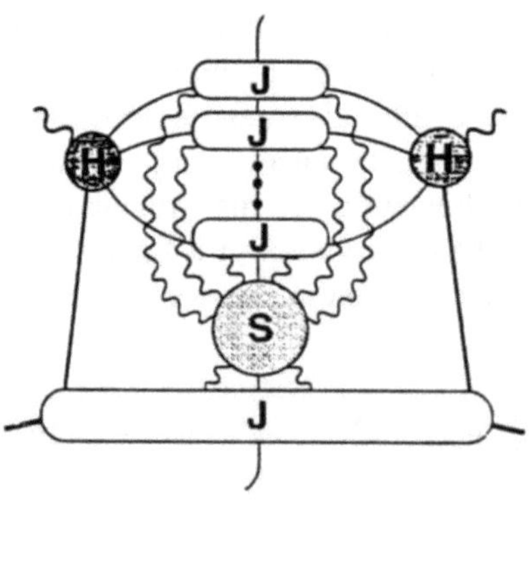

Fig. 5. The structure of the leading singularities of the cut diagrams of the cross section of deeply inelastic electron proton scattering [11]

The Feynman rules for the calculation of the parton distribution function has been developed by Collins and Soper [30]. The operator products in Eqs. (3.17) require renormalization which introduces the dependence on the renormalization scale μ. The renormalization group equation for $f_{a/P}(x,\mu)$ is the Altarelli-Parisi equation

$$\mu\frac{d}{d\mu}f_{a/A}(x,\mu)=\sum_b\int_x^1\frac{dz}{z}P_{a/b}(z,\alpha_s(\mu))f_{b/A}(x/z,\mu)\ . \tag{3.18}$$

The explicit definition of the parton densities could be used to show [29] that the odd and even combination of the moments of the parton densities are precisely the matrix elements of the local opearors appearing in the Wilson operator product expansion. The simplest relations of this type are for example the u-quark number sum rule

$$\int_0^1 dx\left[f_{\mathrm{u}/P}(x,\mu)-f_{\bar{\mathrm{u}}/P}(x,\mu)\right]=\frac{1}{p^+}<p|\bar{u}(0)\gamma^+u(0)|p>=2\ , \tag{3.19}$$

or the momentum sum rule

$$\int_0^1 xdx\sum_a\left[f_{a/P}(x,\mu)+f_{\bar{a}/P}(x,\mu)\right]=\frac{1}{(p^+)^2}<p|\Theta^{++}(0)|p>\ . \tag{3.20}$$

The mass independent $\overline{\mathrm{MS}}$ scheme is a very natural scheme to calculate the parton distribution functions. One can, however, arbitrarily add to and subtract from the parton densities and the hard scattering cross section appearing in Eq. (3.14) some finite terms without changing the value of the physical cross section. A popular example is the DIS scheme [29] where the definition of the parton densities is modified in such a way that the next-to-leading corrections C_{q} in (3.16) appearing in the hard scattering cross section vanish.

3.2 Phenomenology of the Parton Density Functions

The non-perturbatively defined parton densities have to be extracted from the data. This is a very important phenomenological exercise since it provides us with a test of QCD and an increased power in making predictions for future experiments. The significance of a good measurement of the parton densities has been well illustrated by the recent improvements which have been achieved by comparing the BCDMS [31] and EMC [32] data with the SLAC data [33]. Discrepancy has been found between the BCDMS and the EMC data in the measurement of the $F_2(x,Q^2)$ structure function (Fig. 6). A recent reanalysis of the old SLAC data provided an overlap region where the SLAC and BCDMS, and the SLAC and EMC data could be directly compared.

It was found that the consistency of the SLAC and EMC data requires about 7% global upwards renormalization of the EMC data (Fig. 7). An apparent discrepancy has been found in the low- and high-x region also when the SLAC and BCDMS data have been compared. This triggered a reanalysis of the systematic uncertainties of the recent BCDMS data. It has been found that changing the beam calibration

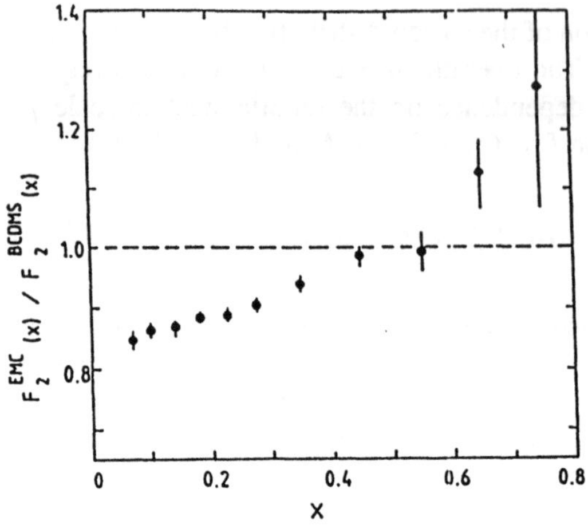

Fig. 6. The ratio of proton structure functions $F_2^{\mathrm{EMC}}(x)/F_2^{\mathrm{BCDMS}}(x)$

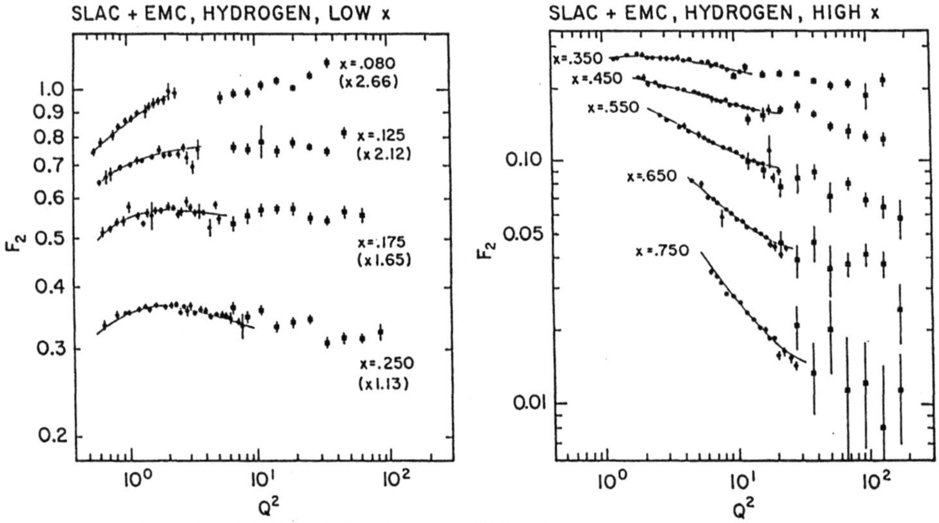

Fig. 7. Comparison of the EMC and SLAC data on F_2^{em} for hydrogen data in the low- and in the high-x region. EMC data are normalized by 1.07 and the curve drawn is the best fit to the SLAC data

and the magnetic field strength within the error bars an agreement could be achieved after a global renormalization of the data by -2% in the low x region. In the large-x region there may still be some small discrepancy (Fig. 8).

The original discrepancy between the EMC and BCDMS data (Fig. 6) led to the largest uncertainty in the prediction of the ratio of the production rate of W and Z at the $p\bar{p}$ colliders which is an important quantity in extracting limits on the top quark mass. It is rewarding that this uncertainty has been eliminated. The BCDMS and EMC data give one of the most reliable measurements of $\Lambda_{\overline{\mathrm{MS}}}$ from the study of the Q^2 behaviour of structure function

$$\begin{aligned} \Lambda^{(4)}_{\overline{MS}} &= (105 \pm 60 + (+90 \text{ or } -45))\ \mathrm{MeV} \quad &: \mathrm{EMC}, \\ \Lambda^{(4)}_{\overline{MS}} &= (220 \pm 15 \pm 50)\ \mathrm{MeV} \quad &: \mathrm{BCDMS} \end{aligned} \tag{3.21}$$

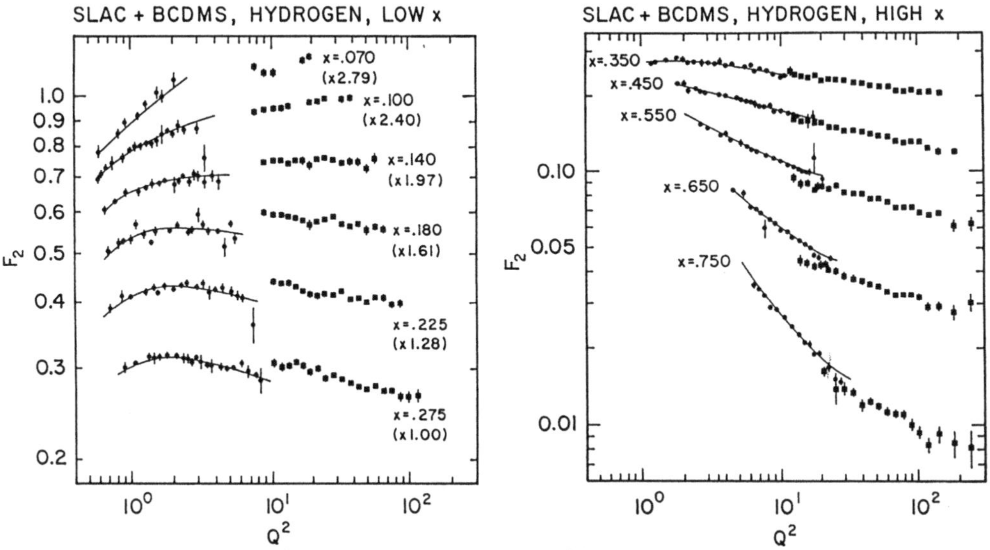

Fig. 8. Comparison of the BCDMS and SLAC data on F_2^{em} for hydrogen data in the low- and in the high-x region. The BCDMS data are normalized by 1.00 and the curve drawn is the best fit to the SLAC data

and the deep inelastic structure functions give the most reliable source of data to extract the quark number densities. A recent global fit to the data together with the next-to-leading order parton evolution equations has been used to obtain recently a new popular set of structure function parametrizations (MRS [34] and DFLM [35]).

3.3 Spin-Dependent Structure Functions

3.3.1 The Puzzling EMC Data

Recently, the EMC collaboration presented results on the polarized structure function of the proton $g_1^p(x)$ in muon-proton scattering [36]. They extended the previous measurement made at SLAC [37] down to smaller values of x (see Fig. 9). The experimental determination of the polarized structure function of the proton $g_1^p(x)$ provides an important information on the spin structure of the nucleon [38, 39]. It appears very difficult to derive the form of nucleon structure functions from first principles (e.g., with the methods of lattice-QCD [40]). Using the physical picture provided by the naive quark parton model various forms of $g_1^p(x)$ have been suggested [39]. The result of the EMC collaboration is surprising since the value of $g_1^p(x)$ at small x is significantly smaller than what has been expected in the naive quark-parton model. This discrepancy has triggered considerable amount of theoretical work.

The main point can be explained by limiting our discussion only to the first moments. Let us introduce the short-hand notation

$$\begin{aligned} \Delta f_i(x,Q^2) &= f_i^{\uparrow}(x,Q^2) - f_i^{\downarrow}(x,Q^2) , \\ \Delta f_i = \Delta f_i(Q^2) &= \int_0^1 dx \Delta f_i(x,Q^2) . \end{aligned} \tag{3.22}$$

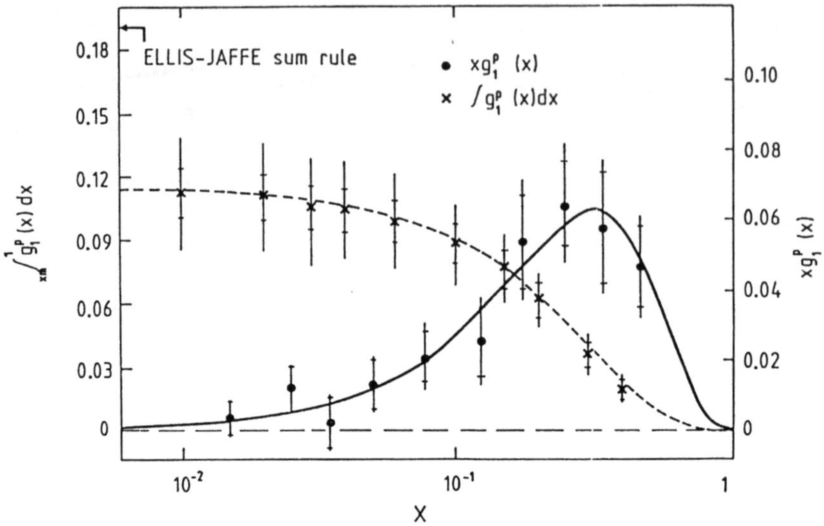

Fig. 9. The quantity $xg_1^p(x)$ (right hand axis and solid circles) versus x. The left hand axis and the crosses show the values $\int_{x_m}^1 g_a^p(x)dx$. The curves represent fits to the data

From Eq. (3.7) we obtain that

$$\begin{aligned} \int_0^1 dx g_1^p(x) &= \frac{1}{2}\left(\frac{4}{9}\Delta u + \frac{1}{9}\Delta d + \frac{1}{9}\Delta s\right) , \\ \int_0^1 dx g_1^n(x) &= \frac{1}{2}\left(\frac{4}{9}\Delta d + \frac{1}{9}\Delta u + \frac{1}{9}\Delta s\right) . \end{aligned} \tag{3.23}$$

It is straightforward to derive the isospin sum rule from the definition of the spin density functions

$$\frac{1}{2}(\Delta u - \Delta d) = \frac{1}{2M} < p\left|J_0^{5,3}(0)\right|p >= \frac{1}{2}\frac{G_A}{G_V} \tag{3.24}$$

where $J_\mu^{5,a}(x)$ denotes the SU(2) triplet or SU(3) octet axial current and G_A, G_V are the weak axial and vector couplings of the proton. From Eqs. (3.22), (3.23) we obtain the Bjorken sum rule

$$\int_0^1 dx \left(g_1^p(x) - g_1^n(x)\right) = \frac{1}{6}\frac{G_A}{G_V} . \tag{3.25}$$

Instead of the moments of definite flavour we can use the SU(3) components

$$I_3 = \Delta u - \Delta d, \quad I_8 = \Delta u + \Delta d - 2\Delta s, \quad I_0 = \Delta u + \Delta d + \Delta s . \tag{3.26}$$

The Bjorken sum rule fixes the value of I_3, and its SU(3) generalization (Ellis-Jaffe [41]) gives the value of I_8

$$I_8 = \frac{F/D - 1}{3(F/D + 1)}\frac{G_A}{G_V} . \tag{3.27}$$

From the weak decay properties of the baryons one can extract the experimental values $G_A/G_V \approx 1.25$ and $F/D \approx 0.631$. Assuming the validity of the sum rules (3.25), (3.27) one can extract the value of the singlet component I_0 from the EMC data. In a recent analysis of the combined EMC and SLAC-Yale data it has been found [42] that the values of the quark moments are

$$\Delta u = 0.73 \pm 0.07, \quad \Delta d = -0.52 \pm 0.07, \quad \Delta s = -0.22 \pm 0.07 \tag{3.28}$$

which tells us that the total spin carried by the quarks is surprisingly small

$$I_0 = \Delta q = \Delta u + \Delta d + \Delta s = -0.01 \pm 0.24. \tag{3.29}$$

Since the total spin of the proton is given by the sum rule

$$\tfrac{1}{2} = \tfrac{1}{2}\Delta q + \Delta G + \Delta L_z\ , \tag{3.30}$$

a small Δq means that the proton spin in the infinite momentum frame is not carried by the quarks. This is against our intuition based on the non-relativistic SU(6)-symmetric wave function of the proton.

One interesting consequence of this result is that the spin dependent neutron structure function is larger than expected [39] and it is negative at smaller values of x. In particular combining the EMC result with the Bjorken sum rule we get

$$\int_0^1 dx\ g_1^n(x) \approx -0.1 \tag{3.31}$$

Hence a direct measurement of g_1^n would give an important independent test of the EMC result.[5]

The puzzling result of Eq. (3.29) has found two alternative theoretical interpretations.

3.3.2 Large Δs Versus Large ΔG

In [45,46] it has been shown that in the Skyrme model [47], in the $1/N_C$ approximation, the matrix element of the singlet axial vector current between proton states, and the first moment of the gluon distribution function as well vanish

$$< p|J_0^{5,0}(0)|p > \approx \Delta u + \Delta d + \Delta s \approx 0\ ; \quad \text{and} \quad \Delta G \approx 0. \tag{3.32}$$

In this model it is not surprising that Δq is small and Δs is large and negative. It has also been shown that such a large strange quark asymmetry is not inconsistent with standard Regge behaviour and the fitted values of the unpolarized quark density functions [46].

Alternatively, it has been recently pointed out that because of the axial anomaly [48,49,50] the QCD corrections [51,52] to the first moment of the singlet part of g_1^p may be enhanced if there is a large gluonic asymmetry which increases with

[5] It should be possible to obtain g_1^n in the near future from experiments using internal polarized gas targets [43] of hydrogen, deuterium and ^{3}He. For a recent phenomenological treatment of g_1^n see [44].

increasing Q^2. Therefore the moments $\Delta u, \Delta d, \Delta s$ appearing in the above analysis have to be replaced by

$$\Delta\tilde{u} = \Delta u - \frac{\alpha_s}{2\pi}\Delta G, \quad \Delta\tilde{d} = \Delta d - \frac{\alpha_s}{2\pi}\Delta G, \quad \Delta\tilde{s} = \Delta s - \frac{\alpha_s}{2\pi}\Delta G. \tag{3.33}$$

If ΔG is large and positive we can explain the experimental value of I_0. Assuming that the spin carried by the strange quark is negligible, $\Delta s = 0$, the EMC data would require

$$\Delta G \approx 8 \pm 2 \quad \text{at} \quad < \mathrm{Q}^2 >= 10.7\,\mathrm{GeV}^2. \tag{3.34}$$

Additional independent measurements of both Δs and ΔG are needed to test whether Δs or ΔG is large. Some information is available on Δs. The cross section of $\bar{\nu}\mathrm{p} \to \bar{\nu}\mathrm{p}$ elastic scattering is proportional [53, 54, 46] to $(\Delta u - \Delta d - \Delta s)$. From the result of a fit to the recent data one can derive [46,54]

$$\Delta s = -0.15 \pm 0.09. \tag{3.35}$$

However, the same problem of interpretation appears here. Δs in Eq. (3.35) has been extracted by fitting the matrix element of the singlet axial current to the data. Therefore actually Δs has to be replaced by $\Delta\tilde{s}$.[6]

3.3.3 The Contribution of the Axial Anomaly

The result given by Eq. (3.34) indicates that in next-to-leading order the singlet part of Δg_1 might have sizable corrections from the gluon content of the proton. Or in other language, the two loop anomalous moment of the singlet axial current is non-vanishing. The corrections appear since the singlet axial current is not conserved. Its divergence equation is the anomaly equation

$$\partial_\mu J_\mu^{5,0} = n_F \frac{\alpha_s}{4\pi} G^a_{\mu\nu}\tilde{G}^a_{\mu\nu} = n_F \partial_\mu K_\mu \tag{3.36}$$

where $\tilde{G}^i_{\mu\nu} = \frac{1}{2}\epsilon_{\mu\nu\sigma\lambda}G_i^{\sigma\lambda}$ and K_μ is a gauge dependent operator

$$K_\mu = \frac{\alpha_s}{2\pi}\epsilon_{\mu\nu\sigma\lambda}\left[\tilde{G}^i_{\lambda\sigma}G_i^\nu - i\frac{1}{3}f_{ijk}G_i^\nu G_j^\sigma G_k^\lambda\right] \tag{3.37}$$

With the help of the anomaly equation we can calculate the corrections appearing in (3.33) rather easily. First we note that although the K_μ operator is gauge dependent it is a useful quantity since its forward matrix elements between massless states are gauge independent. Therefore for our purposes the difference $J_\mu^5 - n_F K_\mu$ can be considered as a conserved current.

In particular for massless partons we have the relations

$$< g|J_\mu^5 - n_F K_\mu|g >= 0\;, \quad < q|K_\mu|q >= 0 \tag{3.38}$$

In next-to-leading order we have the expression (see Fig. 10)

$$\Delta g_1^S \sim \left[\Delta q < q|J_+^{5,0}|q > + \Delta G < g|J_+^{5,0}|g >\right].$$

[6] For elastic scattering the analysis of the QCD corrections is more model dependent.

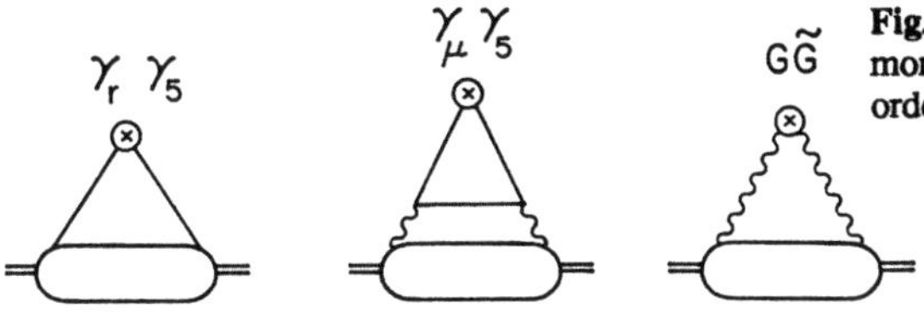

Fig. 10. Diagrammatic representation of the first moment of g_1^p in leading and next-to-leading order

This can be rewritten as

$$\Delta g_1^S = \text{constant}\left[\Delta q < q|J_+^{5,0} - n_F K_+|q> + \Delta G < g|n_F K_+|g>\right]$$

Using Eqs. (3.37),(3.38) after some simple algebra we obtain the desired result

$$\Delta g_1^S \sim \left[\Delta q - n_f \frac{\alpha_s}{2\pi}\Delta G\right]$$

which explains the relations given above in Eq. (3.33). (For more detailed discussions see [55].)

The same result can also be obtained in a completely diagrammatic calculation based on the hard scattering cross section formula discussed in the spin independent case [50,55]

$$\begin{aligned}\Delta\sigma^S_{\gamma^*\mathrm{p}}(q,p) = \int dx \Big[&\Delta q(x,\mu)\Delta\hat{\sigma}^S_{\gamma^*\mathrm{q}}(q,xp,\alpha_s(\mu),Q^2/\mu)\\ &+\Delta G(x,\mu)\Delta\hat{\sigma}^S_{\gamma^*\mathrm{g}}(q,xp,\alpha_s(\mu),Q^2/\mu)\Big]\ .\end{aligned} \tag{3.39}$$

According to the rules of the perturbative calculation $\Delta\hat{\sigma}_{\gamma^*\mathrm{q}}$ and $\Delta\hat{\sigma}_{\gamma^*\mathrm{g}}$ have to be calculated to order $\alpha_s\alpha$. Some representative diagrams are depicted in Fig. 11.

When we project out the first moment only the box diagram contribution remains. An explicit calculation shows that the dominant region of the phase space integral comes from the region $k_\perp \approx Q$, and it reproduces the result given by Eq. (3.33). This is interesting by itself since if the first moment of the gluon density is large then it predicts a large transverse momentum jet production contribution to the spin dependent lepton-hadron scattering which experimentally may be tested. Apparently, the data are most consistent with some amount of strange quark contribution and with a not extremely large gluon contribution.

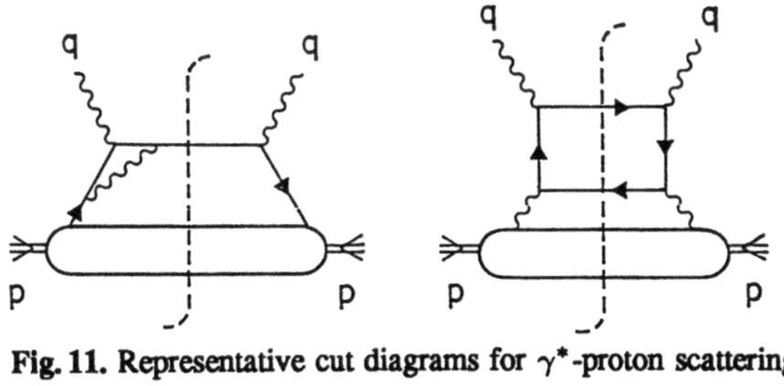

Fig. 11. Representative cut diagrams for γ^*-proton scattering in next-to-leading order

That the first moment of the gluon spin asymmetry distribution

$$\Delta G = \int_0^1 dx (g^{\uparrow}(x) - g^{\downarrow}(x))$$

is large is suggested by the leading order renormalization group equation [56]

$$Q^2 \frac{d}{dQ^2} \left[\alpha_s(Q^2) \Delta G(Q^2)\right] = 0,$$

which means that

$$\alpha_s(Q^2) \Delta G(Q^2) = \alpha_s(Q_0^2) \Delta G(Q_0^2).$$

3.3.4 Parametrization and Measurement of the Spin-Dependent Gluon Density

The direct measurement of the spin asymmetry of the gluon appears to be rather difficult. In principle many reactions depend on ΔG.[7] The most promising one is the observation of the production of large transverse momentum jets in deep inelastic scattering [50,55]. An obvious method to measure gluonic polarization asymmetry is to measure the asymmetry in two-jet production at collider energies (Sp$\bar{\text{p}}$S, Tevatron, LHC, SSC, UNK, HERA) in the regime where gluon-gluon scattering is the dominant production mechanism. Since we do not know how to tag on gluon jets, polarization effects can only be obtained if both the colliding beams (or both the beam and the target) are polarized.[8]

To obtain polarized proton beams at colliding beam machines is a formidable task. However, to my knowledge it has not been proven that it is impossible.[9]

The spin dependent parton densities are not calculable in perturbative QCD. The measured values of $g_{\text{p}}(x)$ have been compared to parametrizations suggested by qualitative features of the naive quark-parton model. The first moments of the polarized quark densities are constrained since they are the matrix element of the axial current of a given quark flavour. They are summarized in the Bjorken [38] and Ellis-Jaffe [41] sum rules. A further relation is the angular momentum conservation (see Eq. 3.30) which is not directly useful since ΔL_z also appears. The Altarelli-Parisi evolution equations combined with this sum rule tell us that a large gluon asymmetry is compensated by large angular momentum [57].

On the gluon wave functions we have only the trivial positivity constraint [58]

$$\left|g^{\uparrow}(x) - g^{\downarrow}(x)\right| < g(x) \tag{3.40}$$

and it has been shown [59] that in the limit $x \to 0$ and $Q^2 \to \infty$, assuming four flavours, the Q^2 evolution [56] leads to the asymptotic form

---

[7] Nine different possibilities are listed, e.g., in [46].

[8] If we could define meaningfully forward-backward asymmetry it would be enough to polarize only one of the beams.

[9] In view of the extreme expenses of building Sp$\bar{\text{p}}$S, Tevatron and HERA no study has been made concerning the feasibility of polarizing the proton beams. We may hope that in the future some progress will be achieved in this direction. We recall that it is feasible to extract polarized beams for fixed target experiments.

$$\frac{\Delta g(x)}{g(x)} \approx x\exp\left[0.82\sqrt{\ln(Q^2/\Lambda^2_{\overline{\rm MS}})\ln(1/x)}\right] \tag{3.41}$$

That is, $\Delta g(x)/g(x)$ remains small at small x, although the effective power of x may be less than one. The same behaviour is expected from Regge analysis [50,55] at moderate values x. Although the first moment of the gluon structure function is infinite, it is important for our discussion that the contribution in the $x = 0.1 - 1$ interval is not overwhelming:

$$\int_{0.1}^{1} g(x, Q^2 = 10\,\mathrm{GeV}^2) \approx 1.5\ . \tag{3.42}$$

Eq. (3.42) and the EMC result, Eq. (3.34), imply that the bulk of the large value of ΔG is at x-values of $x < 0.1$. It is natural to expect large ($\approx 30\% - 50\%$) polarization in the x-region where gluon-gluon scattering is the dominant mechanism for jet production ($x \approx 0.01 - 0.1$).[10]

We illustrate this remark with the curves of Fig. 12 and Fig. 13 [58] where the x-dependence of $\Delta g(x)/g(x)$ and $\Delta g(x)$ are plotted assuming two rather different parametrizations of $\Delta g(x)/g(x)$ satisfying the constraints (3.34), (3.40) and (3.42)

$$\begin{aligned} \Delta g(x)/g(x) &= 2(m+1)x(1-x)^m \qquad &&\text{[curve (a)]} \\ \Delta g(x)/g(x) &= \frac{2nx}{1.0+nx} &&\text{[curve (b)]} \end{aligned} \tag{3.43}$$

In Fig. 12 and Fig. 13 the values $m = 10,\ n = 15$ are assumed. These distributions are significantly different from the ones proposed earlier in the literature [60,61]

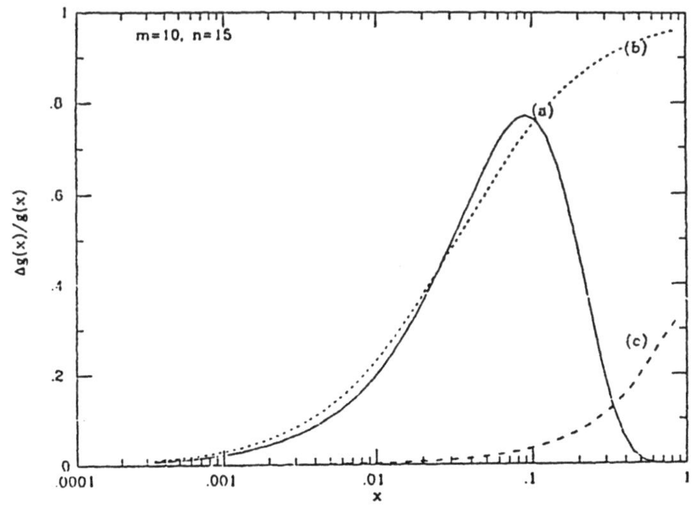

Fig. 12. Relative polarization asymmetry of the gluon number density in the proton $\Delta g(x)/g(x)$ as a function of x for three different parametrizations (a) (b) (c) as defined by Eqs. (3.43) and (3.44)

[10] We note that, e.g., the value of the integral of Eq. (3.42) is slightly smaller at larger values of Q^2. At $Q^2 \approx 100\,\mathrm{GeV}^2$ its value is ≈ 1.1.

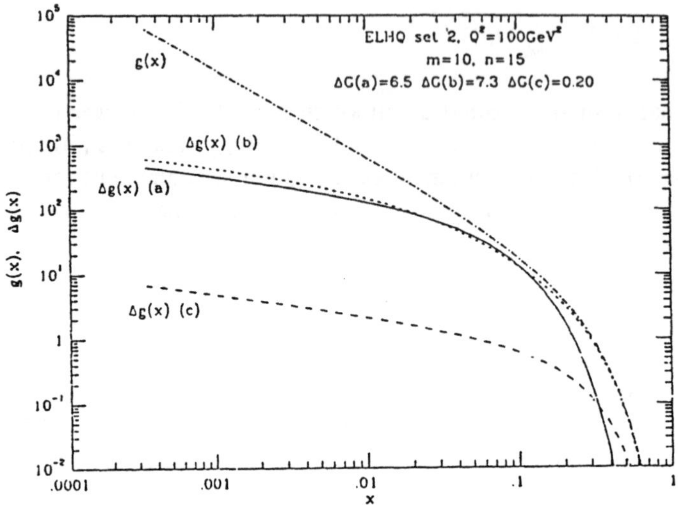

Fig. 13. Polarization asymmetry $\Delta g(x)$ corresponding to the three parametrizations given on Fig. 12. For the unpolarized gluon density function the parametrization EHLQ [62] set 2 at $Q^2 = 100\,\text{GeV}^2$ is used

$$\Delta g(x)/g(x) = \frac{x(2-x)}{3+3(1-x)^2} \qquad \text{[curve (c)]} \tag{3.44}$$

In Fig. 13 the corresponding values of the first moment ΔG for parametrizations (a, b, c) are also indicated.[11] For the unpolarized $g(x, Q^2)$ we have chosen the EHLQ [62] parametrization (set 2, $\Lambda = 200\,\text{MeV}$ and $Q^2 = 100\,\text{GeV}^2$). A different parametrization also consistent with the EMC data has been suggested by Altarelli and Stirling [55]. The corresponding number density functions are depicted in Fig. 14.

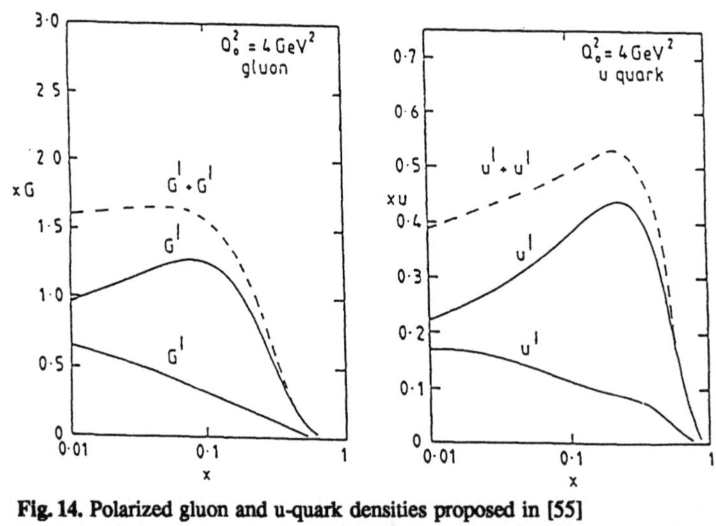

Fig. 14. Polarized gluon and u-quark densities proposed in [55]

[11] Of course we could invent a parametrization consistent with the constraints (3.34), (3.40), (3.42) such that ΔG receives the dominant contribution from the very small x region. Then the gluonic asymmetry becomes small. However, it appears rather difficult to motivate this assumption.

They used their parametrization to make predictions for charm production in spin dependent lepton-hadron scattering.

Here I restrict myself to the discussion of spin asymmetry present in jet production in pp and ep collisions [58] when we have the advantage that the parton scattering subprocesses all have large spin asymmetries at 90°. Defining the asymmetry as

$$A = \frac{|M(++)|^2 - |M(-+)|^2}{|M(++)|^2 + |M(-+)|^2} \tag{3.45}$$

where $|M(\lambda_1\lambda_2)|^2$ denotes the amplitude squared, summed over the final spins, but with fixed helicites λ_1, λ_2 of the incoming partons, it has been found [63], e.g., that

$$\begin{aligned} A(\mathrm{gg} \to \mathrm{gg}) &= \frac{2s^2/tu + ut/s^2 - 3}{3 - ut/s^2 - us/t^2 - st/u^2} \\ A(\mathrm{gg} \to \mathrm{q\bar{q}}) &= -1 . \end{aligned} \tag{3.46}$$

We note that $A(\mathrm{gg} \to \mathrm{gg})$ vanishes in forward and backward directions. Otherwise it is always positive. It has its maximum at 90° when its value is 7/9.

With the suggested large (≈ 30–50%) polarization asymmetry of the gluon number density function in the proton in the kinematical region where gg-scattering is the dominant subprocess, we obtain large asymmetry in the production of two jets at $y_1 = y_2 = 0$

$$\frac{\Delta d\sigma/dp_{1T}dp_{2T}dy_1dy_2}{d\sigma/dp_{1T}dp_{2T}dy_1dy_2} = \frac{\Delta g(x_1,p_T^2)\Delta g(x_2,p_T^2)}{g(x_1,p_T^2)g(x_2,p_T^2)}A(\mathrm{gg} \to \mathrm{gg}) \approx 10 - 20\% . \tag{3.47}$$

Even larger but negative asymmetry is obtained for the photoproduction of two jets in the kinematical region where the $\gamma\mathrm{g} \to \mathrm{q\bar{q}}$ subprocess is dominant. Experimentally, 3000 jet events would allow to establish such a large asymmetry with more than 5 standard deviation significance. I note that at Tevatron $\approx 10^7$ jet events can be produced assuming $5\,\mathrm{pb}^{-1}$ integrated luminosity with $|y_{1,2}| < 1.0$ and $20\,\mathrm{GeV} < p_T < 80\,\mathrm{GeV}$.[12]

In principle one can study spin dependent photoproduction of jet events at HERA. It would be particularly interesting to see the negative asymmetry given by the $\gamma\mathrm{g} \to \mathrm{q\bar{q}}$ subprocess. The physical cross section of electron-proton scattering is obtained by folding the hard scattering cross sections of the contributing subprocesses with the corresponding parton number density functions of the incoming particles. When both incoming beams are polarized, we obtain

$$\begin{aligned} & d\sigma^{\uparrow\uparrow}(s) - d\sigma^{\uparrow\downarrow}(s) \\ &= \sum_{a,b} \int dx_1 dx_2 \Delta F_{a/A}(x_A, Q^2) \Delta F_{b/B}(x_B, Q^2) \\ &\quad \times \left[d\hat{\sigma}^{\uparrow\uparrow}(\hat{s}, \ldots) - d\hat{\sigma}^{\uparrow\downarrow}(\hat{s}, \ldots) \right] \end{aligned} \tag{3.48}$$

[12] Since at 90° all the important subprocesses have large ≈ +50% spin asymmetry [63] the p_T dependence of the measured asymmetry would give the x-dependence of $\Delta g(x)$.

where $\Delta F_{a/A}(x_A, Q^2)$ is the difference of the number densities of polarized partons a with helicity parallel and antiparallel to the helicity of the polarized incoming particle A ($\Delta F = F_+ - F_-$). The summation indices a, b run over all partons which can contribute and $\hat{s} = x_A x_B s$. In two jet production the largest rates are obtained when the scattered electrons in the final states remain undetected. In this case the Weizsäcker-Williams approximation can be used at the electron line. The photon will act as parton of the incoming electron. Only to leading order in α_s and for large photon virtual mass ($Q \gg \Lambda_{\overline{MS}}$) can the photon be considered as a pointlike parton. Important corrections arise from the quark and gluon content of the photon if the photon is near its mass shell. The photon helicity asymmetry function in a polarized electron is the spin dependent Weizsäcker-Williams splitting function

$$\Delta F_e^{\gamma}(x) = \frac{\alpha}{2\pi} \ln(\frac{\hat{s}}{4m_e^2}) \frac{1-(1-x)^2}{x} \tag{3.49}$$

where α is the hyperfine constant and m_e is the electron mass. The photon becomes 100% polarized at $x = 1$.

It has been recently [64] shown that photoproduction of jet events at HERA is dominated by the hadronic component of the photon. The subprocess γ g is dominant only at either relatively large values of the transverse momentum or at negative values of the rapidity. Fortunately, the background coming from the hadronic component of the electron can be suppressed to an acceptable level in the kinematical region $y < 0$, $p_T \geq 10\,\mathrm{GeV}$. In this regime the photon polarization is large. Furthermore, $\gamma g \to q\bar{q}$ is the dominant subprocess. At the relevant x-values the quark polarization asymmetries are small. Therefore, when calculating the contributions of all other subprocesses, it is a good approximation to assume that they have no polarization. With this approximation and assuming again that $x = 0.03 - 0.1$, and for $\Delta g(x)/g(x) \approx 0.5$, we obtain the asymmetry given in Fig. 15 [58].

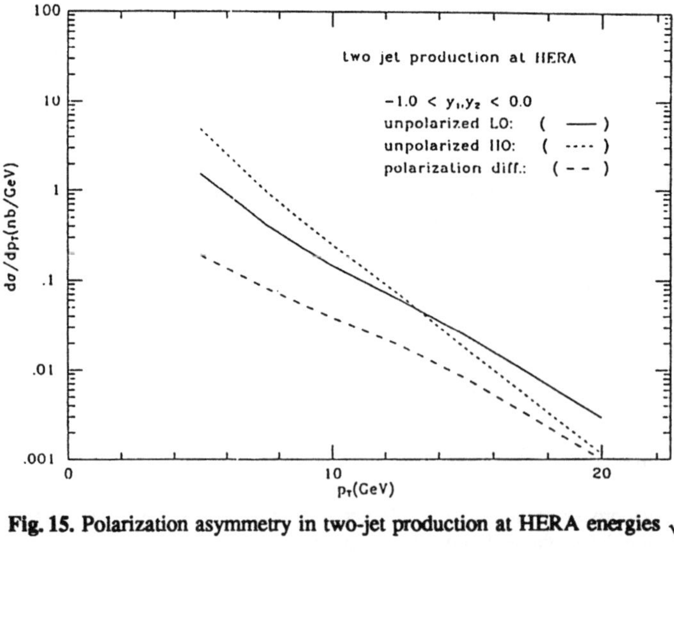

Fig. 15. Polarization asymmetry in two-jet production at HERA energies $\sqrt{s} = 314\,\mathrm{GeV}$

4. Large Transverse Momentum Reactions in Hadron-Hadron Collisions

Many large transverse momentum reactions have beend studied in hadron-hadron collisions. In this Section I shall consider only gauge boson production and heavy flavour production. Jet production will be discussed in the next Section. We do not discuss the production of an inclusive large $p_\perp$-photon since it has been covered by previous lectures [65]. Large angle elastic scattering gives also a very interesting testing ground for QCD ideas. Very recently Bott and Sterman have found further theoretical arguments for the validity of the perturbative QCD formalism in the case of these processes [66].

4.1 γ^*-, W- and Z-Production

The production of a heavy gauge boson in hadron(A)-hadron(B) collisions

$$A(p_A) + B(p_B) \to \mathrm{W},\ \mathrm{Z} \text{ or } \gamma^*(Q) + X \tag{4.1}$$

is the simplest process for large transverse momentum reactions with two colliding hadrons in the initial state. It has provided the theoretical laboratory with an understanding of how factorization works in the case of hadron-hadron collisions [14,11]. With the discovery of the W and Z bosons the theoretical study of this process obtained further phenomenological significance.

4.1.1 Factorization Theorem for the Drell-Yan Process

Let us consider the cross section $d\sigma/dQ^2dy$ for the reaction (4.1) with a virtual photon (Drell-Yan process) when the mass of the virtual photon, Q^2, is large of order s and where y denotes the rapidity

$$y = \frac{1}{2}\log(\frac{Q_0 + Q_3}{Q_0 - Q_3}) \ . \tag{4.2}$$

Let us suppose that the transverse momentum of the gauge boson is not observed. The rapidity can have any finite value. The factorization theorem says that the convoluted form of the cross section and the perturbative algorithm found in the case of deep inelastic scattering (see Eqs. 3.8, 3.9, 3.14 and 2.1) remain valid, except that the convolutions have to be performed over the parton densitites of both initial hadrons

$$\frac{d\sigma(s,Q,y)}{dQ^2dy} = \sum_{a,b}\int_{x_A}^1 dx'_A \int_{x_B}^1 dx'_B f_{a/A}(x'_A,\mu) f_{b/B}(x'_B,\mu)\times$$
$$\times \frac{d\hat\sigma_{a,b}(x_A/x'_A, x_B/x'_B, Q, \mu/Q, \alpha_s(\mu))}{dQ^2dy} \tag{4.3a}$$

where $x_A = e^y\sqrt{Q^2/s}$, $x_B = e^{-y}\sqrt{Q^2/s}$. The hard scattering cross section $d\hat\sigma_{a,b}$ has to be calculated from the singular cross section defined in n dimensions as given in (3.14), except that now we have two convolutions over the singular function $\Gamma_{i/a}$, $\Gamma_{j/b}$ corresponding to the two initial parton lines

$$d\sigma_{ab}^{(B)}(q,p_a,p_b,n) = \sum_{ij} \int_0^1 dx_a \int_0^1 dx_b \Gamma_{i/a}(x_a,\epsilon)\Gamma_{j/b}(x_b,\epsilon)\times \times d\hat{\sigma}_{ij}(q,x_a p_a, x_b p_b, \alpha_s(\mu),\mu) \tag{4.3b}$$

The factorization proof uses power counting arguments to isolate the infrared sensitive regions. In the large Q limit all the masses are negligible and the infrared sensitive regions are given by the singular surfaces at which the denominators of the Feynman amplitudes vanish. According to a theorem by Coleman and Norton these surfaces are defined by configurations which represent classical particle scattering. Contracting these surfaces to a single point we obtain reduced graphs in which all lines carry momenta far from the infrared regions. The reduced graph must depict a classically possible process that can be realized with momentum conservation at each reduced vertex and a classical propagation of the massless particles between vertices. There are two possible classical processes: (a) a massless particle splits into two collinear particles (jet lines) and (b) a soft particle is emitted (soft lines). The proof of the factorization theorem requires showing that the singularites due to soft lines cancel to all orders and that the uncancelled divergent part of the jet lines must be proportional to the Altarelli-Parisi splitting functions and must have the universal process-independent structure (see Eq. (3.14)). In this way any diagram can be decomposed into jet lines, a hard scattering part and soft lines (see Fig. 16).

The structure is rather similar to the case we have seen in deep inelastic scattering. The new feature of the Drell-Yan process is the possibility for soft exchanges between the spectator and the active partons of the initial hadrons long before the hard scattering process takes place. In principle, a soft emission before the hard scattering can change the color of the parton and thus the probability that the subsequent interaction will occur. Fortunately it could be shown that these soft gluon exchanges do not destroy the factorization theorem. The physical picture underlying the factorization theorem is explained in the recent review article of Collins, Soper and Sterman [11]. The proof, however, is rather technical. It is based on semiclassical considerations, carefully introduced subtraction methods, the consideration of pinch singularities and subsequent contour deformation and explicit two-loop calculations [14].

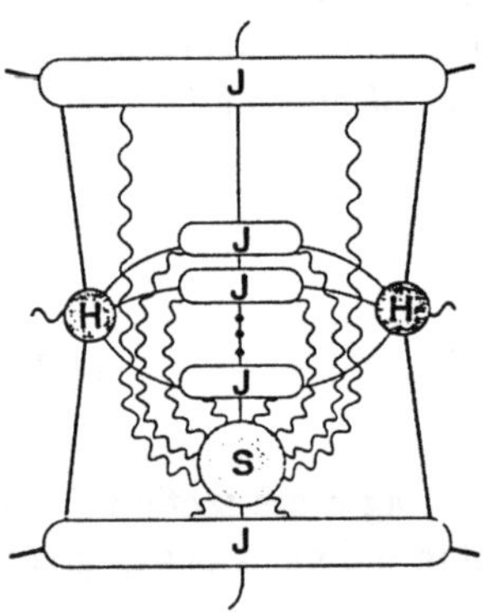

Fig. 16. Leading integration regions for the Drell-Yan process [11]

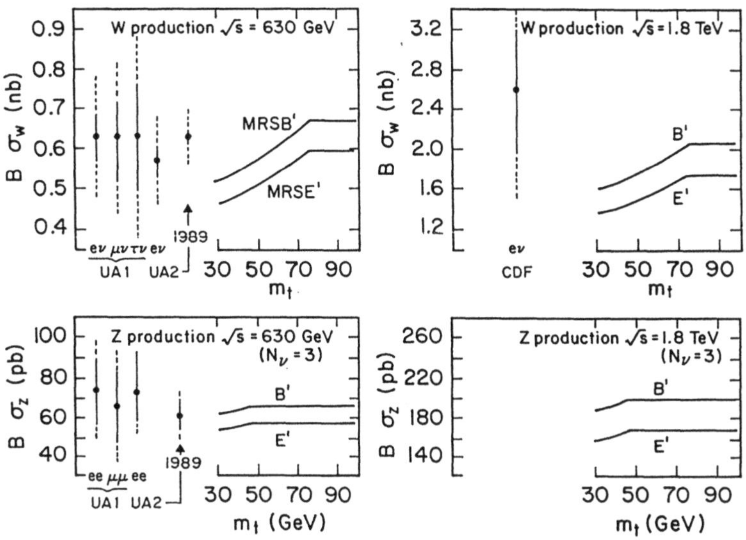

Fig. 17. Cross sections times leptonic branching ratios for W and Z production as a function of the top mass. It is assumed that only three light neutrinos exist $N_\nu = 3$. The upper curves are obtained by the MRSB, the lower curves obtained by the MRSE structure functions [67]. The measured values of the UA1, UA2 and CDF experiments are also indicated

4.1.2 Total Cross Sections and $p_\perp$-Distribution

The theoretical foundation of the validity of the formula (4.3) is rather solid, therefore it is of great interest to compare the predictions with the data. The measured values of total cross sections are in good agreement with the next-to-leading order predictions [67]. This is illustrated with the values depicted in Fig. 17. In view of the rather precise data it is important to reduce the ambiguity in the theoretical predictions. This could be achieved in the future with a better measurement of the deep inelastic structure functions and with the calculation of the full order α_s^2 corrections. Some partial result for corrections of order α_s^2 (soft and virtual gluon corrections) have been calculated recently [68]. This part of the corrections is positive and rather small ($\approx 10\%$). But it is not clear a priori that the remaining corrections are less important than the ones that have been calculated.

The $p_\perp$-distributions of W and Z production will be measured with higher and higher precision in the near future. It is a significant development that recently the next-to-leading corrections have been evaluated by two groups independently [68]. The perturbative calculation is valid provided the order of magnitude of the transverse momentum is not much different from the mass of the W boson. In the small $p_\perp$ region the so called Sudakov logarithms $(\log^2(M_W/p_\perp))^n$ have to be resummed. The result of Altarelli et al. [16] obtained for the low transverse momentum region now could be matched with the the next-to-leading order result of the large $p_\perp$ region. It would be interesting to carry out this improvement.

Studying the sensitivity of the cross section to the value of the renormalization scale [68,69] we can see that the perturbative result still has rather larger ambiguities at $p_\perp$-values below 20 GeV. This is due to the increasing importance of the Sudakov

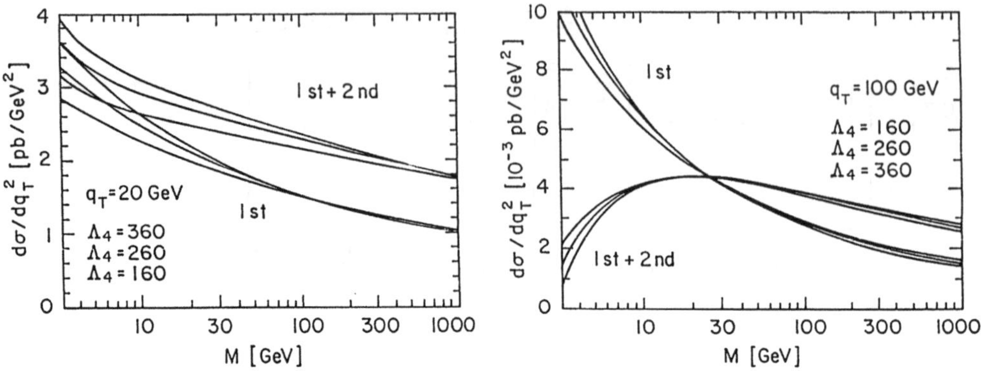

Fig. 18. Scale dependence of the $q_\perp$ distribution of the W production cross section at two different values of the transverse momentum of the W-boson $q_\perp$ =20 and 100 GeV [69]

logarithms (see Figs. 18a–18b). We can see that new improved data from UA2 and CDF [70] are consistent with the theoretical predictions and they favour heavy top mass value $m_t > M_W$ and the existence of only three light neutrinos.

A further interesting feature of the measurement of the $p_\perp$-distribution of the W-boson is that it offers the possibility to measure α_s at hadrons colliders independently from the jet definition ambiguities. The calculation of Arnold et al. has provided the necessary theoretical input. At the moment the experimental error bars are still too large for a quantitative study.

4.1.3 Gauge Boson Production in Association with Multi-jets

The tree-level cross sections are also known for the production of W + 1, 2, 3 jets. The theoretical predictions are in agreement with the recent CDF analysis. The calculation of the process W + 3 jets became feasible thanks to the new interesting technical development in calculating tree diagrams of multibody processes. New recursion relations have been found and the conventional Feynman diagram method could be replaced by a much simpler algorithm [71,72].

It is significant that we have a theoretical estimate for these more complex reactions since they give the dominant background processes to top searches in the (W + jets)-signal [73].

4.2 Heavy Quark Production

Heavy quark production provides interesting new applications of perturbative QCD. It is also of great phenomenological importance. The next-to-leading order corrections have been evaluated recently.

4.2.1 On the Validity of the Factorization Theorem

The proof of the factorization theorem constructed for the Drell-Yan process does not apply directly to the production of a heavy quark-antiquark pair. There are new contributions coming from soft gluon exchanges between the initial partons (active and inactive) and the final heavy quark system, as well as soft and collinear emissions from the heavy quark line. Explicit one-loop calculations [74,75,76] have demonstrated that the factorized form of the parton cross section remains valid since the final soft and collinear contributions cancel. The soft exchanges between the heavy quark line and the spectator lines were also investigated [77] and it was found that they are suppressed by powers of the heavy quark mass. Therefore it appears plausible to assume that the factorization theorem remains valid also for heavy quark production at least if the quark mass value is sufficiently large [78]. Of course, whether the charm and bottom quarks are heavy enough for a valid perturbative treatement can not be decided as long as the non-perturbative mass corrections are not known. Therefore the consistency of the perturbative treatment has to be tested with phenomenological considerations.

4.2.2 Cross Section Formulae

The leading order QCD reactions for heavy quark production are the annihilation processes

$$\begin{aligned} g(p_1) + g(p_2) &\to Q(p_3) + \overline{Q}(p_4) \\ q(p_1) + \bar{q}(p_2) &\to Q(p_3) + \overline{Q}(p_4) \ . \end{aligned} \tag{4.4}$$

The Feynman diagrams are shown in Fig. 19. The total cross sections are given by the convolution

$$\sigma(s, m_Q^2) = \sum_{a,b} \int dx_A dx_B f_{a/A}(x_A) f_{b/B}(x_B) \sigma(\hat{s}, m_Q^2, \mu^2) \tag{4.5}$$

where $\hat{s} = x_A x_B s$, the hard scattering cross section can be calculated according to the algorithm described in the previous Sections to any fixed order of perturbation theory. It can be conveniently parametrized as

$$\begin{aligned} \hat{\sigma}(\hat{s}, \rho) = \frac{\alpha_s^2}{m_Q^2} & \left[F_{ij}^{(0)}(\rho) + \alpha_s(\mu^2) \left(F_{ij}^{(1)}(\rho) + \bar{F}_{ij}^{(1)}(\rho) \log\left(\mu^2/m_Q^2\right) \right) \right] \\ & + \mathcal{O}(\alpha_s^4) \end{aligned} \tag{4.6}$$

where the labels i, j refer to the quark and gluon initial states and $\rho = m_Q^2/\hat{s}$. It is an important new theoretical development that the functions giving the next-to-leading order corrections $F_{i,j}^{(1)}(\rho)$ have been calculated recently [74].[13] The functions $F_{ij}^{(0)}$ and $F_{ij}^{(0)}$ are given in compact form in [74]. It is remarkable that the result of a rather complex calculation can be summarized in such a compact form: it can be given in terms a few one-variable universal functions. Nason, Dawson and Ellis [74] fitted the universal functions $F_{ij}^{(1)}$ with some analytic forms, therefore the result can

[13] In [75] the corrections to the photoproduction of heavy flavour have also been obtained in similar form. The corrections are smaller than in the case of hadron-hadron collisions. This result has great significance for heavy flavour production at HERA.

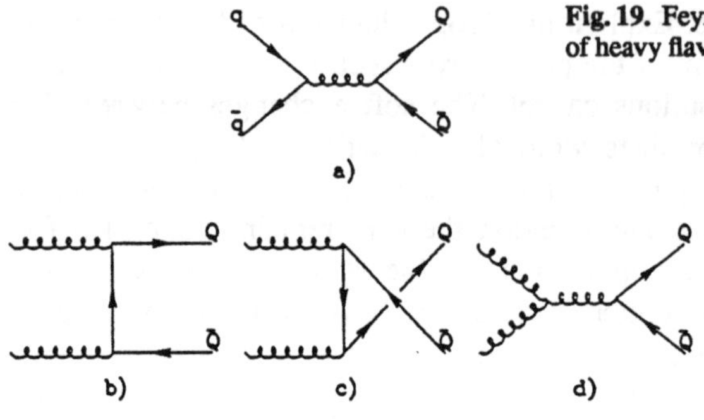

Fig. 19. Feynman diagrams defining the cross section of heavy flavour production in hadron-hadron collisions

be used for further phenomenological considerations [78, 79]. I note that in the case of gluon-gluon initial state the result has been confirmed by a second group [76].

The μ dependence of the next-to-leading order result is dictated by the renormalization group equations. This can be seen explicitly since the functions $\bar{F}^{(1)}_{ij}$ are given in terms of the leading order result $F^{(0)}$ as

$$\bar{F}^{(1)}_{ij}(\rho) = \frac{1}{2\pi}\left[4\pi b_0 F^{(0)}_{ij}(\rho) - \int_\rho^1 dz_1 F^{(0)}_{kj}(\rho/z_1)P_{ki}(z_1) \right.$$

$$\left. - \int_\rho^1 dz_2 F^{(0)}_{kj}(\rho/z_2)P_{kj}(z_2)\right] \tag{4.7}$$

Using the Altarelli-Parisi equations (3.18) we can easily see that the physical cross section (see Eq. (4.7)) satisfies the renormalization equation to order α_s^3

$$\mu\frac{d\sigma}{d\mu} = \mathcal{O}(\alpha_s^4) \ . \tag{4.8}$$

This is the same feature that we have seen in the case of the total cross section of e^+e^- annihilation (see Fig. 3). The physical cross sections calculated to some fixed order of petrurbation theory have some dependence on the unphysical renormalization scale. The scale dependence should, however, be a higher order correction. If the scale dependence is still substantial it either indicates that the higher order corrections are still important or that the perturbative treatment is not valid.

Nason, Dawson and Ellis [80] have also calculated the p_T distribution of an inclusively produced heavy quark which can be compared after some assumption with the data of large $p_\perp$ muon production found by the UA1, UA2 and the CDF experiments.

4.2.3 Phenomenology of Heavy Quark Production

The QCD description of charm, bottom and top production should show improved validity with the increase of the quark mass. In the case of the charm production the question was raised whether one can find at all a charm quark mass value which could give an acceptable, globally consistent result for the description of charm production [74, 75]. The effect due to the ambiguity in the numerical value of the charm quark mass is illustrated in Fig. 20 where the new photoproduction data of

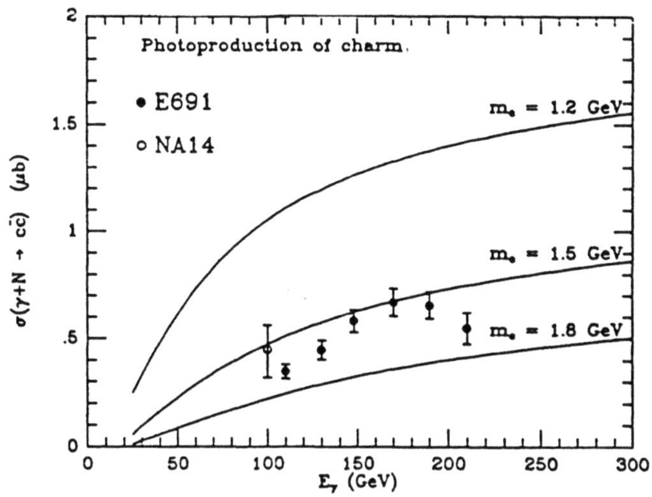

Fig. 20. Next-to-leading order QCD prediction [75] for the photoproduction of charm in comparison with the data [82]

the NA14 and E691 experiments are shown together with the next-to-leading order QCD prediction at three different values of the charm quark mass [82]. Cleary the data favour larger mass values ($\approx 1.8\,\mathrm{GeV}$) if we suppose that the power corrections are negligible.

In the case of bottom production the corrections are very large [74], therefore the next-to-leading order results still show non-negligible scale dependence. Further details are given in F. Pauss' lectures in this volume.

Fortunately the search for top quark is less sensitive to these ambiguities, therefore the CDF experiment could establish the remarkably large limit [83]

$$m_t < 78\,\mathrm{GeV} \quad \text{at} \quad 90\% \text{ confidence level} .$$

The band of the next-to-leading order QCD prediction for the CERN and FERMILAB colliders is given on Fig. 21.

5. Jet Production at Next-to-Leading Order

In this Section we consider the theoretical status of the description of jet production in next-to-leading order of perturbative QCD.

5.1 Jet Production in e^+e^- Annihilation

The smallness of the QCD coupling constant α_s at high energy implies that the hadronic final states in e^+e^- annihilation can be classified into distinct topological signatures characterized by the formation of increased number of jets of hadrons [84, 85].

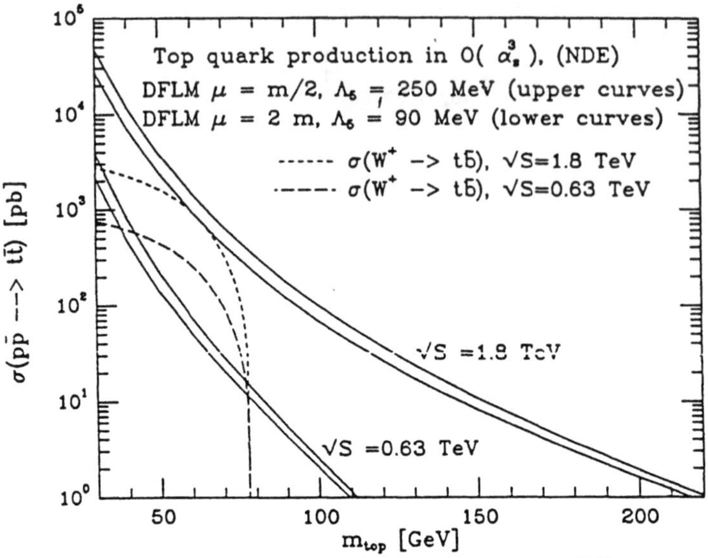

Fig. 21. The production cross section of top quark at Tevatron [74]

The majority of the events have two-jet structure with an angular distribution with respect to the beam axis $(1 + \cos^2\theta)$ typical of spin-$\frac{1}{2}$ quarks [86]. A small fraction of events have planar topology and are dominated by three-jet production. The production of three-jet events at PETRA was predicted by the theory and its observation gave a rather direct evidence for the existence of the gluons. The spin of the gluon has been tested by measuring the angular distributions of the jets. For example, the observed angular distribution [87] in the Ellis-Karliner angle [88] $\overline{\theta}$ is in agreement with the behaviour of the emission of spin-one gluons but it is in disagreement with the emission of spin-zero gluons. To test the details of the three-jet matrix element of perturbative QCD many other topological correlations have been studied.

Infrared-safe quantities defined in terms of quarks and gluons can reliably be calculated in any given fixed order of perturbation theory. According to the fundamental assumption of the QCD-improved parton model the perturbative short-distance partonic distributions of infrared-safe quantities defined in terms of partons can be identified with the corresponding physical quantities defined in terms of hadrons. This assumption might be valid provided we average the result over resolution bins of width large enough for the perturbative result to be justifiable ("bin smearing"). In more physical terms, we assume that for infrared-safe quantities, the confinement effects only limit the angular and energy resolution of our "quark and gluon detectors".

It is expected that the non perturbative effects are power suppressed with some power of 1 GeV/Q where Q is the typical physical scale of the problem in GeV.

In the approach of Sterman and Weinberg [89] the aim was to find an appropriate definition of a jet which is well defined to all orders in perturbation theory. By introducing finite jet resolution parameters the averaging over the non-perturbative effects is taken into account. One need not use the concept of a jet necessarily.

There is a wide variety of infrared shape variables whose distributions are calculable in perturbative QCD and which, after appropriate bin smearing, can be directly compared with the data. The only requirements for such quantities are that they should be infrared and collinear safe.

Theoretically, the most interesting infrared- and collinear-safe quantities are those which are dominated by the three-jet configurations. In leading order of QCD, these quantities are of $\mathcal{O}(\alpha_s)$. Therefore $\Lambda_{\overline{\mathrm{MS}}}$ can be extracted meaningfully from the data using these quantities only if we know the theoretical prediction at least to α_s^2. The next-to-leading order corrections to the production of three partons have been calculated by Ellis, Ross and Terrano (ERT) [90]. Their calculation is valid for quantities which do not depend on the orientation of the final state. The analytic results of ERT are the input in the numerical evaluation of the physical cross sections. The coefficient of the terms of order α_s^2 in the distributions in infrared- and collinear-safe quantities can be calculated once and for all with the required accuracy. An improved treatement is reported in [91].

The ERT calculation has been checked by the completely numerical analysis of Vermaseren et al. [92]. They calculated the thrust distribution and their result has been in agreement with the evaluation of the thrust distribution based on the ERT formulae [93]. More important, the analytic formulae of ERT have been positively confirmed by Fabricius et al. (FKSS) [94]. Hence it can be considered a well established result.

5.1.1 QCD and the Monte Carlo Approach

There is a large variety of distributions of soft and collinear-safe quantities which can be predicted by the theory. It was attempted to implement directly the radiative corrections into Monte Carlo programs [95] (the so-called matrix element (ME) Monte Carlos), and to use appropriate fragmentation models in order to give a complete description of the events. In these studies first appropriate parton clusters were defined, in order to separate two-, three- and four-cluster events. Each cluster is then hadronized according to a certain model. Whatever hadronization model was used, it was tuned to reproduce the bulk of the data, which is in the two-jet regime. This approach has many advantages, especially from an experimental point of view. In view of the many other ambiguities at lower energies this procedure can be justified. Let us discuss in more detail where the weakest point of this method is.

In the ME Monte Carlo programs the multiplicity of the partons is given by the starting configuration. It can only be two, three or four. Therefore the full hadronic final state multiplicity is generated by the hadronization model. Furthermore the virtuality of the initial partons becomes larger and larger as the incoming energy is increased. This is not valid in general. We know that perturbative effects, collinear splitting and soft emission of partons influence the multiplicity at the perturbative level, and non-perturbative hadronization effects come into play only when the virtuality of the partons is as low as the typical hadronic scale. Shower Monte Carlo programs [96, 97] are useful tools to examine this problem. In particular, they also provide a model of the two-, three-, and four-parton production in the pole approximation which is accurate only near the collinear limit (and therefore less accurate at the parton level than ME Monte Carlos). However, they also describe within the same approximation the production of more than four partons which are not included at all in the ME Monte Carlo programs. In the study of [91], it was found that the average number of partons produced perturbatively, i.e., before their virtuality is below 1 GeV, is 6 at PETRA energy, while it is 9 at LEP. If 6 may be considered a good approximation to 4, 10 is not. When fragmentation models are tuned to fit the data in ME Monte Carlos, they are forced to include not only non-perturbative effects, but also some genuine higher order, high transverse momentum effects, which are not contained in the 3- and 4-parton matrix elements.

In the case of the three-jet dominated quantities neither the shower Monte Carlos nor the ME Monte Carlos are good enough to describe the data. Therefore it appears that the best method to test QCD in jet production in e^+e^- annihilation is obtained by comparing directly the parton level QCD prediction with the data. In this case one can carefully estimate also the uncertainties in the QCD predictions.

One can estimate the theoretical errors due to α_s^3 and higher order QCD effects in two ways: either by changing the subtraction scale μ in an appropriate range, or by investigating the predictions obtained with the use of shower Monte Carlo programs [96, 97]. In this last method (see [91]), the higher order corrections are estimated via the study of the the sensitivity to Q_0 which denotes the scale at which the shower development is stopped. In the MC shower programs the sensitivity to hadronization effects, however, is obtained by switching on and off the hadronization which takes place after the shower is stopped. In this way we can relatively well separate the effects of the higher order corrections and the effects of the hadronization.

We shall consider thrust as a typical example of a shape variable. Many other shape variables like energy-energy correlations, broad mass square (with respect also to the thrust axis), oblateness, C-parameter, etc. have been studied.[14] We shall briefly discuss also the jet cluster distributions and inclusive jet production, where the infrared-safe, calorimetric jet definition of [99, 100] is used.

5.1.2 Thrust Distributions

Thrust is defined as sum of the lengths of the longitudinal momenta of the final state particles relative to the axis $\boldsymbol{n}_{\text{thrust}}$ chosen to maximize this sum

$$T = \max \frac{\sum_a |\boldsymbol{p}_a \cdot \boldsymbol{n}_{\text{thrust}}|}{\sum_a |\boldsymbol{p}_a|} \tag{5.1}$$

where a runs over all the final state particles. The weighted differential cross section in thrust, at the renormalization scale μ, is given by

$$\begin{aligned} \frac{1}{\sigma_0}(1-T)\frac{d\sigma}{dT} &= \frac{\alpha_s(\mu)}{2\pi} A_T(T) \\ &+ \left(\frac{\alpha_s(\mu)}{\pi}\right)^2 \left[A_T(T) 2\pi b_0 \log(\mu^2/S) + B_T(T)\right] \end{aligned} \tag{5.2}$$

where σ_0 is the leading order cross section of the reaction $e^+ + e^- \to$ hadrons. A_T and B_T are universal functions, which only depend upon the perturbative structure

[14] For a detailed review of the phenomenology of jet production at PETRA and PEP see the recent review by Ali and Barreiro [98]. Many predictions for LEP are given in [91].

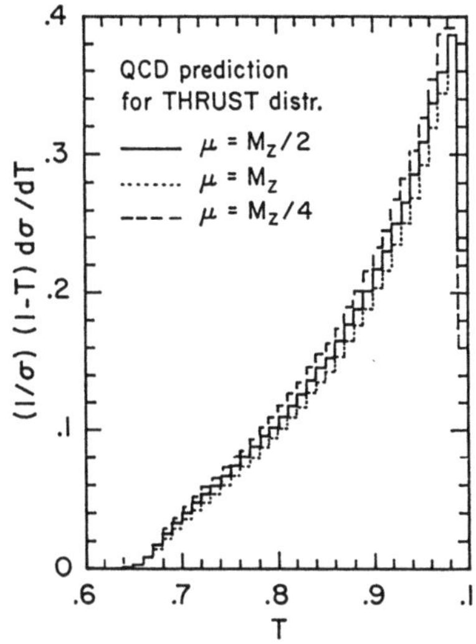

Fig. 22. Thrust distribution at the Z-pole, with three different choices of the renormalization scale [91]

of the theory. The weight $(1-T)$ was introduced for convenience to damp the fast growth of the distribution near the $T \to 1$, two jet region.

The physical quantity expressed by Eq. (5.2) must be independent of μ, within the order of the accuracy of the equation. To order α_s^3 Eq. (5.2) satisfies the renormalization group equation. The structure of the perturbative expansion does not tell us precisely at which scale μ we should evaluate our formula. However, a change of μ by less than an order of magnitude will generate a change of order α_s^3 in the value of the cross section. Various schemes have been proposed, which give recipes for the best choice of the subtraction scale. The only thing we know for sure about the choice of μ is that it should be of the order of the scale involved in the problem. With this choice we avoid the presence of large terms of the order $(\alpha_s \log(\mu/s))^n$ to all orders in the perturbative expansion.

In order to obtain the function B_T, one calculates the thrust of each event, multiplies the weight by $1-T$, and adds it to the appropriate bin in T. The function B_T can be found in [91].

In Fig. 22 we plotted the predictions of QCD at $\sqrt{s} = M_Z = 91\,\text{GeV}$ at three different values of μ ($\mu = M_Z/4$, $M_Z/2$ and M_Z, and $\Lambda_{\overline{\text{MS}}} = 140\,\text{MeV}$, which is the value we will assume throughout our discussion). The three curves for different values of μ give an estimate of the higher order effects. We can see that at LEP energy this corresponds to an ambiguity of roughly 10%. We prefer to choose $M_Z/2$ as our central value, since the energy involved in the strong processes is always somewhat smaller than the total annihilation energy.

The effect of hadronization can be studied using the shower Monte Carlo programs. In order to simplify the discussion we shall only consider the average value of thrust $< 1-T >$. In these programs, from an initial parton with virtuality Q, a shower of partons of decreasing virtuality is developed. The shower is terminated at some small value Q_0 of the virtuality of the partons, and the hadronization is performed by a cluster algorithm or by string fragmentation. In Fig. 23 $< 1-T >$ is plotted as function of Q_0. We can see that at low Q^2 values a plateau develops.

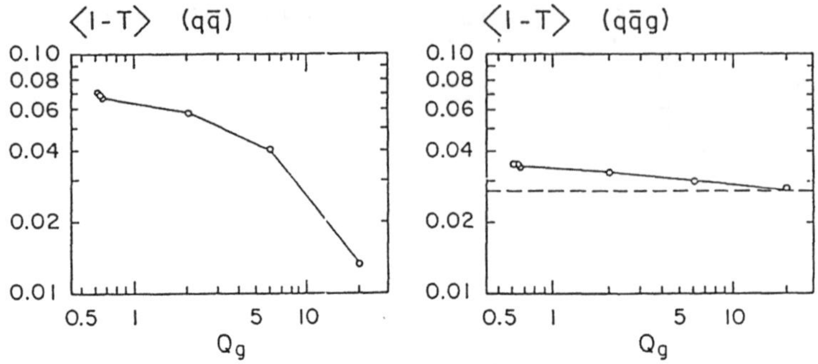

Fig. 23. The value of $< 1 - T >$ as a function of the virtuality of the parton shower starting the evolution either from the $q\bar{q}$ or the $q\bar{q}g$ initial configuration [91]

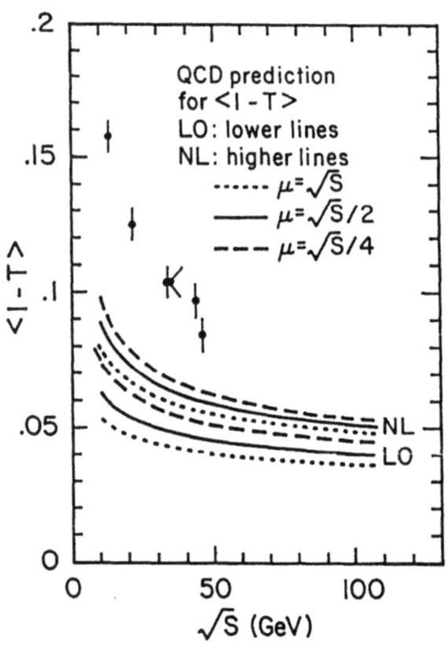

Fig. 24. Comparison of the measured values of $< 1 - T >$ with the next-to-leading order QCD predictions as a function of the annihilation energy. The data points are from [101], and the QCD prediction is plotted for three different values of the renormalization scale [91]

The dashed box at the lowest Q_0 point indicates the effect of the cluster algorithm. It clearly introduces a smaller error (approx. 3%) than the effect of the long shower (approx. 16%) [91].

The perturbative prediction for $< 1 - T >$ is given in Fig. 24 as a function of the annihilation energy, at three different values of the renormalization scale μ. We also plotted in the same figures the data points (which are taken from [101]).

The energy dependence of the data points and the difference between the theoretical and experimental values clearly indicates that at PETRA and PEP energies the higher twist contributions are very important. However, at LEP their effects might become marginal. Assuming $\Lambda^{(5)} = 0.14\,\mathrm{GeV}$, its value at LEP is predicted to be in the interval $< 1 - T > \approx 0.05 - 0.055$. Comparing the leading order (LO) and the next-to-leading order (NL) prediction in the Figure, we note that the K-factor of $< 1 - T >$ is rather large: $K = 1.35 - 1.4$ depending on the choice of scale. We notice that the uncertainty in the theoretical prediction for $< 1 - T >$ is reflected directly in the uncertainty of α_s.

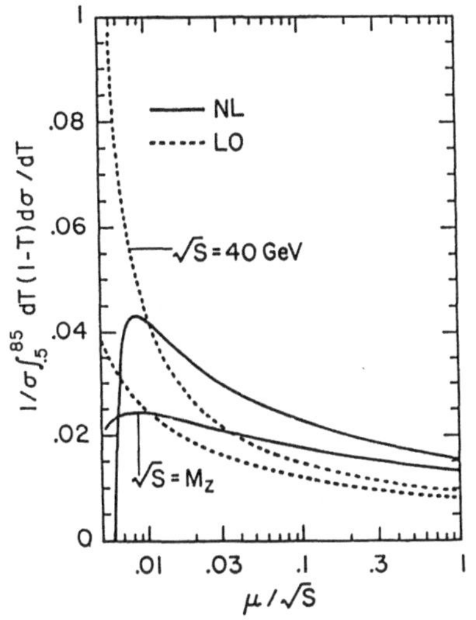

Fig. 25. The scale dependence of the QCD prediction for an integrated value of the thrust distribution (see Eq. 5.4) [91]

The scale ambiguity may be even larger. A rather extreme example is given by the scale dependence of the integrated cross section

$$\Delta\sigma = \int_{0.5}^{0.85} dT(1-T)\frac{1}{\sigma}\frac{d\sigma}{dT} \tag{5.3}$$

In Fig. 25 we plotted its value as a function of $\mu/\sqrt{s}$ for $\sqrt{s}$ = 40 and 91 GeV in leading and next-to-leading order.

The maximum of the next-to-leading order curves corresponds to the optimal scale $\mu_{\rm opt}$ of the minimal sensitivity principle. It may appear somewhat disturbing that the maximum of the next-to-leading order curves is at a scale value less than 1 GeV. This is due to the large ($\approx$ 45%) positive correction. This does not mean that the perturbative result has to be abandoned. Instead we should recall that at the next-to-leading order the MS principle gives a unique correspondence between the size of the correction and the "best" choice of the scale. If a physical cross section is parametrized as

$$d\sigma = C\alpha_s^n\,(1+r\alpha_s) \tag{5.4a}$$

then the MS principle gives the relation

$$\mu_{\rm opt} = \sqrt{s}\,\exp\left(-\frac{r}{2nb_0} - \frac{1}{2(n+1)b_1}\right) \tag{5.4b}$$

where b_0 and b_1 are given by Eq. (2.9).

If the correction r is large and positive the optimal scale is much smaller than Q, if the correction is large and negative the optimal scale is much larger than Q. I recall that for the optimization it is also significant what one chooses to calulate in the first place. Clearly there are always independent higher order effects that cannot be anticipated by intuition, guesswork and prejudice needed for the application of the minimal sensitivity principle and alike.

The considerations of this Section can be repeated for any other shape variable or three-jet dominated quantity. It is important to explore also other shape variables, since in next-to-leading order different quantities might have rather different corrections.

5.1.3 Jet Cluster Distributions

The existence of quark and gluon jets has been proven first in e^+e^- annihilation. In these early studies various shape parameters were used to establish the result. In later applications, various jet resolution parameters have also been used in conjunction with shape parameters. The problem of overlapping jets was avoided by the use of small jet resolution parameters. Clearly, good infrared-safe jet finding algorithms give competitive methods to the shape variables. Moreover, with explicitly defined jets one can ask more detailed questions concerning jet properties.

Any jet analysis should start with some infrared-safe jet definition. Let us consider first the criteria of a good jet definition [99, 100]:

(i) It is defined at any order of perturbation theory;
(ii) it gives finite cross sections at any order of perturbation theory;
(iii) it is insensitive to hadronization;
(iv) it is simple to implement both experimentally and theoretically.

In e^+e^- annihilation simple, infrared-safe jet finding algorithms are provided by the cluster algorithms based on an invariant mass cutoff [102]. There is a certain amount of freedom how we define these algorithms. Consequently, various versions appeared in the literature [94, 101, 102, 103]. Here we shall consider i) the E-cluster algorithm, ii) the P-cluster algorithm and iii) the E_0-cluster algorithm.

These algorithms are all equivalent to order α_s but give different results to order α_s^2. The circumstance that we can use different algorithms is not an ambiguity of the theory. Instead it just reflects the fact that we have a certain amount of freedom in the definition of a jet. Hence a valid comparison between the theory and the experiment requires that the algorithm must be the same in terms of hadrons as in terms of partons.

The E-cluster jet finding algorithm is defined as follows. For all pairs of particles a and b of an event the scaled invariant mass squared $y_{ab} = (p_a + p_b)^2/s$ is calculated. Then the two particles with the smallest value of y_{ab}, denoted as y_{ij}, are replaced by a pseudo-particle of four momentum

$$p_{ij}^{\mu} = p_i^{\mu} + p_j^{\mu} \tag{5.5}$$

if their scaled invariant mass is larger than some given threshold value y_{cut}. This procedure is repeated until all y_{ab} pair invariant masses are greater than y_{cut}. The resulting number of clusters is called the jet or cluster multiplicity of the event.

The P-cluster and E_0-cluster jet finding algorithms are variations of this algorithm. They use different prescriptions for calculating the four momenta of the pseudo-particle. In the case of the P-cluster algorithm instead of Eq. (5.5) the four momenta of the pseudo-particle are calculated as

$$\begin{aligned} \boldsymbol{p}_{ij} &= \boldsymbol{p}_i + \boldsymbol{p}_j \\ E_{ij} &= |\boldsymbol{p}_{ij}| \ , \end{aligned} \tag{5.6}$$

Here we slightly depart from the definition of the P-cluster algorithm which is usually given in the literature. In the usual implementations, one scales all momenta at each step, in order to maintain energy conservation. We emphasize that these different definitions receive different radiative corrections at order α_s^2.

In the case of the E_0-cluster jet finding algorithm we have

$$\begin{aligned} E_{ij} &= E_i + E_j \\ \boldsymbol{p}_{ij} &= \frac{E_{ij}}{|\boldsymbol{p}_i + \boldsymbol{p}_j|} (\boldsymbol{p}_i + \boldsymbol{p}_j) \ . \end{aligned} \tag{5.7}$$

The number of three-jet events with jet resolution $y_{\rm cut}$ can be conveniently parametrized again in the form

$$\begin{aligned} \frac{\sigma_{\rm 3jet}(y_{\rm cut})}{\sigma_0} &= \frac{\alpha_s(\mu)}{2\pi} A_{n3}(y_{\rm cut}) + \left(\frac{\alpha_s(\mu)}{2\pi}\right)^2 \\ &\times \left[A_{n3}(y_{\rm cut}) 2\pi b_0 \log(\mu^2/S) + B_{n3,I}(y_{\rm cut})\right] \ , \end{aligned} \tag{5.8}$$

$I = {\rm E},\ {\rm P}$ and ${\rm E_0}$,

while the number of four-jet events can be given as

$$\frac{\sigma_{\rm 4jet}(y_{\rm cut})}{\sigma_0} = \left(\frac{\alpha_s(\mu)}{2\pi}\right)^2 B_{n4}(y_{\rm cut}) \ . \tag{5.9}$$

The values of the functions A_{n3} and B_{n4} are the same for all the three algorithms since in the leading order no dependence on the algorithm can occur. The values of A_{n3} , $B_{n3,\rm E}$, $B_{n3,\rm P}$ and $B_{n3,\rm E_0}$ are given in [91]. In all the three cases the corrections are rather large (20%–40%). Using Eq. (5.8) (but normalized to $\sigma_{\rm tot}$ and not to σ_0) we can easily calculate the fraction of three-jet clusters as a function of $y_{\rm cut}$. The results obtained for the E-cluster algorithm and the P-cluster algorithm are shown in Fig. 26, at three different values of the renormalization scale. We can see that a change in the algorithm gives a definitely larger change in the prediction than the change in the renormalization scale μ. The former can be as high as 40% while the latter is equal to $\approx \pm 10\%$.

Assuming that in the number of the 3-jet clusters the hadronization effect will not be larger than in the case of the shape parameters, we can see that the consistency of the algorithms at the parton and hadron level is important.

The JADE experiment [102] and the recent CELLO and TASSO experiment [101, 103] used the E_0-cluster jet finding algorithm, with the practical modification that the cluster masses were calculated in terms of the experimentally well defined visible energy, and not of the total energy

$$\tilde{y}_{ab} = \frac{E_a E_b (1 - \cos\theta_{ab})}{E^2_{\rm visible}} \tag{5.10}$$

After correcting for $E_{\rm visible}$, the data on three-jet clusters can be directly compared with the perturbative QCD prediction calculated in the E_0-cluster jet finding algorithm. Therefore, there is no "scheme" ambiguity in the theoretical prediction. There

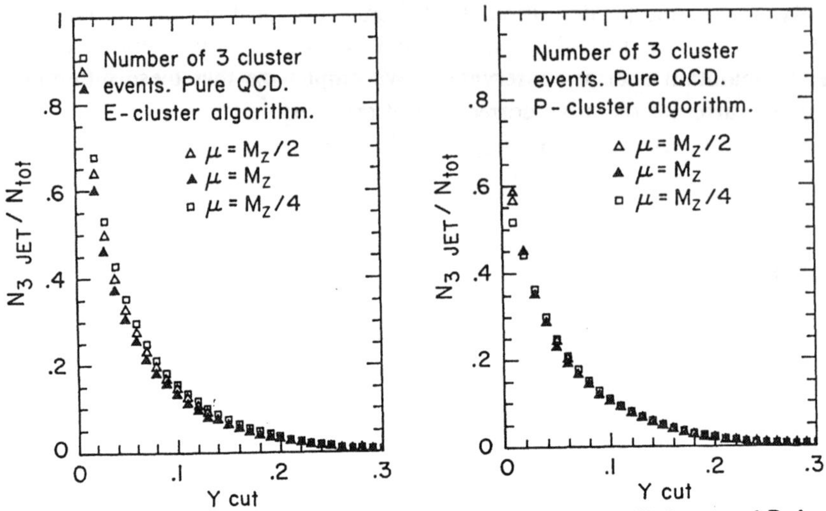

Fig. 26. Three-cluster multiplicities at the Z-pole obtained with the E-cluster and P-cluster algorithms [91]

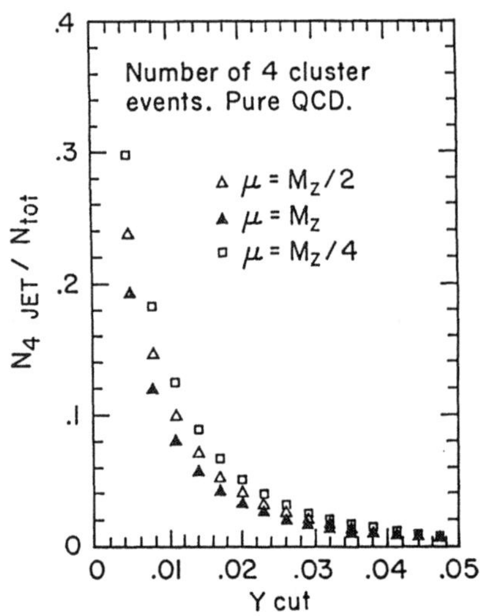

Fig. 27. Four-cluster multiplicities at the Z-pole [91]

are instead different jet finding algorithms, and one has to make sure that the parton and hadron level jet finding algorithms being used are the same.

In Fig. 27 we show the predicted fraction of the production of four-jet clusters at three different values of the renormalization scale M_Z, $M_Z/2$ and $M_Z/4$. This quantity is known only to leading order, therefore the scale ambiguity is large. This unsatisfactory situation can only be improved by calculating the next-to-leading order α_s^3 corrections.

5.1.4 Inclusive Jet Production

The significance of the jet analyses which use some explicit jet definition has considerably increased at the Sp$\bar{\text{p}}$S and the Tevatron experiments [104, 105]. The reasons for this are the reduced topological freedom (the discussion of boost invariant quantities is much simpler), the background of the so-called underlying events and the clarity of the jet signal.

In high energy e^+e^-, ep, pp and $p\bar{p}$ collisions, the discovery of many interesting new-physics-signals depends on jet properties. In many future experiments the background suppression could be considerably improved if the jet properties would be better known. Clearly, it is most important to achieve the best possible description of jet physics and to provide connections between the jet studies carried out at the various colliders. One of the simplest jet measure is the spectrum of inclusively produced jets. It is remarkable that this simple quantity has not been studied in e^+e^- annihilation.

The calorimetric jet definition of [100], after some obvious modifications, can also be applied to the case of e^+e^- annihilation. This definition has been succesfully used by the CDF experiment [105]. One can employ an algorithm [91] in which the total energy of the jets is always equal to the incoming energy and jets within jets are not counted as independent jets.

Since in the soft region the jet energy $E_J \approx 0$ and in the two-jet region it is $E_J \approx \sqrt{s}/2$, the jet spectra are singular because of soft and collinear effects. One should therefore avoid this region. Alternatively, we can suppress the singular behaviour by multiplying the cross section with the weight $e_J(0.5 - e_J)$ where e_J is defined as

$$e_J = E_J/\sqrt{s}\ .$$

The weighted cross section can be conveniently parametrized in the usual form

$$\frac{1}{\sigma}e_J(0.5 - e_J)\frac{d\sigma}{de_J} = \frac{\alpha_s(\mu)}{2\pi}A_J(e_J) + \left(\frac{\alpha_s(\mu)}{2\pi}\right)^2 \times \left[A_J(e_J)2\pi b_0\log(\mu^2/S) + B_J(e_J)\right]\ . \tag{5.11}$$

The scale invariant functions $A_J(e_J)$, $B_J(e_J)$ are tabulated in [91].

In order to exibit the expected size of the uncertaintes in the QCD prediction we plotted the cross section value in Fig. 28 at three different half-opening angles $\delta = 15°$, $30°$ and $45°$ and at three different values of the renormalization scale $\mu = M_Z, M_Z/2, M_Z/4$.

The corresponding Born cross sections are also given for comparison. The K-factor is large, $\approx 30 - 50\%$. We note the expected reduced sensitivity to the choice of scale in the next-to-leading order prediction. The spectrum has a stationary point around $e_J \approx 0.24$ and the corrections change slightly the shape of the spectrum, leading to a more symmetric behaviour. In the region around $e_J \approx 0.24$ the uncertainty due to a change in the renormalization scale is around 13% . This part of the spectrum offers the best opportunity to extract α_s from such a measurement. There is a sizable sensitivity to the change of the jet opening angle.

The shape of the jet spectrum, the dependence on the jet opening angle, the increased size of the jet rate with respect to the Born approximation and the relatively small sensitivity to the choice of the renormalization scale around $e_J \approx 0.24$ are all

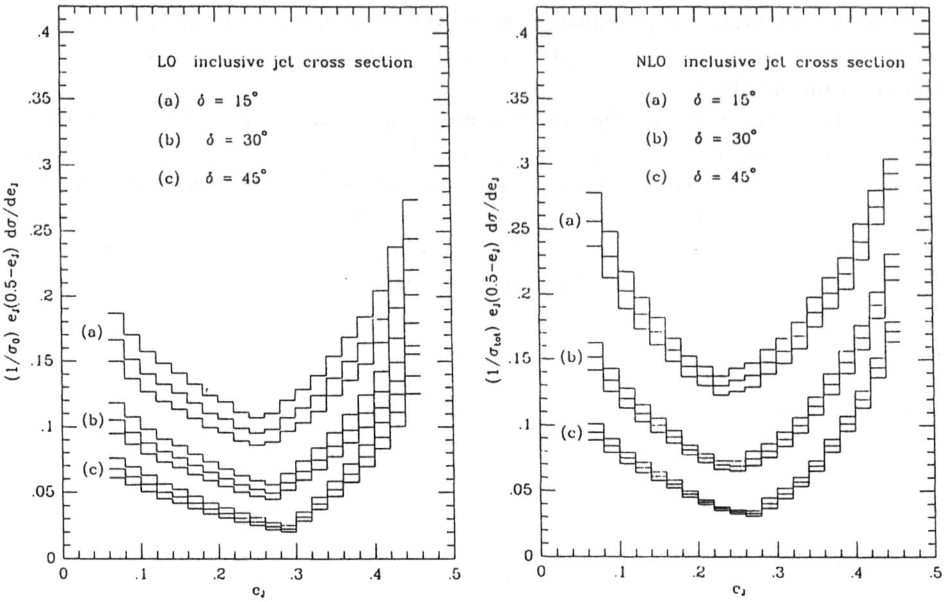

Fig. 28. Leading (LO) and next-to-leading (NLO) order predictions for inclusive cross section at $\sqrt{s} = M_Z$ for three different values of the jet half opening angle δ. The three bands correspond to three different values of the renormalization scale $\mu = M_Z/4,\ M_Z/2,\ M_Z$ and $\Lambda_{\overline{MS}} = 140\,\mathrm{MeV}$ has been used [91]

very interesting features of the QCD prediction. The study of inclusive jet prodution offers a novel method to test QCD at LEP.

5.2 Jet Production in Hadron-Hadron Collisions

The validity of the QCD-improved parton model for the description of the large transverse momentum hadron-hadron interaction got dramatic qualitative confirmation when the clear jet events have been found by the UA1 and UA2 experiments. A startling feature of the data [104, 105] from hadron colliders, both at CERN and Fermilab, is that at above some large transverse energy threshold the events are all jet-like. These jets are an essential tool for organizing and analyzing the data and are potentially important as a test of our quantitative understanding of the underlying strong interaction theory, QCD. This is particularly true if one wants to look [106] for a breakdown of the standard model.

There are significant differences between the jet studies performed at e^+e^- and $p\bar{p}$ colliders. First, the description of a jet at hadron colliders has to be performed in terms of boost-invariant quantities. The reason for this is that the partons in general do not scatter in their centre-of-mass system. But not all boost-invariant quantities are acceptable. For example the cluster algorithm discussed in Sect. 5.1.3 is not a good algorithm for hadron colliders. If we combine pairs of particles and pseudoparticles of a given invariant mass, we would put together low energy back-to-back particles with high energy almost collinear pairs. This is not a problem in e^+e^- collisions where the total annihilation energy is fixed. Another difficulty with jet production at hadron colliders is the background of the so called underlying events given by the hadronization of the spectator partons. This background can only be suppressed if angular and energy cuts are applied. Therefore the jet studies at hadron colliders necessarily require the introduction of some jet finding algorithm based on transverse energy and angular rapidity resolutions. These are the quantities which replace the energy and angle resolution used for jets observed in e^+e^- annihilation.

5.2.1 Partonic Versus Hadronic Jet Cross Sections

The measured jet cross sections obtained with the use of some finite jet resolutions are infrared-safe quantities which can be calculated in any fixed order of perturbation theory. The general formalism, described in some detail in the previous Sections in the case of the Drell-Yan process or in the case of heavy flavour production, appears to remain valid also for jet production. We think so although no all-order proof is available for the validity of the factorization theorem in the case of jet production. However, the same physical arguments which assured the validity of the theorem in the case of the Drell-Yan process also apply here [11]. Furthermore, explicit one-loop calculations have demonstrated the validity of the theorem at least up to this order [100, 107, 108].

As we have seen in Sect. 3 the factorization theorem gives a well defined algorithm to calculate finite partonic cross sections at any fixed orders of perturbation theory. Therefore, similarly to the case of gauge boson production or heavy flavour production discussed in Sect. 3, the cross sections of the production of partonic jets, folding with properly defined initial parton number densities give, up to negligible power corrections, directly the cross sections of the production of hadronic jets.

For jet production the next-to-leading order corrections have special significance, since in leading order of perturbation theory the parton cross section is independent of the jet resolution parameters. Strictly speaking, the cross section at this level is not a jet cross section but is rather just the parton cross section. It has no jet definition dependence to match that in the experimental cross section at higher orders in perturbation theory and it is infinite unless a finite jet size is introduced.

At order α_s^3, graphs like those depicted in Fig. 29 are allowed and there is an explicit dependence on the jet definition. In the calculation, one must decide when two partons count as two jets and when they count as one. The calculation at this order allows us to account for the power of the "experimental microscope" to resolve

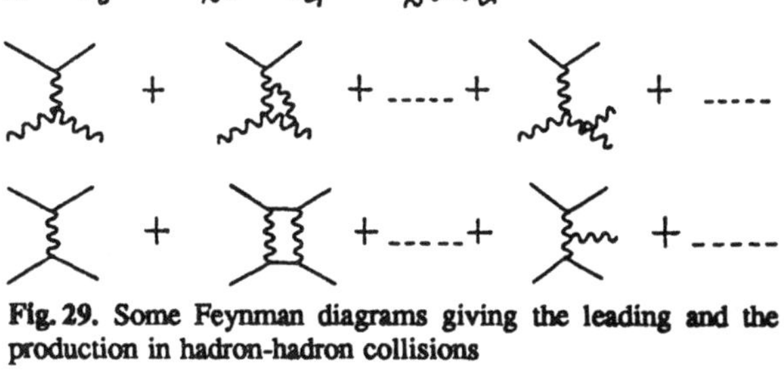

Fig. 29. Some Feynman diagrams giving the leading and the next-to-leading order corrections to jet production in hadron-hadron collisions

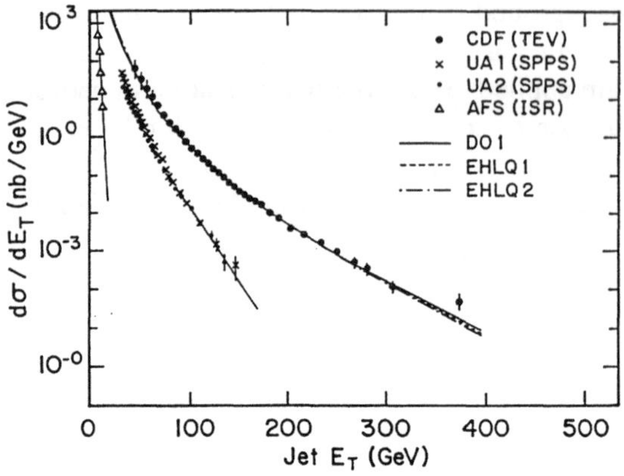

Fig. 30. Compilation of the results on inclusive jet cross sections obtained by UA1, UA2, CDF and AFS experiments [110]

one parton into two. It is exactly this careful treatment of the finite size of the jet which renders the jet cross section finite at all orders in perturbation theory in analogy to what we have seen above for similar quantities in e^+e^- physics. Also, when two partons do count as one jet, one must define the resulting jet axis and jet-E_T. In the experimental measurement, the differences between jet definition algorithms are now expected to matter, in that they can change the measured cross section at the same level as the α_s^3 corrections in the theory.

Unfortunately, the theoretical cross section for jet production is known only to within a factor of two or so, because it has only been calculated at the Born level so far. Although the absolute normalization is ambigous, the shape of the $p_\perp$ distribution is in very good agreement with the data (see Fig. 30).

A new theoretical development is that recently all the important theoretical ingredients of the next-to-leading order calculations have been provided. The full order α_s^3 matrix elements in $(4-2\epsilon)$ dimensions have been calculated by Ellis and Sexton [109]. A method to cancel the singularities analytically with finite jet resolution parameters has been given in [100]. In the same paper jet cross sections have been calculated using these full matrix elements applied to the simplified case of gluons only. Note that the general qualitative and even quantitative results are not expected to be very different from the gluon-only case, since the essential physics of asymptotic freedom, collinear parton splitting, and soft gluon emission is already present in the gluon-only case. Related calculations have recently reported [108] by Aversa, Chiappetta, Greco and Guillet. These calculations include quarks as well as gluons, but so far are limited to very small sizes of the cone that is used to define a jet.

Therefore in the near future the data can be compared with complete calculations of one order beyond the Born approximation, i.e., at order α_s^3). It is expected that to order α_s^3 the theoretical uncertainty will be smaller than the current experimental error.

5.2.2 The Cross Section of Inclusive Jet Production

The quantity that we wish to study is the inclusive cross section $d\sigma/dy_J dp_J$ for production of a jet with pseudo-rapidity y_J and transverse energy E_T plus anything. Let us suppose that gluon 2 makes the observed jet. We shall use light-cone coordinates in which vectors are given by components $p^\mu = (p^+, p^-, \boldsymbol{p}_\perp)$, with $p^\pm = (p^0 \pm p^3)/\sqrt{2}$ and $\boldsymbol{p}_\perp = (p^1, p^2)$. Let us take momentum components in the hadron-hadron center-of-mass frame, with the z-axis aligned with the beam direction. With this notation, we denote the outgoing gluon momenta by

$$
\begin{aligned}
p_1^\mu &= (\frac{1}{\sqrt{2}} p_J e^{y_1}, \frac{1}{\sqrt{2}} p_J e^{-y_1}, -\boldsymbol{p}_{\perp,J}) \\
p_2^\mu &= (\frac{1}{\sqrt{2}} p_J e^{y_J}, \frac{1}{\sqrt{2}} p_J e^{-y_J}, \boldsymbol{p}_{\perp,J}),
\end{aligned}
\tag{5.12}
$$

where p_i denotes the absolute value of the transverse part, $\boldsymbol{p}_{\perp,i}$, of p_i^μ ($i = 1,2$). We have used the fact, which follows from transverse momentum conservation, that

$$
\boldsymbol{p}_{\perp,1} = -\boldsymbol{p}_{\perp,2} \,. \tag{5.13}
$$

Note that y_1 is not fixed. We will express the other variables in the problem in terms of the observed jet variables, the hadron-hadron center-of-mass energy s, and also y_1, which we will retain as an integration variable.

Let us now look at the momenta of the incoming partons. They may be written in terms of momentum fraction variables x_A, x_B as

$$
\begin{aligned}
p_A^\mu &= (x_A\sqrt{s/2}, 0, \boldsymbol{0}) \\
p_B^\mu &= (0, x_B\sqrt{s/2}, \boldsymbol{0}).
\end{aligned}
\tag{5.14}
$$

Let us define another set of momentum fraction variables X_A, X_B as functions of y_1, y_J and p_J, so that $X_A\sqrt{s/2}$ is the plus-component of the momentum of the system formed by gluon 1 and the jet and $X_B\sqrt{s/2}$ is the minus-component of the momentum of this system. Thus

$$
X_A = \frac{p_J}{\sqrt{s}} \left(e^{y_1} + e^{y_J}\right) \quad , \quad X_B = \frac{p_J}{\sqrt{s}} \left(e^{-y_1} + e^{-y_J}\right). \tag{5.15}
$$

In the $2 \to 2$ hard scattering, momentum conservation implies that X_A and X_B are the momentum fractions of the respective incoming gluons:

$$
x_A = X_A \quad , \quad x_B = X_B. \tag{5.16}
$$

However, this relation is modified when there is an additional gluon in the final state.

We also introduce variables $\hat{S}, \hat{T}, \hat{U}$ formed from y_1, y_J and p_J in such a way that, when we are discussing the $2 \to 2$ cross section, these are the Mandelstam variables $\hat{s}, \hat{t}, \hat{u}$ of the elementary gluon scattering

$$
\begin{aligned}
\hat{S} &= 2p_J^2\left[1 + \cosh(y_J - y_1)\right] \\
\hat{T} &= -p_J^2\left(1 + e^{y_J - y_1}\right) \\
\hat{U} &= -p_J^2\left(1 + e^{y_1 - y_J}\right).
\end{aligned}
\tag{5.17}
$$

(We use upper case letters for $\hat{S}, \hat{T}, \hat{U}$ in order to avoid confusion with, for instance, the center of mass energy squared $\hat{s}$ of the two incoming gluons in the $2 \to 3$ process.)

The differential cross section in leading order $n = (4-\epsilon)/2$ dimensions can be written as a one dimensional integral

$$\frac{d\sigma}{dp_J dy_J d^{1-2\epsilon}\phi_J} = \frac{p_J}{(4\pi)^2}\left(\frac{p_J^2}{4\pi^2\mu^2}\right)^{-\epsilon} \times \int dy_1\, X_A f_A(X_A)\, X_B f_B(X_B)\, \frac{1}{\hat{S}^2}\, \langle|\mathcal{M}|^2\rangle. \tag{5.18a}$$

where the kinematical relations between the transverse momenta, longitudinal momentum fraction and rapidities can be seen from the light-cone component of the four momenta. At the Born level, the contribution of gluon-gluon scattering is computed using graphs like the first one shown in Fig. 31. The invariant function $|\mathcal{M}|^2(\hat{S},\hat{T},\hat{U};\epsilon)$ is

$$\begin{aligned}|\mathcal{M}|^2(\hat{S},\hat{T},\hat{U};\epsilon) &= \text{const}\ \alpha_s^2 16(N_C^2-1)N_C^2\,(1-\epsilon)^2 \\ &\quad\times\left(3-\frac{\hat{U}\hat{T}}{\hat{S}^2}-\frac{\hat{U}\hat{S}}{\hat{T}^2}-\frac{\hat{S}\hat{T}}{\hat{U}^2}\right) \\ &= \text{const}\ \alpha_s^2\, 4(N_C^2-1)N_C^2\,(1-\epsilon)^2 \\ &\quad\times\frac{(\hat{S}^2+\hat{T}^2+\hat{U}^2)\,(\hat{S}^4+\hat{T}^4+\hat{U}^4)}{\hat{S}^2\hat{T}^2\hat{U}^2}.\end{aligned} \tag{5.18b}$$

The first expression is the form given by Ellis and Sexton [109]. The second form is equivalent (using $\hat{S}+\hat{T}+\hat{U} = 0$) and is more convenient for algebraic manipulations.

In the next-to-leading order, the loop contributions (see Fig. 29) give terms that are singular as $\epsilon \to 0$ together with some simple non-singular terms. They are all given by Ellis and Sexton [109]. The ultraviolet singularities are cancelled by the standard UV renormalization counter terms while the infrared and collinear singularities are cancelled and factored out according to the algorithm of the factorization theorem (see Eq. (3.14)).

When the $2 \to 3$ contributions are considered one must decide when two partons count as two jets and when they count as one. When two partons count as one jet, one must then define the jet axis and the jet-p_J. If the experimental measurement is to be compared with a higher order calculation, then the differences between jet definition algorithms can matter. The experimental definition and the theoretical definition must match (at least to order α_s^3).

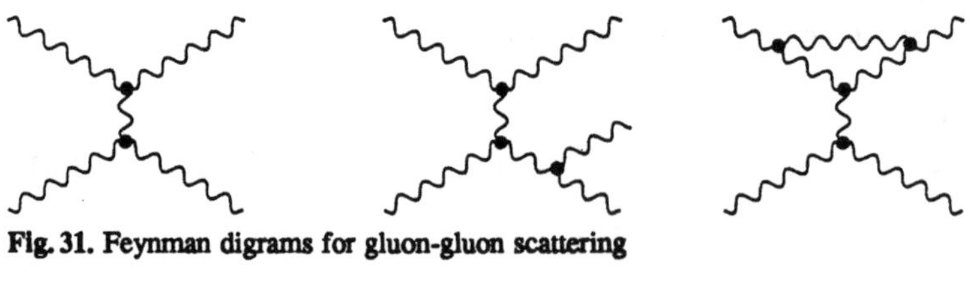

Fig. 31. Feynman digrams for gluon-gluon scattering

We thus need a good jet definition. The criteria for a good jet definition for e^+e^- have been listed in Sec. 5.1. They remain valid also in the case of the jets produced in hadronic collisions. I describe the definition which has been proposed in [99]. Let the calorimeter consist of cells i, in which transverse energies $E_{T,i}$ are measured. Define a jet cone of radius R in y-ϕ space, $R = \sqrt{\Delta y^2 + \Delta\phi^2}$. The $E_T \equiv p_J$ of the (trial) jet is then

$$E_{T,J} = \sum_{i \text{ in cone}} E_{T,i} \, . \tag{5.19}$$

The jet axis is defined by the weighted averages:

$$\begin{aligned} y_J &= \frac{1}{E_{T,J}} \sum_{i \text{ in cone}} E_{T,i} \, y_i \, , \\ \phi_J &= \frac{1}{E_{T,J}} \sum_{i \text{ in cone}} E_{T,i} \, \phi_i \, . \end{aligned} \tag{5.20}$$

Finally, the cone axis must agree with the jet axis determined by Eq. (5.20). If it does not, one simply iterates the process until there is agreement.

Let us examine how this works in our calculation of order α_s^3. A single isolated parton with parameters (p, y, ϕ) will be "reconstructed" as a jet with these same parameters. Two partons with parameters (p_1, y_1, ϕ_1) and (p_2, y_2, ϕ_2) may be combined into a single jet.

When two partons are combined into one jet, the jet transverse energy is

$$p_J = p_1 + p_2 \, , \tag{5.21}$$

while the jet angles are

$$y_J = \frac{1}{p_J}[p_1 y_1 + p_2 y_2], \tag{5.22a}$$

$$\phi_J = \frac{1}{p_J}[p_1 \phi_1 + p_2 \phi_2]. \tag{5.22b}$$

One must then examine whether the two partons fit in a cone of radius R about the jet axis. The condition that parton 1 fits into the cone is (denoting a two dimensional vector $\Omega = (y, \phi)$)

$$|\Omega_1 - \Omega_J|^2 = [y_1 - y_J]^2 + [\phi_1 - \phi_J]^2 < R^2, \tag{5.23}$$

or

$$\frac{p_2}{p_1 + p_2}|\Omega_1 - \Omega_2| < R.$$

Similarly for parton 2. Thus the combined condition is

$$|\Omega_1 - \Omega_2| < \frac{p_1 + p_2}{\max(p_1, p_2)} R. \tag{5.24}$$

If the parton angles satisfy this condition, then we count one combined jet, but not the two smaller jets.

Note that in the case

$$R < |\Omega_1 - \Omega_2| < \frac{p_1 + p_2}{\max(p_1, p_2)} R, \tag{5.25}$$

the two partons might, with some logic, count also as separate jets, as it does in Furman's original definition [107], but they do *not* in the calculation of [100]. This ambigous situation does not generally arise for unequal jets but it can arise for jets of nearly equal p_J. According to the algorithm of [100] only the combined jet is counted when the combined jet is composed of subjets, ignoring the subjets. This somewhat arbitrary choice is taken because one wants to "see" the highest-E_T jet, which reflects the most interesting physics.

There is a potential problem in recognizing experimentally a combined jet that is made of distinct subjets, since some jet finding algorithms used by the experimental groups may not locate the combined jet because there is no "seed" near the jet axis. Another concern is the allocation of energy in cases when the cones of two jets which cannot be merged into a single jet still overlap. One could allocate the E_T of the cells in the overlap region to the nearest jet or the largest jet or just allocate it to both jets. This problem does not occur at order α_s^3, so we need not specify how it is to be resolved in our calculation. The differences resulting from different experimental choices should be small, $\mathcal{O}(\alpha_s^4)$.

There is a simple algorithm [99] for finding the jet axis as defined above. Define a smeared version $\Psi(y, \phi)$ of the E_T-distribution in the calorimeter by

$$\Psi(y, \phi) = \sum_i E_{T,i} \; \theta\big(R^2 - (y_i - y)^2 - (\phi_i - \phi)^2\big) \\ \times \big[R^2 - (y_i - y)^2 - (\phi_i - \phi)^2\big].$$

Then a cone axis (y_J, ϕ_J) that is a stable solution of the jet condition (5.20) can be easily located because it is a local maximum of the function $\Psi(y, \phi)$.

I would now like to describe the calculational method. The differential cross section for (2 partons $\to$ 3 partons) is singular when two of the partons become collinear (or when one becomes soft, if the soft parton is a gluon). When one integrates over the collinear or soft configuration, one gets a divergence, which is cancelled by a divergence in the order α_s^3 virtual graphs for (2 partons $\to$ 2 partons) (or is absorbed into parton distribution functions). The whole calculation is regulated by working in $(4 - 2\epsilon)$ space-time dimensions, so that the divergences appear as terms proportional to $1/\epsilon^2$ or $1/\epsilon$ after the integrations are performed. One must extract the divergent terms, show that they cancel against the divergences in the virtual graphs, then keep the remaining part that is finite as $\epsilon \to 0$. The integrand is very complicated, so that in the end one must perform the integrations numerically. The problem is that numerical integration is not suited to extracting divergent contributions and keeping the finite part: for small ϵ, the finite part is only a tiny fraction of the $1/\epsilon^2$ and $1/\epsilon$ parts, and is smaller than the error in the numerical integration.

Let us consider now some of the results obtained in [100] in a gluon-only world. I should refer the reader to the original papers concerning some specific choices that

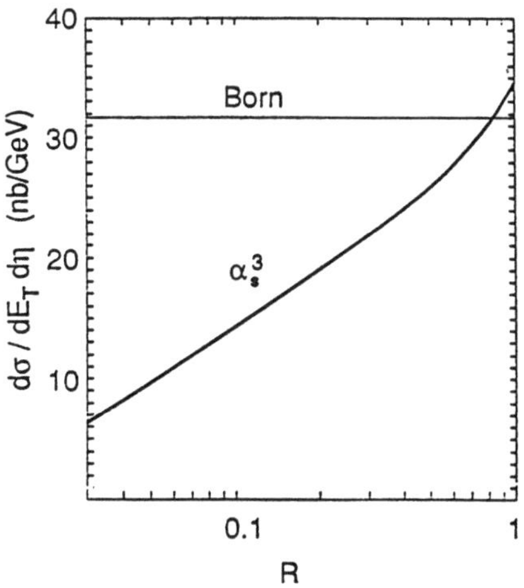

Fig. 32. Dependence of the jet cross section on the jet resolution parameter R [100]

were made in this calculation. I only mention that the value of $\Lambda_{\overline{MS}}$ was fixed so that $\alpha_s = 0.13$, as in real world. This gives $\Lambda_{\overline{MS}} = 1600\,\text{MeV}$. Furthermore a gluon distribution that begins with

$$G(x, \mu_0^2) = \frac{6.0}{x}(1 - x)^5 \tag{5.26}$$

at $\mu_0^2 = 5$ GeV has been used. The gluon distribution then was evolved using the computer code of Wu-Ki Tung [110] to second order. To get real parton distributions that are based on data at values of μ much different from the values of μ at which they are used, one should use second order evolution.

Fig. 32 shows the dependence of the cross section on the radius R of the jet cone for $E_T = 50\,\text{GeV}$ and $\sqrt{s} = 1800\,\text{GeV}$ using $\mu = E_T$. The Born result does not depend on R, but the α_s^3 result does.

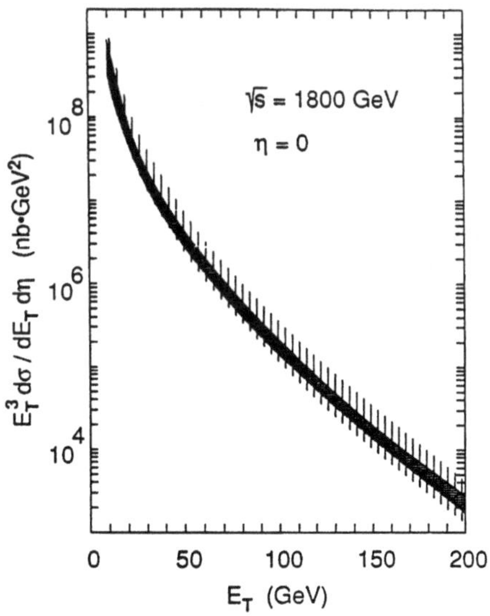

Fig. 33. Dependence on the scale parameter μ. The dashed area shows the leading order, the shaded area is the next-to-leading order result [100]

One should note that the R dependence is logarithmic for small R. As $R \to 0$, the calculated cross section tends to $-\infty$. This illustrates that one cannot use fixed order perturbation theory for small R. Perturbation theory is effectively an expansion in $\alpha_s \log R$, and when R is small, the expansion parameter is no longer small.

This R dependence for moderate values of R should be reliably given by perturbation theory. It is thus a prediction of QCD that can be tested by experiment.

In Fig. 33, the cross section versus the transverse energy of the jet is shown at $\sqrt{s} = 1800\,\mathrm{GeV}$ and $R = 0.6$. The bands indicate the range of results obtained at the Born level and at order α_s^3 as μ varies in range $E_T/2 < \mu < 2E_T$. When this curve is available for the full QCD including quarks, one will have to include the uncertainties in the parton distribution functions in determining the error bands. Then a deviation of the experiment from the error band (most likely at high E_T) would be a signal for a breakdown of the standard model.

References

1 M. Gell-Mann: Acta Physica Austriaca Suppl. **IX** (1972) 733;
H. Fritzsch and M. Gell-Mann: In *XVI International Conference on High Energy Physics, Batavia*, Vol. II, p.135 (1972);
H. Fritzsch, M. Gell-Mann and H. Leutwyler: Phys. Lett. **47B** (1973) 365

2 D. J. Gross and F. Wilczek: Phys. Rev. Lett. **30** (1973) 1343; Phys. Rev. **D8** (1973) 3633;
H. D. Politzer: Phys. Rev. Lett. **30** (1973) 1346

3 F. E. Close: *An Introduction to Quarks and Partons* (Academic Press, London 1979)

4 K. G. Wilson: Phys. Rev. **D10** (1974) 2445, and references therein

5 M. Creutz: *Quarks, Gluons and Lattices* (Cambridge Univ. Press 1983)

6 D. Weingarten: Phys. Rev. Lett. **51** (1983) 1830;
E. Witten: ibid. **51** (1983) 2351;
C. Vafa and E. Witten: Nucl. Phys. **B234** (1984) 173; Phys. Rev. Lett. **53** (1984) 353; Commun. Math. Phys. **95** (1984) 257

7 P. Langacker and H. Pagels: Phys. Rev. **D8** (1973) 4620;
S. Weinberg: Physica **96A** (1979) 327;
J. Gasser and H.Leutwyler: Ann. Phys. (NY) **158** (1984) 142

8 G. Altarelli: Phys. Rep. **81** (1982) 1;
A. H. Mueller: Phys. Rep. **83** (1981) 237

9 R. K. Ellis: FERMILAB-CONF-88/60-T, May 1988 (Lectures given at 1987 Theoretical Advanced Study Inst. in Elementary Particle Physics, Santa Fe, NM, Jul. 5–Aug. 1, 1987)

10 F. J. Yndurain: *QCD: An Introduction to the Theory of Quarks and Gluons* (Springer-Verlag, New York 1983);
T. Muta: *Foundation of QCD*, Lecture Notes in Physics (World Scientific 1987)

11 J. C. Collins and D. E. Soper: Ann. Rev. Nucl. Part. Sci. **37** (1987) 383;
J. C. Collins, D. E. Soper and G. Sterman: In *Pertubative QCD*, ed. by A.H. Mueller (World Scientific 1989)

12 A. Bassetto, M. Ciafaloni and G. Marchesini: Phys. Rep. **100** (1983) 201;
Yu.L. Dokshitzer et al.: Rev. Mod. Phys. **60** (1988) 373

13 D. Amati and G. Veneziano: Nucl. Phys. **B140** (1978) 54;
R .K. Ellis et al.: Nucl. Phys. **B152** (1979) 285;
A. V. Efremov and A. V. Radyushkin: Teor. Mat. Fiz. **44** (1980) 17;
S. Libby and G. Sterman: Phys. Rev. **D18** (1978) 3252; ibid. **D18** (1978) 4737;
A. Mueller: Phys. Rev. **D18** (1978) 3705

14 G. Bodwin: Phys. Rev. **D31** (1985) 2616; ibid. **D34** (1986) 3932;
J. C. Collins, D. E. Soper and G. Sterman: Nucl. Phys. **B261** (1985) 104; ibid. **B308** (1988) 833

15 Yu. L. Dokshitzer, D. I. Dyakonov and S. I. Troyan: Phys. Rev. **58** (1980) 269

16 G. Altarelli et al.: Nucl. Phys. **B246** (1984) 12;
G. Parisi and R. Petronzio: Nucl. Phys. **427** (1979) 154;
C. H. T. Davies and W. J. Stirling: Nucl. Phys. **B244** (1984) 337
17 J. C. Collins, D. E. Soper and G. Sterman: Nucl. Phys. **B250** (1985) 199
18 L. V. Gribov, E. M. Levin and M.'G.'Ryskin: Phys. Rev. **100** (1983) 1;
A. H. Mueller and J. W. Qiu: Nucl. Phys. **B268** (1986) 427;
L. V. Lipatov: In *Perturbative QCD*, ed. by A.H. Mueller (World Scientific 1989)
19 G. 't Hooft: Nucl. Phys. **B61** (1973) 455
20 W. A. Bardeen et al.: Phys. Rev. **D18** (1978) 3998
21 P. V. Landshoff: Phys. Lett. **169B** (1986) 69;
P. J. Doust and J. C. Taylor: Phys. Lett. **197B** (1988) 232;
G. Leibrandt: In *Proceedings of the XXIV International Conference on High Energy Physics, Munich*, ed. by R. Kotthaus and J. H. Kühn (Springer-Verlag, 1988) p. 782
22 W. Caswell: Phys. Rev. Lett. **33** (1974) 244;
D. R. T. Jones: Nucl. Phys. **B75** (1974) 531
23 O. V. Tarasov, A. A. Vladimirov and A. Yu. Zharkov: Phys. Lett. **93B** (1980) 429
24 R. K. Ellis: In *Proceedings of the XXIV International Conference on High Energy Physics, Munich*, ed. by R. Kotthaus and J. H. Kühn (Springer-Verlag, 1988) p. 48
25 K. G. Cheterkyn, A. L. Kateev and F. V. Tkachov: Phys. Lett. **85B** (1979) 277;
M. Dine and J. Sarpinstein: Phys. Rev. Lett. **43** (1979) 668;
W. Celemaster and R. J. Gonsalves: Phys. Rev. Lett. **44** (1979) 560; Phys. Rev. **D21** (1980) 3112;
S. G. Gorishny, A. L. Kateev and S. A. Larin: Phys. Lett. **212B** (1988) 238
26 B. A. Kniehl and J. H. Kühn: MPI-PAE/PTh 12/89, and references therein
27 R. P. Feynman: *Photon Hadron Interactions* (Benjamin, Reading, MA 1972)
28 K. Wilson: Phys. Rev. **179** (1969) 1699;
W. Zimmermann: Commun. Math. Phys. **15** (1969) 208; Ann. Phys. (NY) 77 (1970) 536, 570
29 G. Altarelli, R. K. Ellis and G. Martinelli: Nucl. Phys. **B157** (1979) 461
30 J. C. Collins and D. E. Soper: Nucl. Phys. **B194** (1982) 445
31 A. C. Benvenuti et al.: Phys. Lett. **233B** (1989) 485,490
32 EM Collaboration J. Ashman et al.: Phys. Lett. **202B** (1988) 603
33 L. W. Whitelow et al.: SLAC-PUB-5100 (1989)
34 M. Diemoz et al.: Zeit. Phys. **C39** (1988) 21
35 A. D. Martin, R. G. Roberts and W. J. Stirling: Mod. Phys. Lett. **A4** (1989) 1135
36 EM Collaboration J. Ashman et al.: Phys. Lett. **206B** (1988) 364
37 G. Baum et al.: Phys. Rev. Lett. **51** (1983) 1135
38 J. D. Bjorken: Phys. Rev. **148** (1966) 1467;
J. Kuti and V. Weisskopf: Phys. Rev. **D4** (1971) 3418;
R. P. Feynman: *Photon-Hadron Interactions* (Benjamin, New York 1972)
39 V. W. Hughes and J. Kuti: Ann. Rev. Nucl. Part. Sci. **33** (1983) 611
40 G. Martinelli and C. Sachrajda: Southampton preprint SHEP-87/88-16 (1988)
41 J. Ellis and R. Jaffe: Phys. Rev. **D9** (1974) 1444
42 G. Baum et al.: Yale preprint (July 1988)
43 K. Rith: private communication;
S. J. Friedman et al.: preprint MPI H-1988 V24 (August 1988)
44 A. Schaefer: Phys. Lett. **208B** (1988) 175
45 S. J. Brodsky, J. Ellis and M. Karliner: Phys. Lett. **206B** (1988) 309
46 J. Ellis and M. Karliner: Phys. Lett. **213B** (1988) 73
47 E. Witten: Nucl. Phys. **B223** (1983) 422 ;
G. Adkins, C. Nappi and E. Witten: Nucl. Phys. **B228** (1983) 552;
E. Guadagnini: Nucl. Phys. **B236** (1984) 35;
P. O. Mazur, M. A. Novak and M. Praszalowicz: Phys. Lett. **147B** (1988) 137
48 G. Altarelli and G. G. Ross: Phys. Lett. **212B** (1988) 391
49 A. V. Efremov and O. Teryaev: Dubna prepint JINR-E2-86-287 (1988)
50 R. D. Carlitz, J. C. Collins and A. H. Muller: Pittsburgh preprint 40423-11-P8 (1988)
51 J. Kodaira: Nucl. Phys. **B165** (1980) 129
52 R. L. Jaffe: Phys. Lett. **123B** (1987) 101
53 L. Wolfenstein: Phys. Rev. **D19** (1979) 3450;
H. Georgi, D. B. Kaplan and L. Randall: Phys. Lett. **169B** (1986) 73
54 D. B. Kaplan, and A. Manohar: Harvard preprint HUTP-88/A024, May(1988)
55 G. Altarelli and W. J. Stirling: CERN preprint CERN-TH 5249/88 (1988)

56 G. Altarelli and G. Parisi: Nucl. Phys. **B126** (1977) 298
57 P. G. Ratcliffe: Phys. Lett. **192B** (1987) 180
58 Z. Kunszt: Phys. Lett. **218B** (1989) 243
59 M. B. Einhorn and J. Soffer: Nucl. Phys. **B274** (1986) 714
60 R. Carlitz and J. Kaur: Phys. Rev. Lett. **38** (1976) 673
61 P. Chiappetta, J. Ph. Guillet and J. Soffer: Phys. Lett. **183B** (1987) 214
62 E. Eichten et al.: Rev. Mod. Phys. **56** (1984) 579
63 J. Babcock, E. Monsay and D. Sivers: Phys. Rev. **D19** (1979) 1483
64 Z. Kunszt and W. J. Stirling: "Photoproduction of Large Transverse Momentum Jets at HERA", in *Proceeding of the HERA Workshop, Hamburg, October 1987* (see also RAL-88-65 (1988)); Z. Kunszt: ETH-PT/88-8 (1988); M. Drees and R. M. Godbole: Phys. Rev. Lett. **61** (1988) 682
65 R. Baier: Talk given at the Advanced Research Workshop on QCD Hard Hadronic Processes, St.Croix, Virgin Island (1987)
66 J. Bott and G. Sterman: Phys. Lett. **B224** (1989) 201, and references therein
67 For a recent review see W. J. Stirling: Invited talk given at the EPS Conference on High Energy Physics, Madrid (1989)
68 P. B. Arnold and M. H. Reno: Nucl. Phys. **B319** (1989) 37; R. J. Gonsaves, J. Pawlowski and C.-F. Wai: Buffalo Univ. preprint UB-HET-89/2 (1989)
69 P. Arnold, R. K. Ellis and H. Reno: Fermilab-Pub-89/60-T (1989)
70 CDF Collaboration I. Abe et al.: Phys. Rev. Lett. **62** (1989) 1005; UA2 Collaboration H. Plotow-Besch: paper presented at the EPS Conference on High Energy Physics, Madrid (1989)
71 F. A. Berends, W. T. Giele and H. Kuijf: Nucl. Phys. **B321** (1989) 595; R. Kleiss and W. J. Stirling: Nucl. Phys. **B262** (1985) 235; J. F. Gunion and Z. Kunszt: Phys. Lett. **161B** (1985) 333
72 V. Barger et al.: Phys. Rev. Lett. **62** (1989) 1971; F. A. Berends et al.: Phys. Lett. **244B** (1989) 237
73 W. T. Giele and W. J. Stirling: Leiden preprint 89-0634 (1989)
74 P. Nason, S. Dawson and R. K. Ellis: Nucl. Phys. **B303** (1989) 607
75 R. K. Ellis and P. Nason: Nucl. Phys. **B312** (1989) 551
76 W. Beenacker et al.: Phys. Rev. **D40** (1989) 54
77 S. Brodsky, J. F. Gunion and D. E. Soper: Phys. Rev. **D36** (1987) 2710
78 J. C. Collins, D. E. Soper and G. Sterman: Nucl. Phys. **B263** (1986) 37
79 G. Altarelli et al.: Nucl. Phys. **B308** (1988) 724
80 E. Berger: In *Proceedings of the XXIV International Conference on High Energy Physics, Munich*, ed. by R. Kotthaus and J. H. Kühn (Springer-Verlag 1988)
81 P. Nason, S. Dawson and R. K. Ellis: Nucl. Phys. **B303** (1989) 607
82 J. C. Anjos et al.: Phys. Rev. Lett. **62** (1989) 513; R. Forty: In *Proceedings of the XXIV International Conference on High Energy Physics, Munich*, ed. by R. Kotthaus and J. H. Kühn (Springer-Verlag 1988)
83 CDF Collaboration: paper presented at the EPS Conference on High Energy Physics, Madrid (1989)
84 J. Ellis, M. K. Gaillard and G. G. Ross: Nucl. Phys. **B111** (1976) 253; Erratum Nucl. Phys. **B130** (1977) 516; A. M. Polyakov: JETP Sov. Phys. **32** (1971) 296; ibid. **33** (1971) 850
85 A. De Rujula et al.: Nucl. Phys. **B138** (1978) 387
86 R. F. Schwitters et al.: Phys. Rev. Lett. **35** (1975) 1320; G. G. Hanson et al.: Phys. Rev. Lett. **35** (1975) 1609
87 R. Brandelik et al.: Phys. Lett. **97B** (1980) 453
88 J. Ellis and I. Karliner: Nucl. Phys. **148B** (1979) 141
89 G. Sterman and S. Weinberg: Phys. Rev. Lett. **39** (1977) 1436
90 R. K. Ellis, D. A. Ross and A. E. Terrano: Nucl. Phys. **B178** (1981) 421
91 Z. Kunszt et al.: ETH Zürich preprint ETH-PT/89-39
92 J. A. M. Vermaseren, K. J. F. Gaemers and S. J. Oldham: Nucl. Phys. **B187** (1981) 301
93 R .K. Ellis and D. A. Ross: Phys. Lett. **106B** (1981) 88; Z. Kunszt: Phys. Lett. **107B** (1981) 123
94 K. Fabricius et al.: Zeit. Phys. **C11** (1981) 315
95 P. Hoyer et al.: Nucl. Phys. **28** (1983) 243; A. Ali, E. Pietarinen and J. Willrodt: DESY T-80-01, (1980); A. Ali et al.: Phys. Lett. **93B** (1980) 155;

T. Sjöstrand: Comput. Phys. Commun. **28** (1983) 243;
R. D. Field and S. Wolfram: Nucl. Phys. **B213** (1983) 65;
G. C. Fox and S. Wolfram: Nucl. Phys. **B168** (1980) 285;
T. D. Gottschalk: Nucl. Phys. **B239** (1984) 349

96 G. Marchesini and B. R. Webber: Nucl. Phys. **B238** (1984) 1;
G. Marchesini and B. R. Webber: Cavendish-HEP-88/7 (1988) (Updated version)

97 T. Sjöstrand and M. Bengsston: Comput. Phys. Commun. **43** (1987) 367

98 A. Ali and F. Barreiro: In *Advanced Series on Directions in High Energy Physics*, Vol. 1, ed. by A. Ali and P. Soding (World Scientific 1988)

99 D. E. Soper: In *Proceedings of the XIXth International Symposium on Multiparticle Dynamics, Arles, France, June, 1988*

100 S. D. Ellis, Z. Kunszt and D. E. Soper: Phys. Rev. Lett. **62** (1989) 726; Oregon preprint OITS-395 (1989)

101 CELLO Collaboration H. J. Behrend et al.: DESY report DESY 89-019 (1989)

102 W. Bartel et al.: Zeit. Phys. **C33** (1986) 23;
N. Magnussen: PhD Thesis, Internal Report DESY F22-89-01 (1989)

103 TASSO Collaboration W. Braunschweig et al.: Phys. Lett. **B214** (1988) 286

104 G. Arnison et al.: Phys. Lett. **123B** (1983) 115;
M. Banner et al.: Phys. Lett. **118B** (1982) 203

105 J. Huth: In *Proceedings of the SSC Workshop on Calorimetry for SuperCollider, March 13-17, 1989*

106 See, for example, the discussion in V. Barnes et al.: In *Proceedings of the Workshop on Experiments, Detectors and Experimental Areas for the Supercollider, Berkeley, July, 1987*, ed. by R. Donaldson and M. G. D. Gilchriese (World Scientific, Singapore 1988) p. 235

107 R. K. Ellis et al.: Nucl. Phys. **B173** (1980) 397;
M. A. Furman: Nucl. Phys. **B197** (1982) 413;
W. Furmański and W. Slomiński: Jagellonian preprint TPJU-11/81 (1981)

108 F. Aversa et al.: Phys. Lett. **210B** (1988) 225;
P. Chiappetta: In *Proceedings of the XXIV International Conference on High Energy Physics, Munich*, ed. by R. Kotthaus and J. H. Kühn (Springer-Verlag 1988)

109 R. K. Ellis and J. C. Sexton: Nucl. Phys. **B269** (1986) 445

110 J. C. Garfinkel: Contribution given to the EPS Conference on High Energy Physics, Madrid (1989)

111 J. C. Collins and Wu-Ki Tung: Nucl. Phys. **B278** (1986) 934

MIX
Papier aus verantwortungsvollen Quellen
Paper from responsible sources
FSC® C105338

If you have any concerns about our products,
you can contact us on
ProductSafety@springernature.com

In case Publisher is established outside the EU,
the EU authorized representative is:
Springer Nature Customer Service Center GmbH
Europaplatz 3, 69115 Heidelberg, Germany

Printed by Libri Plureos GmbH
in Hamburg, Germany